Y0-DWP-091

ADVANCES IN MOLECULAR AND CELL BIOLOGY

Volume 7 • 1993

BIOLOGY OF THE CANCER CELL

ADVANCES IN MOLECULAR AND CELL BIOLOGY

BIOLOGY OF THE CANCER CELL

Series Editor: E. EDWARD BITTAR
Department of Physiology
University of Wisconsin
Madison, Wisconsin

Guest Editor: GLORIA H. HEPPNER
Breast Cancer Biology Program
Michigan Cancer Foundation
Detroit, Michigan

VOLUME 7 • 1993

Greenwich, Connecticut *London, England*

55 Old Post Road, No. 2
Greenwich, Connecticut 06836

JAI PRESS LTD.
The Courtyard
28 High Street
Hampton Hill, Middlesex TW12 1PD
England

ISBN: 1-55938-624-X

Manufactured in the United States of America

CONTENTS

LIST OF CONTRIBUTORS

Mina J. Bissell
Life Sciences Division
Lawrence Berkeley Laboratory
Berkeley, California

Michael G. Brattain
Department of Biochemistry and
Molecular Biology
Medical College of Ohio
Toledo, Ohio

Donald S. Coffey
Department of Urology
The Johns Hopkins Oncology Center
Johns Hopkins School of Medicine
Baltimore, Maryland

Laura J. Counterman
Department of Physiology
Michigan State University
East Lansing, Michigan

Robert H. Getzenberg
Department of Urology
The Johns Hopkins Oncology Center
Johns Hopkins School of Medicine
Baltimore, Maryland

Elieser Gorelik
Pittsburgh Cancer Institute and
Department of Pathology
University of Pittsburgh
Pittsburgh, Pennsylvania

Sandra Z. Haslam
Department of Physiology
Michigan State University
East Lansing, Michigan

Erik J. Hegmann
Cancer Research Division
Sunnybrook Health Science Centre
Reichmann Research Institute
Toronto, Ontario, Canada

Gloria H. Heppner — Breast Cancer Biology Program
Michigan Cancer Foundation
Detroit, Michigan

G. Howell — Department of Pharmacology
Baylor College of Medicine
Houston, Texas

Constantin G. Ioannides — Department of Gynecology
University of Texas
Houston, Texas

Robert S. Kerbel — Cancer Research Division
Sunnybrook Health Science Centre
Reichmann Research Institute
Toronto, Ontario, Canada

Misoon Kim — Pittsburgh Cancer Institute and Department of Pathology
Pittsburgh, Pennsylvania

Bonnie E. Miller — Breast Cancer Biology Program
Michigan Cancer Foundation
Detroit, Michigan

Fred R. Miller — Breast Cancer Biology Program
Michigan Cancer Foundation
Detroit, Michigan

Kathleen M. Mulder — Department of Pharmacology
Pennsylvania State University
College of Medicine
Hershey, Pennsylvania

Brian C. Murphy — Department of Urology
The Johns Hopkins Oncology Center
Johns Hopkins School of Medicine
Baltimore, Maryland

Katherine A. Nummy — Department of Physiology
Michigan State University
East Lansing, Michigan

Catherine A. O'Brian — Department of Cell Biology
University of Texas
Houston, Texas

Ole W. Petersen — Life Science Division
Lawrence Berkeley Laboratory
Berkeley, California

Kenneth J. Pienta — Wayne State University School of Medicine
Detroit, Michigan

Janusz W. Rak — Cancer Research Division
Sunnybrook Health Science Centre
Reichmann Research Institute
Toronto, Ontario, Canada

Calvin D. Roskelley — Life Sciences Division
Lawrence Berkeley Laboratory
Berkeley, California

L. Sun — Department of Pharmacology
Baylor College of Medicine
Houston, Texas

Daniel W. Visscher — Department of Pathology
Wayne State University School of Medicine
Detroit, Michigan

Nancy E. Ward — Department of Cell Biology
University of Texas
Houston, Texas

J. K. V. Willson — Department of Medicine
Case Western Reserve University
Cleveland, Ohio

Sandra R. Wolman — Breast Cancer Biology Program
Michigan Cancer Foundation
Detroit, Michigan

S. P. Wu — Department of Pharmacology
Baylor College of Medicine
Houston, Texas

B. L. Ziober — Department of Pharmacology
Baylor College of Medicine
Houston, Texas

PREFACE

Even though I readily agreed to be the Guest Editor for *The Biology of the Cancer Cell*, and have (more-or-less) enjoyed it, I have to admit to a basic incompatibility with the job: I do not believe in "the cancer cell." In fact, I do not believe that cancer is a cellular, much less molecular, disease, an opinion which would seem to disqualify me from editing a volume for *Advances in Molecular and Cell Biology*. In my view, cancers are diseases of tissues, of multicellular organizations and interactions, and of "cellular societies." The cancer cell does not exist without the complicity of neighboring cells, both normal cells and other cancer cells, of their products and of the cells and products of other organ systems that collectively comprise the "host." With the current emphasis in cancer research on oncogenes and suppressor genes, and on autocrine factors and their receptors, it is easy to conclude that the "whole story" can be read in the cancer cell *per se*. No one would deny that the tremendous recent advances in describing the molecular and cellular alterations in cancer cells have greatly added to our understanding of neoplasia. But, learning how to translate the meaning of these alterations into ways of treating and, better yet, preventing cancer will require at least as deep an understanding of the context in which it develops. The topics of this volume were selected to lead the reader through the complex series of events by which cancer cells and their "environment" interact to produce malignant disease. Underlying themes are the diversity in the pathways that can lead to malignancy and the basic heterogeneity of neoplastic cell populations. As has often been said, in cancer no generalizations are always true (including this one). Thus, the various chapters must be viewed as examples of possible processes and mechanisms, not as universally applicable laws.

Chapter 1 focuses on the many alterations that may be detected in the genome of cancers. Using breast and colon cancers as models, the authors address the difficult issue of which changes are causally related to the development and behavior of cancer cells and which are "simply" by-products of genetic instability. Are the critical changes sequential or cumulative? How are the multiplicity of genetic changes reflected in the phenotypic heterogeneity between putatively similar cancers and even within individual cancers?

Chapters 2 and 3 begin the shift in focus from changes within cancer cells themselves to how these cells are able to either perceive or ignore their environment. Both of these chapters deal with colon cancer. Chapter 2 is concerned with autocrine mechanisms by which cancer cells may circumvent the need for external sources of growth factors, and Chapter 3 considers oncogene-associated alterations in signal transduction which may confer selective advantage on cancer cells, both during the course of their development and, later, during treatment.

Chapters 4 through 7 continue the theme of how cancer cells interact with their immediate environment, but put emphasis on the active roles of that environment in affecting or controlling the expression of the cancerous phenotype. Chapter 4 describes the roles of the extracellular matrix in the regulation of neoplastic gene expression, and Chapter 5 brings in the activities of other cells, the normal stromal cells, in growth and differentiation during cancer development. Both these chapters deal with breast cancer whereas Chapter 6 moves to another epithelial malignancy, cancer of the prostate. Chapter 6 focuses on the cytoskeletal framework to describe how signals from the outside can trigger the cellular scaffolding to effect changes in the nuclear matrix which, in turn, regulate gene expression. Thus, with this chapter the concept of the *unity* of cancer cells and their environment comes full circle from Chapter 1.

The objective of the remaining chapters is to broaden the horizon of cancer development from that of the emerging cancer cell and its most immediate environment to considerations of the dynamics of cancer and normal cell populations and how these dynamics contribute to malignant behavior. Chapters 7 through 9 share a common vision of cancers as evolutionary systems. Chapter 7 builds on the theme of cancer cell diversity, first introduced in Chapter 1, to discuss how interactions among differing cancer cell clones within the same neoplasm modulate the growth and behavior of the cancer as a whole, with a special emphasis on metastasis, the process by which cancer cells colonize tissues away from the site of their primary development. This chapter, as well as Chapter 9, depends upon the concept of the inherent genetic instability of cancer cells which allows for new phenotypes and combinations of phenotypes to be introduced into the population, and which contributes to the aggressiveness of the cancer, including the development of resistance to therapy. These chapters show cancer as a dynamic phenomenon with the capacity for change throughout its course. Thus neoplastic heterogeneity has a time dimension, referred to as "progression," which adds further complexity to cancer cell biology.

Chapters 8 and 9 present examples of how a cancer can subvert the host into becoming a partner in progression. Chapter 8 is concerned with the ways in which the immune system may be neutralized as a defense against cancer, even perhaps be a force in its growth and, by selecting against less aggressive subpopulations, contribute to progression. Chapter 9 discusses how cancer cells utilize host angiogenesis to provide the structural and nutritional framework necessary for growth and metastatic progression.

Chapters 7 through 9 use a spectrum of cancer types, including breast cancer and melanoma, as illustrations.

Cancer research, like other types of research tends to be reductionist. Individual researchers study particular systems and particular mechanisms. This approach is certainly productive, but needs to be balanced by an appreciation of a larger picture. In selecting and grouping the topics for this volume, I have tried to give the reader a sense of the complexity of cancer biology. I want to thank all the authors for their very fine presentations and Dr. E. Edward Bittar for the opportunity to bring them together.

Gloria H. Heppner
Guest Editor

CELLULAR GENETIC ALTERATIONS: MODELS OF BREAST AND COLON CANCER

Sandra R. Wolman and Daniel W. Visscher

Advances in Molecular and Cell Biology
Volume 7, pages 1–34.

ISBN: 1-55938-624-X

I. INTRODUCTION

Human epithelial tumors are characterized by a histologically defined sequence that corresponds to clinical disease progression. This morphologic continuum begins with abnormal proliferation and differentiation, termed dysplasia, followed by intra-epithelial (*in situ*) carcinoma, which progresses to invasive growth into host tissues and culminates in dissemination to distant organs (metastasis). These morphologic stages are accompanied by, and indeed may be determined by, specific genetic alterations. In colorectal neoplasia, histologic progression from normal, to proliferative changes, to increasingly severe forms of dysplasia, to invasive adenocarcinoma is marked by the acquisition of 5q* deletion, *ras* mutation, and losses or mutations of genes on 17p and 18q, respectively (Fearon and Vogelstein, 1990). In general, human tumors also display a spectrum of clinical aggressiveness, once invasion has occurred, which correlates with a collection of morphologic features defined as "grades." Tumor grade reflects degree of loss of differentiation (the extent to which the neoplasm resembles or recapitulates histologic patterns of the organ from which it arose). In glial tumors, for example, low-grade astrocytomas display losses of chromosomes 13, 17, and/or 22; losses from 9p are found in anaplastic (less differentiated) astrocytomas, and high-grade tumors (i.e., glioblastoma multiforme) also show chromosome 10 loss and amplification of the epidermal growth factor receptor (James et al., 1990).

Sequential or "stepwise" accumulation of genetic aberrations implies a consistent sequence, but there is little evidence that genetic alterations are *acquired* in a predictable order in these or other tumor systems. Rather, each event is believed to induce phenotypic changes which, in aggregate, reflect capability for invasion and metastasis. Although the colon cancer model has become a paradigm for solid tumor development, definition of critical genetic events in other malignancies has been hampered by several inherent properties of human neoplasias, especially their lengthy natural history, architectural complexity, and relative inaccessibility for sampling (particularly at the small, early stages). Marked intertumoral heterogeneity of etiologic factors, as well as of mechanisms of disease progression, represents another major obstacle to a unified formulation of the genetic events responsible for neoplastic progression.

Genetic alterations at the cellular level (aggregate DNA content, karyotype) resulting in a malignant phenotype are often less precise but more accessible to analysis than those at the molecular level (gene structure, regulation, and expression) that are a step closer to the mechanistic basis for the disease. Quantitative DNA and cytogenetic analyses are widely applied in diagnostic and prognostic evaluation of many human tumors and provide a necessary context for detection and interpretation of molecular data. We will explore evidence relating cellular genetic changes to biological and clinical aspects of two common human cancers, one for which precancerous lesions are well defined (colorectal neoplasia) and another, adenocarcinoma of the breast, which lacks a generally accepted morphologic basis for definition of a genetic pathway. The methodological approaches that permit assessment of cellular genetic changes may serve as a bridge between traditional morphologic diagnosis and molecular analyses.

II. TISSUE MODELS

Normal epithelial surfaces are characterized by exquisite regulation of proliferation, functional differentiation, and intercellular linkage. Virtually all epithelial tumors are preceded by a pathologic entity termed "dysplasia" in which each of these parameters is altered. Compared to the corresponding normal epithelium, dysplastic surfaces display thickening (i.e., more cell layers) and increased architectural complexity, with relative cell crowding and absent or inappropriate differentiation. Dysplasias also exhibit an array of morphologic nuclear changes, including enlargement, increased chromatin staining, and abnormal, highly variable shapes, indicative of cellular genetic changes. The morphologic spectrum of dysplasia can range from subtle abnormalities (marginally distinguishable from reactive or inflammatory changes) to complete loss of cellular differentiation (sometimes seen in carcinoma *in situ*). Functional cellular changes, particularly those involving increased cell cycling, may precede objective histologic evidence of dysplasia. In carefully studied tumors amenable to serial sampling (such as squamous cell carcinoma of the uterine cervix), the degree or severity of dysplasia correlates with the propensity for subsequent clinical development of invasive cancer. Progression to invasion is not, however, inevitable. Approximately 30–50% of cervical carcinomas *in situ* remain clinically stable for prolonged periods. Further, evolution from milder forms of dyplasia is not demonstrable in many examples of carcinoma *in situ*. Finally, degrees of morphologic heterogeneity are recognizable even in carcinoma *in situ*. In other organs where sampling is more limited, the relationships between precursor lesions, such as dysplasia, and invasive cancer are less clear and based largely on circumstantial data.

There is widespread acceptance that dysplastic (adenomatous) lesions of the large bowel represent direct precursors of colorectal adenocarcinoma, although, as with cervical dysplasia, not all of these lesions will necessarily progress to frank malignancy. An adenoma-to-carcinoma sequence is supported by observations of the contiguity of benign polyps to malignant lesions, by correlations of increasing adenoma size with increasing risk of malignancy, and by statistical associations of the prevalence of both diseases (Fenoglio and Lane, 1974; Muto et al., 1975; Morson, 1976). The inherited polyposis syndromes and their clear association with increased risk of colorectal cancer affirm the importance of polyps as precursor lesions (Wolman et al., 1990). The observed co-segregation and high incidence of both polyps and cancers of the colon within families without such syndromes (Burt et al., 1985; Cannon-Albright et al., 1988) lend additional support to interpretation of these lesions as an adenoma-to-carcinoma sequence. The existence of such a sequence would reinforce the potential diagnostic value of markers identified in the antecedent lesions.

Yet even for colorectal tumors, a model based on progression via adenomatous polyps is incomplete. Recent observers have called attention to the origin of many colorectal cancers from small, so-called "flat adenomas" (Muto et al., 1985; Adachi et al., 1988; Wolber and Owen, 1991). Examples of these lesions in both sporadic (Desigan et al., 1985) and hereditary forms (Lynch et al., 1988; 1991) have been documented, as have small, flat carcinomas without associated polyps (Hunt and Cherian, 1990). In the hereditary form, they commonly appear as poorly differentiated mucinous tumors (Lynch et al., 1991). These neoplasms may constitute a second morphologic pathway of cancer formation in the colon, possibly determined by a different set of genetic alterations from those connected with a polyp-dependent progression. Moreover, evidence that colorectal cancers are associated with mucosal alterations adjacent to tumors (Filipe, 1984; Viola et al., 1991) seems to support the thesis of a widespread "field" of premalignant alteration. From such a field many atypical clones could arise and even merge to form an apparently polyclonal tumor.

In the breast, a spectrum of intraepithelial changes, described in diagnostic surgical biopsy material, has been termed proliferative breast disease (PBD). Recently, the histopathologic features of PBD have been shown to convey differential risk for subsequent development of clinical invasive neoplasia in the same or contralateral breast tissue (Page and Dupont, 1990). Patients with the most severe form of this process, atypical hyperplasia (AH), have a 4- to 10-fold relative risk of progression to malignancy. The incidence of malignancy is augmented in patients with AH and a family history of breast carcinoma, suggesting that AH is a reflection of inherited alterations or susceptibility (ibid). The focal and microscopic nature of PBD and, thus, the limited availability of precancerous breast tissue for study have hampered epidemiologic and genetic analyses of PBD as a possible premalignant or preinvasive condition of mammary adenocarcinoma. Although it is dogma that invasive breast neoplasia traces its origins to areas of ductal (or lobular) *in situ* carcinoma, there is little agreement on morphologic evidence of progression from earlier stages. The frequent co-existence of atypical ductal hyperplasia with intraductal and invasive ductal disease suggests a sequence in breast cancer equivalent to that of colorectal tumors. However, the protracted intervals to detection of clinical malignancy in patients with AH (8–12 years) and the complexity of the lobular-ductal system in the human breast add to the subjectivity of direct morphologic observations (Rosai, 1991), and all these factors limit the value of traditional morphologic studies.

III. APPROACHES TO ANALYSIS

A. DNA Content: Flow Cytometry, Image Analysis, Replication

The presence of abnormal DNA content in malignant tumor cells was inferred from their bizarre nuclear shapes and abnormal chromatin distributions in conventional histologic preparations. Semi-quantitative (Feulgen) staining methods confirmed these observations, and recent developments in image and flow cytometry have made DNA assessment routinely available, even from archival specimens. However, despite widespread clinical application of DNA analysis to human tumors, little is known about the mechanism(s) of DNA aneuploid clone generation or the specific relationship of this anomaly to neoplastic progression. In viral transformation experiments, induction of tetraploidy appears to be the first visible event (Lehman and Defendi, 1970; Hirai et al., 1974), and there is some evidence for tetraploidy as an intermediate stage in the evolution of a few human tumors (Tribukait et al., 1991). Subsequent loss of genetic material secondary to abnormal chromosome segregation would then be required to produce hyperdiploidy, which is the most common DNA content anomaly in human tumors. The far less common hypertetraploid clones in human tumors are probably generated by a second tetraploidization event.

An overwhelming body of published literature now indicates that for most adult epithelial malignancies including colorectal and breast cancers, diploid-range lesions have a survival advantage over aneuploid cases. This is by no means true for all human tumors: Aneuploid pediatric neuroblastomas have a more favorable prognosis than diploid-range cases (Look et al., 1984). In endocrine organ neoplasia, no prognostic relevance has been found for DNA aneuploidy, which is in fact regularly demonstrated in clinically benign lesions. Thus, determination of ploidy is both common and useful for certain groups of human tumors, but its mechanistic relation to prognosis is not understood. Presumably, an excess or unbalanced complement of cellular DNA materially alters gene expression or regulation, as well as proliferation potential. In this respect, it is noteworthy that DNA aneuploidy is reliably and strongly correlated with poor differentiation in most human tumors of epithelial derivation (Koss et al., 1989). The association between abnormal DNA content and high proliferative fraction in epithelial tumors is also reported with remarkable consistency. However, in relation to prognosis there is some evidence in head and neck cancers that aneuploid cells may be more susceptible to chemotherapy, presumably because of high proliferative index, than are diploid cells. In such a case aneuploidy is advantageous prognostically in indicating potential response to treatment. It seems odd that an excessive DNA replication burden would constitute a cell-cycling advantage; however, abnormal cycling may represent a cause rather than an effect of DNA aneuploidy, possibly resulting from altered genomic stability.

B. Cytogenetics: Metaphase, Interphase

One of the most productive routes to detection of genes that are causal of or contributory to cancer is recognition of frequent or specific chromosome aberrations associated with particular tumors or tumor-prone individuals. Furthermore, cytogenetic specificity has contributed to the perception that tumors are clonal proliferations and that the cytogenetic alteration marking the tumor also confers a selective growth advantage. Cytogenetic classification of leukemic patients has demonstrated the value of karyotypic aberrations as independent prognostic factors capable of predicting duration of remission and survival (Fourth International Workshop on Chromosomes in Leukemia, 1984). Correlation between cytogenetic breakpoint sites and the loci of human cellular oncogenes, most notably the t(9;22) of chronic myeloid leukemia and the t(8;14) of Burkitt's lymphoma, has resulted in recognition of new cellular oncogenes, important in cellular growth regulation (Croce, 1987). We expect that identification of non-random chromosomal changes in solid tumors should also contribute to better understanding of tumor biology and form the basis for prognostic and therapeutic strategies, since non-random or specific cytogenetic changes may point to causal or functionally relevant cellular genes.

On the other hand, an increase in apparently random alterations could also provide biologic relevance by reflecting increased genetic instability. Distinguishing random from non-random events, however, is far more difficult in solid tumors than in the leukemias. Extensive and complex karyotypic changes, heterogeneity of chromosome pattern for cells within a tumor, and variability among tumors of the same histotype are attributes of many solid tumors. The problems of defining tumor-specific aberrations are further compounded by technical difficulties of cell separation, low intrinsic mitotic rates, and contamination by normal cellular elements. Despite the many problems that plague solid tumor cytogenetic studies, a growing body of data suggests that consistent karyotypic alterations do characterize solid tumors (Sandberg, 1990; Mitelman, 1991).

Once a potentially important locus is suspected, investigations can be extended beyond metaphase analysis to explore the extent and clinical representation of its aberrations. The advent of new methods based on *in situ* hybridization for identification of chromosomal components in interphase nuclei makes possible the study of previously inaccessible populations of tumor cells and permits expansion to direct and population-based assessment of chromosome patterns (Pinkel et al., 1988; Herrington and McGee, 1990). Thus, critical questions pertaining to chromosome instability and heterogeneity can be explored. Probes that can target repetitive centromeric regions or unique sequences within chromosomes by *in situ* hybridization techniques allow definition of individual genetic alterations in the contexts of whole nuclear architecture and whole cell and tissue organization.

C. Theoretical Concerns: Methodologic Pros and Cons

The terms "diploid" and "aneuploid" are routinely employed in the cytophotometry literature, implying analogy to cytogenetic studies. Image and flow cytometric DNA analysis, however, represent estimates of total DNA content derived from fluorescence or optical density measurements of dyes which bind DNA stoichiometrically. DNA content determinations are made relative to an endogenous or admixed normal reference population, generally peripheral blood lymphocytes, and are only able to detect a 2–5% net gain or loss (i.e., one to several chromosomes) of total DNA by present standard technology. This level of sensitivity is materially affected by numerous factors, especially the efficiency with which tumors may be dissociated into single-cell suspensions or smears. The extent to which cytophotometric DNA aneuploidy reflects the degree of cytogenetic aberration of a given neoplasm is unclear. Strong correlations have been observed between cytophotometric DNA content and total chromosome counts in some tumor systems, including adenocarcinomas of the breast and colon (Remvikos et al., 1988a, 1988b). However, karyotypic abnormalities which do not result in significant net quantitative changes of total DNA (e.g., reciprocal translocations, balanced chromosomal duplications/losses) clearly should yield "normal" (i.e., diploid-range) DNA histograms.

About two-thirds to three-fourths of human breast and colorectal carcinomas contain neoplastic populations with abnormal modal DNA contents. Quantitative DNA analyses are represented by histograms which plot cell number (vertical axis) against DNA content (horizontal axis) and provide cell cycle data (synthesis phase fraction [SPF] estimates) (see Figure 1). Most commonly, the abnormal mode falls between diploid and tetraploid (i.e., hyperdiploid), although hypertetraploid and hypodiploid stemlines are detected occasionally (5–10% each). At present, no method reliably distinguishes near-diploid tumor populations from contaminating stromal or inflammatory cells in flow-cytometric DNA histograms. Thus, it is unclear what proportion of flow-cytometrically DNA-aneuploid tumors also contain diploid-range neoplastic populations. Studies employing multiple sampling techniques or image analysis, however, have demonstrated that several populations within a single tumor are cytophotometrically distinguishable in 15–50% of breast and colorectal adenocarcinomas (Beerman et al., 1991; Russo et al., 1991). These data imply clonal heterogeneity or genetic evolution in such tumors. In contrast, comparison of DNA histograms from primary neoplasms and their synchronous nodal metastases or recurrences reveals similarity of DNA indices in most cases. The dominant stemline, whether diploid or aneuploid, within a given tumor appears to be the one most likely to metastasize, and its dominance probably depends upon aggressive growth potential. (See Chapters 7 and 9 for further discussion of "clonal dominance" within primary tumors and its relation to metastasis.)

It should be noted that image and flow cytophotometry are both significantly limited in their ability to detect intratumoral clonal heterogeneity. Image analysis

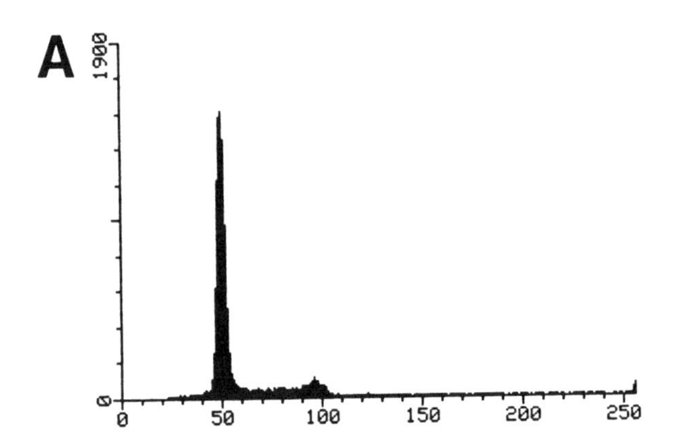
A
1900
0
0
50
100
150
200
250

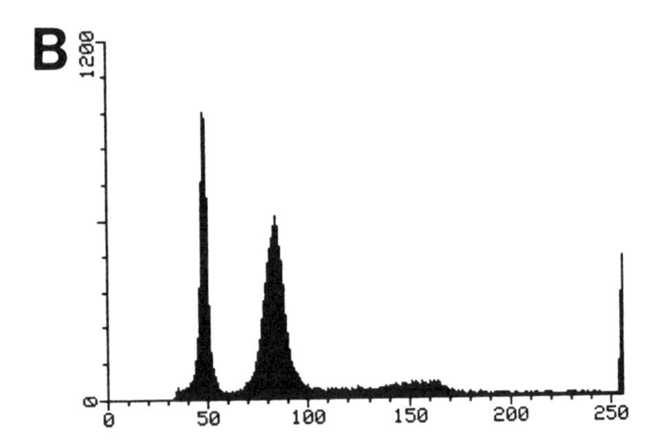
B
1200
0
0
50
100
150
200
250

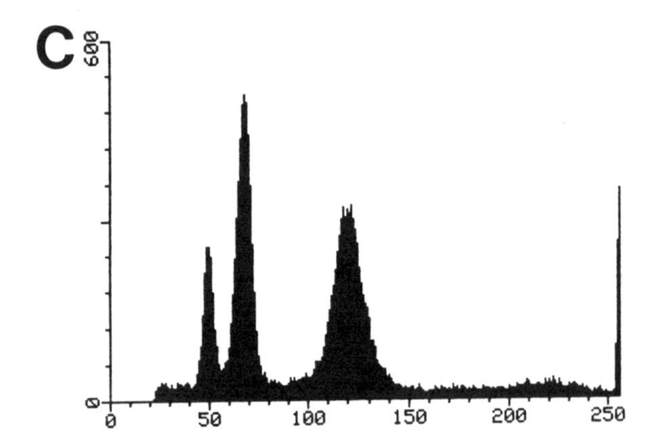
C
600
0
0
50
100
150
200
250

requires manual visualization of neoplastic cells for DNA measurements, and the resulting histograms are derived from only 100–300 cells. Reliable flow-cytometric analysis, on the other hand, is rarely based on histograms containing fewer than 10,000–20,000 counts. It is well established that some dissociation protocols, particularly those employing enzymatic digestions, result in differential loss (i.e., cell lysis) of aneuploid cells. Therefore, relatively small population subsets may be obscured or under-represented using either method. However, small regions, such as small flat colonic lesions or apparently uninvolved mucosa adjacent to tumor, are accessible to study by image but not by flow-cytometric methods (and can be further investigated for genetic alterations with *in situ* hybridization techniques).

Documentation of specific cytogenetic aberrations in solid tumors has been difficult because even the best-recognized chromosome-specific associations are relatively inconstant. For example, only 10% of a series of Wilms' tumors showed the "characteristic" deletion in 11p13 (Solis et al., 1988); the deletion of 13q14 in retinoblastoma is found in approximately 21% of the tumors (Potluri et al., 1986). Detection of a specific aberration is often obscured by the presence of extensive chromosome aberrations in the tumor cells, with considerable variability from cell to cell. Some aberrations are common to many tumor types and probably represent common pathways of oncogene or tumor suppressor gene involvement that may well be independent of the tissue of origin. For example, multiple copies of 1q have

Figure 1. Flow-cytometric DNA histograms of breast carcinomas. Histograms represent quantitative DNA analyses by plotting cell number (vertical axis) against DNA content (horizontal axis). The cell cycle data (synthesis phase fraction [SPF]) can be estimated from the extent of the area between the G1 and G2 peaks. a) *Diploid Range*—Normal—This histogram shows two peaks: the first consists of cells in the resting and presynthetic phase of the cell cycle (G_0 and G_1) that exhibit a mean fluorescence intensity similar to control cells (not shown), with a normal (2N) complement of DNA. A second, smaller peak is present at twice normal DNA content, representing premitotic and mitotic phase cells (G_2/M). Intervening cells are in the process of synthesizing DNA (the synthesis-phase fraction [SPF]). Note that it is not possible to distinguish neoplastic cells from admixed stromal or inflammatory cells in suspensions of diploid-range tumors. b) *DNA Aneuploid —Hyperdiploid*—A more complex pattern of two distinct G_0/G_1 peaks is present, one abnormal corresponding to a clonal population of neoplastic cells and a second, normal peak representing host-derived stromal cells (alone or possibly with a second diploid-range neoplastic clone). G_2/M cells of the DNA aneuploid population are evident, but those corresponding to diploid-range cells are obscured by the synthesis-phase events of the abnormal cells. c) *DNA Aneuploid-Multiploid*—Two DNA aneuploid G_0/G_1 stemlines are present, one in the hyperdiploid range, the second in the hypertetraploid region.

been observed in breast cancers, myeloid leukemias, and testicular tumors. On the other hand, aberrations such as trisomy 7, which has been reported in renal tumors, glial tumors, bladder tumors, prostate cancers, and melanomas (Balaban et al., 1986; Bigner et al., 1986, Rey et al., 1987; Wolman et al., 1988; Babu et al., 1990; Berrozpe et al., 1990), may favor growth in culture.

Solid-tumor cytogenetic studies, because of limits imposed by slow spontaneous rates of cell division, often depend upon cells which have been cultured for differing periods of time, and the cultured cells may represent only a subset of the original tumor population. Diploid metaphases may originate from stromal or inflammatory cells, although in certain tumors it is difficult to rule out the possibility that they represent a component of the tumor cell population, partly because of prognostic associations. Major problems in cell preservation and disaggregation, in bacterial contamination, and in selectivity of culture conditions, and emergence of new aberrations with time in culture, have contributed to the many difficulties in interpretation of culture results (Leibovitz, 1989). Much of our knowledge of the biology of breast epithelial tumors, for example, derives from cell cultures that are heterogeneous (both with respect to the tumor *per se* and to possible admixture with non-tumor cells), as demonstrated by differences in proliferation, morphology, antibody reactivity, or protein synthesis of growing clonal populations (Schmidt-Ullrich et al., 1986). Thus, *in vitro* as well as *in vivo* results are unlikely to be fully representative of the primary tumors.

The most important problem, however, is that of true tumor heterogeneity, as contrasted with the heterogeneity arising in culture as described above. Recurrent observations of variability or heterogeneity of both the neoplastic phenotype and genotype confound definition and assessment of genetic alterations in cancer. Heterogeneity is evident among histologically similar cancers from different patients (intertumor heterogeneity) and among different cells of the same cancer at a single time (intratumor heterogeneity), as well as at different points in time (progression). All these aspects of tumor heterogeneity pertain to karyotypic heterogeneity as well, although in many tumors the chromosome aberrations are related and serve to trace clonal evolution and tumor progression (Nowell, 1976). In addition to this genetic basis, other underlying mechanisms of neoplastic heterogeneity include adaptive (Farber and Rubin, 1991) and environmentally determined events (Nowell, 1976; Heppner, 1984)

Clonality, the property that the cells within a tumor are derived from a single parent cell, is often indicated by uniformity or relative uniformity of genetic aberrations contained within many or all cells of the tumor. The aberration(s) are assumed to confer or reflect biological distinctions relevant to tumor behavior, and thus to be relevant to tumor initiation and clonal expansion. However, the assumption of clonality in tumors based on genetic markers may not sufficiently take into account the prolonged periods of growth and the powerful selective influences that could result in dominance of a single pattern despite many different cells of origin. Arguments favoring a single cell of origin depend on the identification of markers that are relatively uniform within a tumor cell population and that differ from the host (cytogenetic aberrations, uniformity of X chromosome inactivation) or, more powerfully, the identification of a presumptive *causal* relation between the marker and altered behavior in the tumor cell population. The well known examples of the t(8;14) of Burkitt's and the t(9;22) of CML both result in repositioning of cellular oncogenes and produce quantitative or qualitative changes in important cellular proteins.

Multiclonality appears to be the norm for some of the most ordinary of human tumors—the squamous and basal tumors of the skin (Heim et al., 1989), and squamous and other tumors of the respiratory tract, as well as some leukemias (Heim and Mitelman, 1989; Kobayashi et al., 1990), gliomas (Shapiro et al., 1981), and possibly some AIDS-related tumors. It may prove to be common in other solid tumors as our ability to study them improves. Thus the appearance of clonality does not necessarily indicate clonal origin: Many cells could respond to the initiating event(s) at the time of onset to form a tumor; however, a subpopulation within the mass that is more resistant to local growth-limiting conditions, such as low oxygen supply, or is dividing more rapidly, could easily become the dominant and apparently sole population of the tumor. Conversely, true clonal origin could be masked by differing physical conditions and selective influences within the primary tumor that foster survival of a mixed cell population (see Chapter 7). Clonal evolution has been studied in the course of tumor growth by incorporating cells with different genetic markers into experimental tumors (Kerbel et al., 1988; see Chapter IX), and it is clear that clonal selection can rapidly result in homogeneous populations that appear to have originated from single cells. Conversely, the term "polyclonal" may be descriptive, but not necessarily reflective of tumor origin. If the cell of origin were diploid but genetically unstable, one would expect not only multiple aberrant clones but also an array of non-clonal aberrations (which are frequently seen and usually ignored in tumor cell analysis). A few recent studies have emphasized that frequent non-clonal numerical and structural aberrations serve to distinguish tumor-derived from non-tumor-derived cultures (Geleick et al., 1990; Wurster-Hill et al., 1990; Micale et al., 1992).

IV. COLORECTAL NEOPLASIA

A. DNA Content and Cell Proliferation

The frequent demonstration of DNA aneuploidy in various colorectal adenomatous lesions (Goh and Jass, 1986; McKinley et al., 1987) is not surprising, given the obvious cytologic anomalies that characterize these lesions. Aneuploidy has been associated with large size and with increasing degrees of dysplasia (Goh and Jass, 1986; Sciallero et al., 1988). These observations suggest that abnormal DNA content could represent a relevant marker in the proposed adenoma-to-carcinoma sequence; however, because polyps are routinely extirpated, clinical or epidemiologic approaches to the question of natural progression are of limited feasibility. As noted above, aneuploidy by no means signifies inevitability of disease progression (Auer et al., 1982), and the diploid and aneuploid histograms from adenomas cannot be distinguished readily from those generated by carcinomas. Thus the earliest point at which DNA aneuploidy occurs in tumor development is poorly defined. Sensitivity levels of current cytometric technology limit detection of aneuploidy mainly to lesions which unequivocally demonstrate abnormal nuclear features. Further, abnormalities of proliferation regularly *precede* the cytologic changes of malignancy in many cancers, including colorectal tumors (Deschner and Lipkin, 1975). These observations also support the possibility that alterations in cell cycling predispose to DNA content aberrations, as noted earlier.

Numerous reports address the clinical prognostic value of DNA measurements in colorectal cancer, with promising if not dramatic results (Wolley et al., 1982; Banner et al., 1985; Kokal et al., 1986; Emdin et al., 1987). However, a consensus on the clinical utility of DNA measurements has been slow to evolve in colorectal neoplasia for several reasons. First, although aneuploidy is correlated, albeit weakly, with clinical parameters such as higher Dukes' Stages (Crissman et al., 1989), the careful statistical analysis of follow-up that is essential is lacking from much of the published data, as is clear separation of colonic vs rectal tumors within individual series. Tumors arising from the distal bowel are clinically, epidemiologically, and possibly biologically different from right-sided lesions, as partially evidenced by their higher incidence of DNA aneuploidy (see below). Individual studies are extremely difficult to compare (as is apparent from Table 1) because of variations in selection of cases, particularly by stage, in follow-up intervals, and in flow-cytometric techniques. The definition of DNA aneuploidy employed is not uniform. Although some define aneuploidy as any measurable deviation from the control population, others require a minimum of 20% increase in DNA content. Estimates of S-phase fraction (SPF) are even more subject to methodologic variability, which is unfortunate given the potentially great prognostic impact of this parameter. The significance of ploidy is often eliminated following statistical correction for its association with higher proliferative fraction (Bauer et al., 1987).

Recent work (see Table 1) has consistently demonstrated a 10–30% survival advantage for diploid-range cases in stage-similar patients. This difference, however, becomes only marginally significant after statistical correction for correlation with factors such as proliferative fraction. In fact, these studies as a group strongly suggest that the biological aggressiveness attributed to DNA aneuploidy is largely mediated by increased proliferative activity. Apart from cell cycle analysis, relatively few attempts have been made to correlate cytometric DNA content of neoplastic cell populations with other genetic, functional, or molecular properties that might be expected to facilitate neoplastic progression. Exceptions to this generalization are found in the work of Remvikos and colleagues, who have provided significant correlations between cytogenetic and flow-cytometric find-

Table 1. Flow Cytometric Variables versus Outcome (Recurrence or Survival) in Colorectal Adenocarcinoma

(N)	*Dukes' Stage*	*% Aneuploid*	*Mean SPF*	*% Difference in Survival by:* Ploidy	SPF	*F/U*	*Author*
(60)	A + B⁺	78.3%	16.7%	NS*	39%	5 yr	Bauer
(86)	C⁺	64%	NA	32%	33%	52 mo mean	Schutte
(694)	B + C⁺	49%	10.8% diploid 19.1% aneuploid	12%	15%	5 yr	Witzig
(69)	C⁺	87%	17.6%	18%	50%	5 yr	Harlow
(45)	A – C⁺	78%	NA	47%	NA	30 mo mean	Russo
(236)	A⁺	52%	NA	1%	NA	2 yr minimum	Armitage
	B	59%	NA	20%	NA		
	C	68%	NA	10%	NA		
(115)	A – D⁺	62%	NA	38%	NA	75 mo	Giaretti
(100)	A – D⁺	63%	NA	33%	NA	6 – 10 yr	Rognum

Notes: ⁺A = partial thickness invasion
B = transmural invasion
C = regional node metastasis
D = hepatic metastasis
*NS = not significant, data not shown

ings and more recently between fractional allelic loss and DNA aneuploidy (1988a and b; 1991). These studies link abnormalities at the cellular and molecular levels, providing a functional basis for the apparent clinical importance of DNA content.

B. Chromosome Aberrations

Chromosome studies in adenomas of the colon support the correlation of advancing stages of malignancy with progressive increase in chromosome aberrations (Levin and Reichmann, 1986). Of 18 lesions classified as adenomas, 5 were cytogenetically normal, 5 had only numerical aberrations of which 3 were mosaic for normal cells, and 7 showed both numerical and structural abnormalities; all were in the diploid range. More recent studies of adenomas (Petersen et al., 1991) revealed non-clonal hyperploid cells in many lesions, although all but one had diploid or near-diploid modes. In another series of 25 polyps recurrent aberrations, consisting of trisomy 7 and 13, have been noted, and there was limited correlation of increasing dysplasia with the presence of chromosome aberrations (Longy et al., 1990).

Relatively few detailed banding analyses of colorectal carcinomas have been published (Becher et al., 1983; Ochi et al., 1983; Muleris et al., 1985, 1987; Levin and Reichmann, 1986), in many of which pertinent patient information (history, pathology, therapy, etc.) is lacking. Direct tumor analyses have not found consistent abnormalities but did show frequent trisomies for chromosomes #7, 8, 13, 19, 20 and 21, loss of #17, and structural aberrations of 1, 5 and 12 (Becher et al., 1983; Ochi et al., 1983; Levin and Reichmann, 1986). The largest series of adenocarcinomas included 9% of cases that were chromosomally normal, and another 12% showed only simple gains or losses of whole chromosomes (Levin and Reichmann, 1986). Preferential involvement of #17 in chromosomal recombination and monosomy for 17p and 18 were the most frequent karyotypic changes in 18 near-diploid colorectal tumors (Muleris et al., 1987, 1988).

The chromosomal observations in near-diploid colorectal tumors have been critical in the eventual molecular detection of several genes. The localization of the Familial Polyposis Coli (FPC) gene to chromosome 5 was stimulated by identification of an individual with FPC and multiple developmental anomalies who had an interstitial deletion of chromosome 5q (Herrera et al., 1986). Studies of polymorphic genomic probes assigned to chromosome 5 in FPC families demonstrated linkage with the C11p11 probe (Bodmer et al., 1987; Leppert et al., 1987), and a gene for FPC was localized to 5q21-q22, which is consistent with the chromosomal deletion originally described. More recently, a second colon cancer gene located nearby on 5q has been identified (Nishisho et al., 1991). The 17p deletion has been linked with loss of the TP53 tumor suppressor gene, and the 18q gene called DCC (deleted in colon cancer) has also been isolated recently (Fearon et al., 1990).

The significance of other molecular findings remains obscure. RAS mutations appear relatively early in tumor progression, more often in well-differentiated tumors that are clearly associated with adenomatous polyps (Laurent-Puig et al., 1991), and somewhat less commonly in more highly malignant tumors. Evidence for gene amplification in colorectal cancers is sporadic in primary tumors, but when permanent lines derived from colorectal cancer are examined, more than half have shown chromosomal evidence of amplification (Bruderlein et al., 1990) either as double minutes (dmin) or, in one case, as a homogeneously stained region (HSR). Three of 18 cultured lines were amplified for the ERBB2 oncogene, but no examples of amplification for *MYC*, *MYB*, *FOS* or *RAS* were found.

There is a developing awareness of at least two major subsets of colorectal malignancy that are characterized by rather different arrays of chromosome aberrations (Bardot et al., 1991). Moreover, the two subsets arise preferentially in different parts of the large bowel (Muleris et al., 1988). The first group, described as monosomic in type [MT], is found most commonly in the left colon and right-sided lesions are rare; it is associated with chromosome losses that most often involve chromosome 18, but also include loss of 17p and relative gain of 17q. Observations of endoreduplication are frequent. The other type, denoted TT or trisomic type, is rare in the left colon, with chromosome gains predominating, and a less uniform pattern of losses. Gains affecting chromosomes 5, 6, 7, and 12 are prevalent in TT tumors whereas duplications of 8q, 13, 17q, 20 and the early replicating X are common to both types (Muleris et al., 1990b). Another preferential region of chromosome aberration in MT tumors is 1p (where deletions were noted in approximately half the tumors analyzed) (Leister et al., 1990; Bravard et al., 1991). These alterations are assumed not to involve random segments; the lost or duplicated regions carry, among others, genes important in nucleotide synthesis, and the gains may affect salvage pathways that are important to cell survival.

Additional evidence of the non-random nature of chromosome losses and gains suggests that such distribution, independent of sites that are subject to recombination and selection, is similar among different epithelial tumors (Atkin and Baker, 1991). Colon tumors showed overrepresentation of chromosomes 7, 13, 20, and X and particularly low levels of chromosomes 17 and 18 relative to other tumor types examined, possibly because of their involvement in structural rearrangements. The preferential involvement in colon cancer of chromosomes 1, 8, 12, and 13 noted in several laboratories may mark sequences of tumor progression other than that based on the polyp-centered model, and that may be represented by different morphologic lesions, such as progression from flat adenomas.

V. BREAST CANCER

A. DNA Content and Cell Proliferation

In no other tumor system has the clinical and biological significance of cytometric DNA content been as thoroughly evaluated as in breast cancer. The large numbers of tumors studied have permitted classification of DNA content beyond the level of "normal" (i.e., diploid range) or "abnormal" (aneuploid) to distinguish tetraploid, hypertetraploid, and polyploid (multimodal) cases. The literature on breast carcinoma also includes a generous sampling of image analyses which provide different information, complementary to that from flow cytometry. A major difference between flow and image histograms is the relative frequency with which hypertetraploid cells are observed in the latter, particularly in otherwise normal histograms from malignant tumors (Auer et al., 1987). A significant number of image DNA histograms are characterized by diffuse DNA content abnormalities containing numerous hypertetraploid events, often without a detectable mode (see Figure 2) (ibid; Bocking et al., 1989). This diffuse, non-modal pattern of DNA distribution, also observed in tumors other than breast, has not been identified by flow-cytometric analysis for reasons which are unclear.

Numerous impressively detailed studies have compared DNA content with clinical outcome for invasive breast neoplasia. Breast cancers in the diploid range have been associated with better histologic differentiation and with positive hormone receptor status, which is itself evidence of differentiation. Large retrospective studies have shown marked differences in survival between diploid and aneuploid tumors (Auer et al., 1980; Levack et al., 1987; Dressler et al., 1988; Fallenius et al., 1988). These series regularly demonstrate a 15–35% survival advantage among

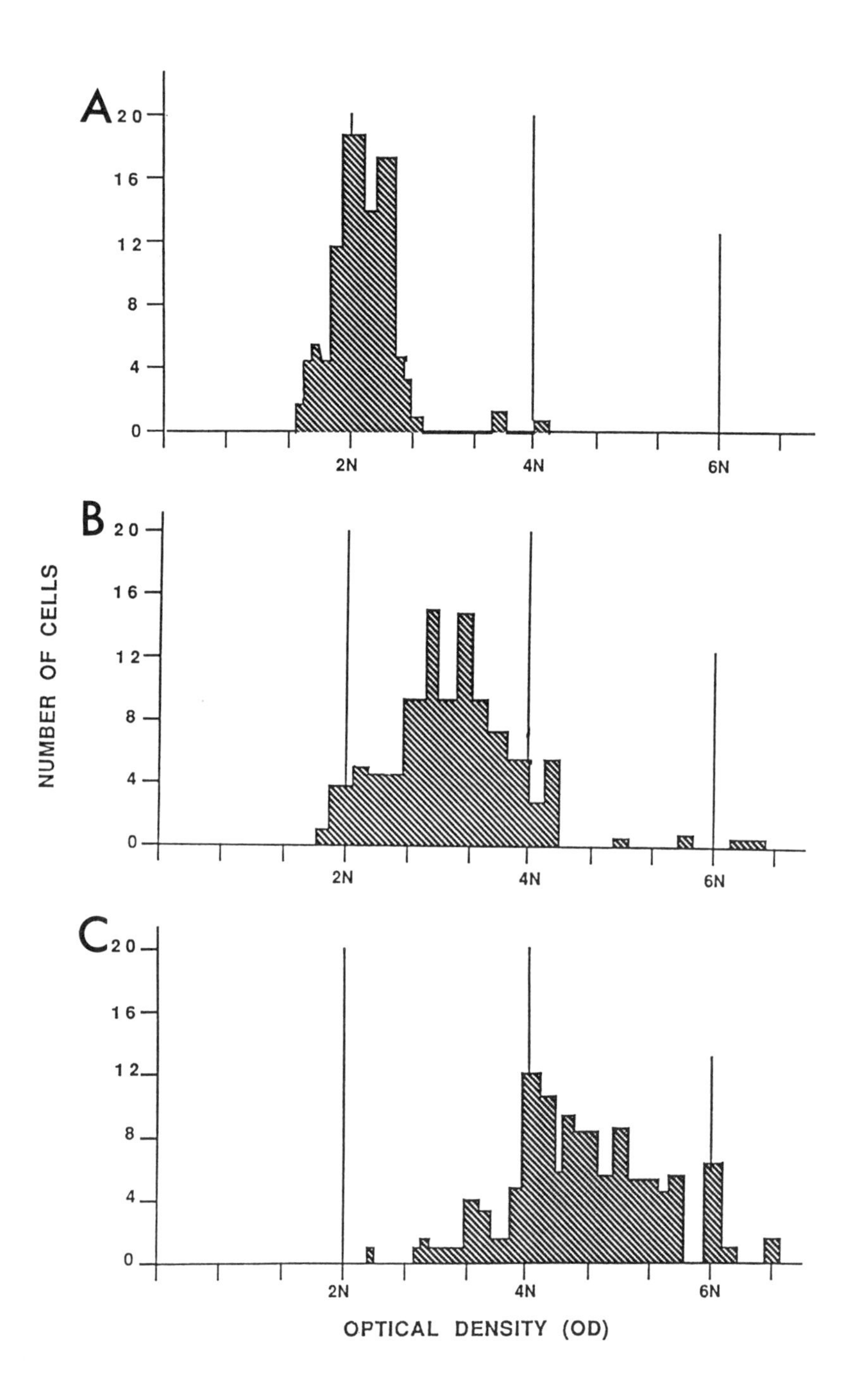
A
B
C
20
16
12
8
4
0
2N
4N
6N
NUMBER OF CELLS
OPTICAL DENSITY (OD)

diploid-range as contrasted with aneuploid, stage-matched cases. However, a careful review has emphasized that published data are contradictory with regard to correlation of ploidy patterns with disease staging, lymph-node status, and long-term survival, and that DNA measurement is not yet fully accepted as an independent prognostic tool in breast cancer (Koss et al., 1989).

As in the case of colon cancer cytometry, multivariate analyses have revealed that apparent differences between diploid and aneuploid cases are largely accounted for by the strong relationship between abnormal DNA content and proliferative fraction. Although diploid tumors generally show a far lower median value for SPF than do aneuploid tumors (Dressler et al., 1988), the diploid tumors with high SPF values are associated with survival reduced to levels approximating those of aneuploid lesions (Clark et al., 1989). Overall, although there is considerable overlap, the mean SPF of DNA-aneuploid breast carcinomas is approximately twice that of diploid-range tumors.

In contrast to the case of colorectal neoplasia, the association of poor prognosis in breast cancer with increased proliferation is well documented by various measures (mitotic rate, radioactive thymidine uptake), whereas other measures of proliferation, such as proliferating cell nuclear antigen (PCNA) or silver-stained nucleolar organizing regions (AgNOR), have not yet been correlated with survival for either tumor. It is unclear whether there are meaningful biologic differences between the two tumor systems or merely technical difficulties in estimating the growth fractions of colonic tumors. The clinical relevance of proliferative activity in neoplastic populations is reinforced by correlations with time to relapse and probability of metastasis. Rapidly growing breast tumors metastasize at a smaller size than those with lower kinetic indices (Tubiana and Koscielny, 1988), suggesting a profound influence of cell-cycling rate on biologic behavior.

Another difference between breast and colorectal DNA analyses is the relative sophistication of the interpretations of DNA histograms in relation to clinical

Figure 2. Image cytophotometric DNA analysis—Histograms derived from image analysis of Feulgen-stained neoplastic cells demonstrate significant differences from those from flow cytometric studies. Peaks are less well resolved with wider coefffi-cients of variation (CV), in part reflecting the smaller number of events (100–200) collected. Thus SPF estimates are generally not attempted. a) The G_0/G_1 peak mode is at 2N (the diploid DNA content established by measurements of normal lymphocytes) and this tumor is therefore in the diploid range. b) The G_0/G_1 peak is visibly shifted to a position between 2N (diploid) and 4N (tetraploid) that is characteristic of a hyperdiploid population. c) The population shown contains numerous tetraploid range and hypertetraploid events, without a well-defined peak mode. Hypertetraploid events are rare in image analytic measurements, but are not observed by flow-cytometric analyses for reasons which are not understood.

outcome. Breast cancers with near-tetraploid histograms are associated either with survival rates similar to those with near-diploid tumors or intermediate between those for diploid-range and aneuploid tumors (Coulson et al., 1984). In contrast, cases with cytometrically hypertetraploid populations have, on the whole, worse survival than cases with hyperdiploid modal DNA indices (ibid). This finding is corroborated by image analysis studies in which groups of patients whose samples show numerous hypertetraploid events also demonstrate poorer survival than those with DNA-aneuploid tumors (Auer et al., 1980; Fallenius et al., 1988; Bocking et al., 1989). These data are particularly interesting in that they provide empiric support for the notion that large-scale genetic events detectable at the cellular level are associated with and may facilitate aggressive behavior.

The inaccessibility and microanatomic complexity of breast tissues have severely restricted the application of DNA analysis to early events in the evolution of breast tumors. Similarly to other epithelial neoplasms, ductal carcinoma *in situ* clearly demonstrates an incidence of aneuploidy similar to that of invasive carcinoma. DNA contents of epithelial proliferations with some, but not all, features of neoplasia, such as atypical hyperplasias (AH), have been analyzed in a few studies, with aneuploidy detected in 15–30% of lesions (Uccelli et al., 1986; Carpenter et al., 1987). These data, in conjunction with studies demonstrating abnormal oncogene expression, suggest that AH represents a form of dysplasia, albeit low grade and clinically heterogeneous. It is noteworthy that reported DNA-content abnormalities in AH are generally subtle (consisting of small hyperdiploid subpopulations), which is not surprising in light of the relatively minor nuclear cytologic changes that characterize the lesions. Unfortunately, no studies have yet correlated DNA-content abnormalities with risk of subsequent malignancy for these proliferative lesions.

B. Chromosome Aberrations

The search for specificity of chromosomal aberrations in breast cancer has yielded many candidates but is still highly controversial. A review of the 1988 edition of Mitelman's *Catalog of Chromosome Aberrations in Cancer* revealed that band 1p13 was the locus most frequently involved in structural changes (Mitchell and Santibanez-Koref, 1990). That edition of the Catalog includes over 100 examples of breast cancer, but fewer than 70 represented solid tumors rather than effusion fluids, and many of the solid tumors were metastatic rather than primary. Lesions at several other locations on chromosome 1 (1p11, 1q23, 1q21) also appeared relatively common in the pooled data.

The emphasis on chromosome 1 aberration was reinforced by the addition of 14 new cases of *in situ* and invasive primary lesions in which the most common finding was rearrangement at 1p13, although alterations in 16q22 and deletions of 17p were also significant (Mitchell and Santibanez-Koref, 1990). However, a comparable recent study yielded meaningful data on 26 cases in which the most common

Table 2. Prognostic Significance of DNA Content and Proliferative Fraction in Breast Carcinoma

			% Difference in Survival by:			
(N) Stage(s)	*% Aneuploid*	*Median SPF*	*Ploidy*	*SPF*	*F/U*	*Author*
(101) N*–	54%	12.5%	8%	15%	4.3 yr median	Muss
(345) N–	68%	5.2%	14%	20%+	39 mo median	Clark
(472) multiple	63%	7.7%	9%	33%	7 yr	Stal
(222) multiple	74%	12% (mean)	19%	18%	22 yr	Joensuu
(198) N–	64%	4.5% (diploid)	10%	32%+ (aneuploid)	80 mo median	Winchester
(398) multiple	57%	3%(diploid) 6% (aneuploid)	0%	14%	10 yr	Fisher
(197) multiple	52%	8%	8%	3%	3.6 yr median	Kute

Notes: *N– = no axillary node metastases
+ diploid range cases

breakpoint was 1p36 (Hainsworth et al., 1991), considerably distant from that of previous reports. Additional and equally confusing support for frequent aberrations of chromosome 1 is evident from other cytogenetic (Gerbault-Seureau et al., 1987; Dutrillaux et al., 1990; Devilee et al., 1991) and molecular studies. Molecular evidence of loss of heterogeneity (the demonstration that two alleles of a locus are polymorphic in normal host tissues and monomorphic in the tumor) has become the most widely used approach to identification of tumor suppressor genes. Although several locations on chromosome 1 have shown loss of heterozygosity (LOH) for particular genes or genetic loci, the losses reported in the 1p32-36 region (Genuardi et al., 1989; Gendler et al., 1990; Bieche et al., 1990), and at 1q21 and distal 1q sites (Chen et al., 1989; Gendler et al., 1990), do not correspond to the chromosomal sites of greatest recombination.

The prevalence of normal diploid cells in cultured primary breast cancers is a persistent problem, and almost every series of tumors studied by direct analysis includes both cases that are exclusively diploid and cases with diploid cells as well as cytogenetically aberrant cells (Hainsworth et al., 1991; Pandis et al., 1992). Clonal structural aberrations are infrequent in short-term culture studies (Wolman et al., 1985; Zhang et al., 1989; Geleick et al., 1990), and when present usually compose a subpopulation within the culture. In contrast, non-clonal aneuploid cells appear generally more common in primary cancer cultures (Geleick et al., 1990), as they do in direct preparations where it is often difficult to find two karyotypically identical cells within the same tumor (Mitchell and Santibanez-Koref, 1990; Hainsworth et al., 1991).

Heterogeneity of a different type is displayed by the detection of non-overlapping patterns of chromosomal aberrations within the same tumor (Pandis et al., 1992).

Both clonal and non-clonal aberrant cells were observed, with a substantial number of cases demonstrating polyclonality. From other reports (Mitchell and Santibanez-Koref, 1990; Hainsworth et al., 1991), it has not been clear whether the tumors with multiple aberrations demonstrated clonality, or polyclonality with several non-overlapping stemlines, except for one case where clonal evolution was traced in detail (Mitchell and Santibanez-Koref, 1990). However, polyclonality was not observed in another series of 52 cases; and in the 40 cases with abnormal karyotypes, although there was great variation among individual cells of a case, overlap of aberrations among cells within individual tumors was consistent with clonal tumor origin (Slovak et al., in press).

Another recurrent "hotspot" for chromosome rearrangement, also buttressed by molecular LOH at loci on 11p, is chromosome 11. A recent report focused on non-random abnormalities of chromosome 11, and found rearrangements at several loci in 20 of 30 cases examined after direct harvest of primary tumors (Ferti-Passantonopoulou et al., 1991). However, it is difficult to evaluate this work: No indications were given of the frequency of all aberrations or of other presumptively specific aberrations; the ploidy levels are not stated; nor is there information on clonality. Thus, the preferential involvement of chromosome 11 could, in theory, be less than that of several other chromosomes.

Lesions involving chromosome 16 have been identified frequently in near-diploid metaphases, including rearrangements of 16q or loss of 16 (Rodgers et al., 1984; Dutrillaux et al., 1990; Mitchell and Santibanez-Koref, 1990). Moreover, molecular studies have shown LOH at several chromosome 16 loci in breast tumors (Devilee et al., 1989; Larsson et al., 1990). Because few or no other chromosome aberrations were present, it is tempting to think that the 16q alteration is a "primary" change, specific to breast cancer. In any event, it is likely to represent either a very early (initiating?) change or to mark a biologically distinct subset of breast cancers.

Chromosomal evidence of gene amplification in the forms of double minutes (dmin) or homogeneously stained regions (HSR) has been reported with varying frequency in breast cancer. Some observers have reported high incidences of dmin in primary tumors (48%) (Gebhart et al., 1986) and others have found HSRs or abnormally banded regions (ABRs) in relatively large numbers of cases (Dutrillaux et al., 1990). Chromosome aberrations indicative of gene amplification are of particular concern because of evidence of genomic amplification in breast cancer, particularly of the *ERBB2*, *INT2*, and *MYC* oncogenes. However, examination for amplification of these and other oncogenes in a series of tumors with HSRs failed to reveal either degrees of amplification consistent with the size of the HSRs or localization to the HSR (Saint-Ruf et al., 1990). Even a mammary cancer cell line with multiple copies of the HST gene (a member of the fibroblast growth factor family) and an HSR at the appropriate locus, 11q13, did not reveal transcripts of HST or the related INT2 genes (Lafage et al., 1990)

The new probes that enable identification and localization of chromosomes within interphase nuclei by hybridization to regions of homology within DNA have been and will continue to prove a rich source of information in breast and other cancers. The chromosome-specific repetitive centromeric probes are a powerful adjunct to metaphase analysis, permitting assessment of aneuploidy in poorly mitotic tumors and of cytogenetic heterogeneity within tumor cell populations. Probes for the centromeric regions of chromosome 18 and the X, when applied to the study of breast cancer, revealed both aneuploidy and heterogeneity of the tumor (Devilee et al., 1988). Such probes will also react with structurally altered chromosomes in tumor cells and could potentially be used to determine the component chromosomes of origin in recombinant markers. These probes are not confined to the highly repetitive centromeric regions but can also be utilized to examine specific genes. For example, a probe for the ERBB2 gene demonstrated levels of amplification consistent with Southern-blot results in breast cancer cell lines (Kallioniemi et al., 1992). Moreover, suitable labeling has demonstrated not only heterogeneity of chromosome composition but also that, in some breast cancers, proliferation may be selectively associated with the aberrant cells (Waldman et al., 1992). In this study, 6 chromosomal centromere-specific probes were used and 22 breast cancers were examined. Although monosomy was infrequent, and many of the tumors examined were DNA-diploid, the extent of cytogenetic heterogeneity found was impressive. Each tumor showed heterogeneity of chromosome copy number (defined as 2 or more different subpopulations each comprising at least 20% of the total cells) for at least one of the 6 chromosomes. Individual tumors displayed cell clusters that were uniformly aberrant, corresponding to the regions from which touch preparations (representative of the cut surfaces of the tumor) were made. A more rigorous demonstration of isolated, chromosomally variant domains within a tumor is possible by application of probes to tissue sections (Wolman et al., 1992).

Partial or complete loss of chromosome 17 has been associated with increased labeling index, indicative of cell replication (Waldman et al., 1992) and with tumors with high proliferation indices (Merlo et al., 1992). There is ample evidence that selective loss of loci on 17p is the most common molecular aberration in breast and many other tumors (Sato et al., 1991; Coles et al., 1990). It now appears that several tumor suppressor loci are present in this region (Coles et al., 1990; Sato et al., 1991; Biegel et al., 1992) in addition to the p53 gene. Some of these may not act as recessive tumor suppressor genes in that a normal allele is still present. The linkage of loss with altered cell proliferation is likely to constitute an important mechanistic basis for the near-ubiquity of chromosome 17 aberrations in tumors.

Fluorescence *in situ* hybridization [FISH] (Figure 3) with chromosome-specific probes can be used to study paraffin-embedded tumors, either after disaggregation (Wolman et al., 1993) or in thin sections which permit focused analysis and precise correlation of genetic alterations with tumor extent and morphology (Wolman et al., 1992). The problems of tumor heterogeneity and admixture with benign epithelium, stroma, or inflammatory cells can be circumvented, and small localized lesions become accessible to genetic evaluation. This approach will probably

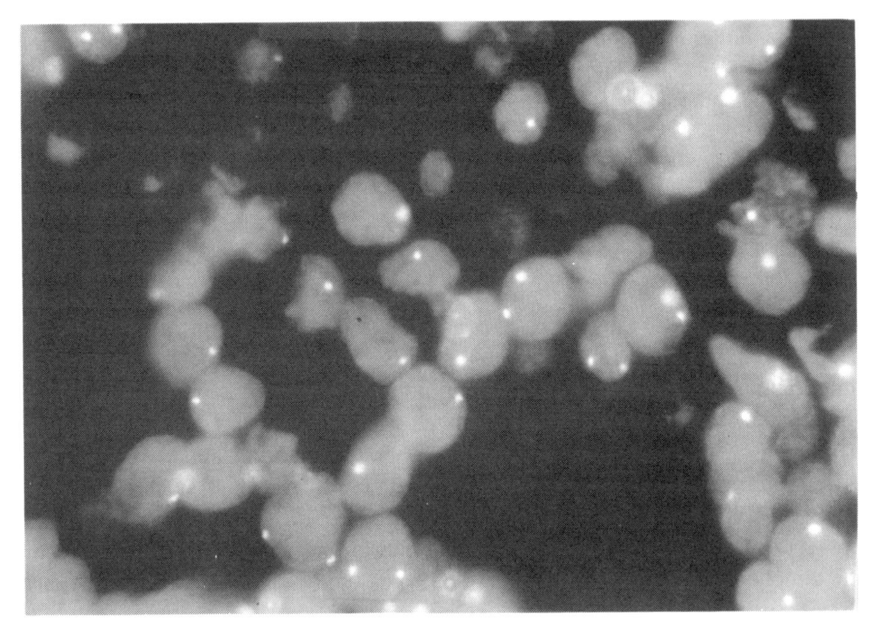

Figure 3. Fluorescence *in situ* hybridization—Detection of chromosome copy number aberrations in a 4 μ section of breast tissue with atypical adenomatous hyperplasia. The section has been hybridized with a centromeric probe for chromosome 17 (D17Z1, Oncor). The cluster of cells at center [R] shows 2 spots per nucleus, but an acinar array [L center] shows a single signal per nucleus indicating monosomy for this chromosome. (A few signals that are below the plane of section show halos.) The expected frequency of disomic signal in this type of preparation is in the range of 80–90%.

provide more compelling evidence of the extent of genetic heterogeneity than is currently available.

VI. COMMENTS AND CONCLUSIONS

At present, the accepted model for colon carcinogenesis is a series of molecular alterations linked with recognizable histologic stages in tumor development taking place over many years in a given cell lineage. There is also a suggested alternative pathway to that of the adenomatous-polyp-to-carcinoma molecular model, manifested by another set of morphogenetic steps. In human breast cancer development, the morphologic evidence for similar genetic pathways is less clearly defined, and for many other tumors it is entirely lacking. Nevertheless, accumulation of more-or-less specific sets of genetic aberrations is perceived to be responsible for individual tumor types, such as breast (Sato et al., 1991). The earliest events in tumorigenesis may not be specific, however. Genetic instability remains a hallmark of cancer and its precursor lesions, reflected in changes in chromosome number, due to reduplication of chromosomes, tetraploidy, and considerable variation in chromosome number from cell to cell within a given tumor. Its role is essential in permitting stepwise tumor evolution.

It is assumed, as well as indicated in the morphologic progression models cited above, that critical steps in tumor progression (e.g., host invasion, metastasis) are a direct consequence of specific genetic events. The extent to which these critical steps are reflected in cytophotometrically detectable DNA content shifts has been largely unexplored, principally because the available technology is not well adapted to a microdissection approach (necessary to compare *in situ* versus invasive neoplasia). Neither the mechanism(s) nor the sequence(s) of DNA-aneuploid clone generation are well understood. Viral transformation experiments suggest that induction of tetraploidy via interference with the mitotic apparatus or process may be an important initiating event. Subsequent loss of genetic material would then produce clones with hyperdiploid DNA content. Cytogenetic studies employing marker chromosome analysis often indicate that different DNA stemlines within a given tumor have a common predecessor, or ancestral clone (Slovak et al., in press). Thus, genetic evolution, as opposed to multicentric origin, may contribute to the intratumoral heterogeneity of some tumors. The presumption of evolution in tumors would be that of increasing deviation from the norm, i.e., diploid → multiples of haploid (e.g., tetraploid) → aneuploid. However, an alternative pattern of evolution, i.e., diploid → hypoploid (aneuploid) → polyploid, has also been proposed based on the duplication of rearranged chromosomes observed in breast and colorectal tumors (Muleris et al., 1990a; Dutrillaux et al., 1991).

Evidence of progressive deviation of DNA content is weakened, however, by observations that preinvasive conditions (dysplasias, *in situ* carcinomas) from a variety of sites (bladder, uterine cervix, larynx) display the full spectrum of DNA content anomalies, including polyploidy and hypertetraploidy, found in cancers (Bjelkenkrantz et al., 1983; Bocking et al., 1985; Hofstädter et al., 1986; Bibbo et al., 1989). Further, regression or stability of cytophotometrically aneuploid pre-invasive lesions has been documented in studies of bronchial and cervical epithelium (Auer et al., 1982). Thus, DNA aneuploidy is not necessarily irreversible in neoplastic progression. By contrast, cytogenetic studies of premalignant lesions are usually characterized by absent or relatively minor degrees of cytogenetic aberration in comparison with the more prevalent and more extensive aberrations of invasive tumors.

The premise that human solid tumors progress via functional aberrations, acquired through a number of discrete genetic events, implies that intratumoral DNA-content heterogeneity is an important property of neoplasms. To the extent that significant shifts in DNA content accompany phenotypic changes, DNA analysis represents a means to study tumor progression at the cellular level.

Approaches to the question of progression have entailed multisite sampling of primary neoplasms and comparisons of histograms from primary tumors with synchronous nodal metastases or recurrences. In both breast and colorectal neoplasia, ploidy determinations from various locations within a tumor, or between the primary and its nodal metastasis, are concordant in 70–85% of cases. This does not imply, however, that most tumors of breast or colon are cytometrically monoclonal. Multiple DNA stemlines are detectable in approximately 50–60% of these tumors, with 10–15% having multiple aneuploid stemlines and the remainder showing an aneuploid mode in combination with diploid- or tetraploid-range populations.

The coexistence of differing populations in most colon and breast tumors could reflect emergence of one from another, or independent origins. Current dogma in tumor biology claims that a more aggressive population should become numerically dominant as the result of clonal expansion (Nowell, 1976), but in view of the concordances described above, it is more likely that coexistence persists over time (see Chapter VII). Several other caveats to generalizations about neoplastic progression are suggested by data on clonal DNA content abnormalities. First, some tumors, notably adenocarcinoma of the prostate, exhibit heterogeneity of DNA content in virtually all cases, with low concordance of modal DNA content between primary tumor and nodal metastasis (Peters et al., 1991). Prostate tumors are slow-growing, multifocal, and notoriously heterogeneous in morphology. Undifferentiated, presumably aggressive foci are localizable geographically within these tumors but could be interpreted to represent lesions of independent origin rather than the result of clonal evolution. Second, non-diploid stemlines do not necessarily represent the "most aggressive" population within a tumor. In carcinoma of the breast and other tumors, diploid-range recurrences of nodal metastases have been observed with DNA-aneuploid primary tumors, sometimes attributed to differential susceptibility of the aneuploid clones to radiation or cytotoxic therapy (Ensley et al., 1989). Location may be a more important determinant of nodal metastasis or recurrence than aggressiveness. Recent reports have correlated the modal DNA index of regional lymph node metastases with that of superficially located cells from the primary colon cancer (Kim et al., 1991). By contrast, the DNA indices of hepatic metastases correlated best with those of samples from the deepest point of penetration of the primary (ibid).

To appreciate aggregate DNA-content shifts (a form of clonal evolution) adequately, it is probably necessary to define the various components of a neoplasm by careful microdissection. Flow cytometry, which requires a relative abundance of tissue, and karyotype analysis, which is essentially morphologically blind, are less than optimal for such assessment. The conditions described may be met for image cytophotometry studies of carcinoma of the uterine cervix, where the lesions are small and readily sampled and the analytic approach permits isolation of small cell groups. Analyses have shown not only DNA content shifts between early invasive and corresponding surface components in the majority of tumors, but also that, in most cases, the microinvasive component shows a lower modal DNA index than the overlying *in-situ* disease (Fu et al., 1980). Thus, neither the existence nor the direction of clonal evolution is well supported by this and other studies cited above.

In other respects, the data now emerging present a picture of far greater heterogeneity than was appreciated with the previously available tools for investigation: polyclonality for many epithelial tumors has been reported. The application of FISH to investigation of aberrations in thin tissue sections as well as whole cell direct preparations and touch preparations has revealed extensive intratumoral regional heterogeneity for chromosome copy number and a remarkable diversity of chromosome aberrations among histologically similar lesions. The evidence of marked disparities in chromosome number from FISH studies on touch preparations of breast cancers (Waldman et al., 1993) is reinforced by recent metaphase studies that have found multiple cytogenetically unrelated clones in many breast cancers (Pandis et al., 1993). If, as has been suggested (Wolman and Heppner, 1992), the baseline frequencies of LOH reflect levels of genetic instability that are considerably greater for other tumors such as colon or lung than is generally true for breast cancer, then we should expect that these tumors, when better characterized cytogenetically, will also display great heterogeneity and polyclonality. The assumption, when an aberration is manifest in a large fraction of the tumor cell population, is that there is an associated growth or survival advantage. However, diversity itself may be advantageous to tumor survival since it provides flexibility to meet the challenges of different microenvironments within the disparate tissues of the host.

Instability should not be assumed from observations of heterogeneity alone; but there is ample evidence of emergence of new populations within tumors, both *in vivo* and *in vitro*, to justify assumptions of inherent genomic instability. There is evidence that colonic cancers show mucosal alterations distant from the tumor site, possibly supporting a widespread "field" of premalignant alteration from which many atypical clones could arise and merge. It is possible that the clonality that appears to characterize many tumors, and that has dominated our thinking with respect to tumor origin, may instead be largely a result of tumor evolution. If that is the case, then tumorigenic insult commonly results in a wide variety of genetic disruptions at the genic and chromosomal levels, and selection is the critical factor in the *apparent* clonality of tumors.

In summary, although few studies have dealt with the relationship between karyotypic and cytometric changes directly, most of the available data are mutually supportive (Tribukait et al., 1986; Remvikos et al., 1988a,b). However, the phenotypic correlates of these genetic measures are speculative and currently often appear inconsistent. Preinvasive lesions appear to be characterized by a full spectrum of DNA content alterations, but the extent of cytogenetic aberration is more clearly linked with malignant progression. Moreover, the different primary sites examined exhibit similarities of quantitative DNA abnormalities but have shown substantial differences in their accessibility to karyotypic assessment and in the probable localization of critical defects. The quantitative alterations at both sites appear closely linked to abnormalities of proliferation, and the probable underlying similarity of mechanism is likely to depend on increased genetic instability. The chromosomal and molecular tumor-specific changes then result from selection, based on host- and tissue-related factors. The heterogeneity resulting from genetic instability is a prominent and consistent attribute of malignant and premalignant lesions, and this attribute may be better perceived by analytic approaches at the cellular than at the molecular level. These observations place important constraints upon interpretations of molecular alterations found in tumors. Further, they suggest that the critical tumor-initiating events are those that result in destabilization of the genome.

NOTE

*The cytogenetic nomenclature used is based on an International System for Human Cytogenetic Nomenclature (1985) published in collaboration with the March of Dimes Birth Defects Foundation and Cytogenetics and Cell Genetics by S. Karger, Basel, Switzerland, and modified for Tumor Cytogenetics, 1991.]

REFERENCES

Adachi, M, Muto, T., Morioka, Y., Ikenaga, T., & Hara, M. (1988). Flat adenoma and flat mucosal carcinoma (?ii b type). A new precursor of colorectal carcinoma? Report of two cases. Dis. Colon Rectum 31, 236–243.

An International System for Human Cytogenetic Nomenclature (1985). (Harnden D.G., & Klinger, H.P., eds.) S. Karger A.G., Basel and New York. Supplement—Guidelines for Cancer Cytogenetics (1991). (Mitelman, F., ed.).

Armitage, N.C.M., Ballantyne, K.C., Sheffield, J.P., Clarke, P., Evans, D.F., & Hardcastle, J.D. (1991). A prospective evaluation of the effect of tumor cell DNA content on recurrence in colorectal cancer. Cancer 67, 2599–2604.

Atkin, N.B. & Baker, M.C. (1991). Numerical chromosome changes in 165 malignant tumors. Evidence for a nonrandom distribution of normal chromosomes. Cancer Genet. Cytogenet. 52, 113–121.

Auer, G.U., Askensten, U., Erhardt K., Fallenius A., & Zetterberg, A. (1987). Comparison between slide and flow cytophotometric DNA measurements in breast tumors. Anal. Quant. Cyt. and Histol. 9, (2)138–146.

Auer, G.U., Caspersson, T.O., & Wallgren, A.S. (1980). DNA content and survival in mammary carcinoma. Anal. Quant. Cytol. 2, 161–165.

Auer, G.U., Ono, J., Nasiell, M.J., Caspersson, T., Kato, H., Konaka, C., & Hayata, Y. (1982). Reversibility of bronchial cell atypia. Can. Res. 42, 4241–4247.

Babu, V.R., Miles, B.J., Cerny, J.C., Weiss, L., & Van Dyke, D.L. (1990). Cytogenetic study of four cancers of the prostate, Cancer Genet. Cytogenet. 48, 83–87.

Balaban, G.B., Herlyn, M., Clark, W.H., & Nowell, P.C. (1986). Karyotypic evolution in human malignant melanoma. Cancer Genet. Cytogenet. 19, 113–122.

Banner, B.F., Tomas-De La Vega, J.E., Roseman, D.L., & Coon, J.S. (1985). Should flow cytometric DNA analysis precede definitive surgery for colon carcinoma? Ann. Surg. 202, 740–744.

Bardot, V., Luccioni, C., LeFrancois, D., Muleris, M., & Dutrillaux, B. (1991). Activity of thymidylate synthetase, thymidine kinase and galactokinase in primary and xenografted human colorectal cancers in relation to their chromosomal patterns. Int. J. Cancer, 47, 670–674.

Bauer, K.D., Lincoln, S.T., Vera-Roman, J.M., Wallemark, C.B., Chmiel, J.S., Madurski, M.L., Murad, T., & Scarpelli, D.G. (1987). Prognostic implications of proliferative activity and DNA aneuploidy in colonic adenocarcinomas. Lab. Invest. 57(3), 329–335.

Becher, R., Gibas, Z., & Sandberg, A.A. (1983). Involvement of chromosomes 7 and 12 in a large bowel cancer: Trisomy 7 and 12q. Cancer Genet. Cytogenet. 9, 329–332.

Beerman, H., Bonsing, B.A., van de Vijver, M.J., Hermans, J., Kluin, P.H., Caspers, R.J., van de Velde, C.J.H., & Cornelisse, C.J. (1991). DNA ploidy of primary breast cancer and local recurrence after breast-conserving therapy. Br. J. Cancer 64, 139–143.

Berrozpe, G., Miro, R., Caballin, M.R., Salvador, J., & Egozcue, J. (1990). Trisomy 7 may be a primary change in non-invasive transitional cell carcinoma of the bladder. Cancer Genet. Cytogenet. 11, 429–439.

Bibbo, M., Dytch, H.E., Alenghat, E., Bartels, P.H., & Wied, G.L. (1989). DNA ploidy profiles as prognostic indicators in CIN lesions. Am. J. Clin. Pathol. 92, 261–265.

Bieche, I., Champeme, M.H., Merlo, G., Larsen, C.J., Callahan, R., & Lidereau, R. (1990). Loss of heterozygosity of the L-myc oncogene in human breast tumors. Human Genet. 85, 101–105.

Biegel, J.A., Burk, C.D., Barr, F.G., & Emanuel, B.S. (1992). Evidence for a 17p tumor related locus distinct from p53 in pediatric primitive neuroectodermal tumors. Can. Res. 52, 3391–3395.

Bigner, S.H., Mark, J., Bullard, D.E., Mahaley, M.S., & Bigner, D.D. (1986). Chromosomal evolution in malignant human gliomas starts with specific and usually numerical deviations. Canc. Genet. Cytogenet. 22, 121–135.

Bjelkenkrantz, K., Lundgren, J., & Olofsson, J. (1983). Single-cell DNA measurements in hyperplastic., dysplastic and carcinomatous laryngeal epithelia, with special reference to the occurrence of hypertetraploid cell nuclei. Anal. Quant. Cytol. 5, 184–188.

Bocking, A., Auffermann, W., Vogel, H., Schlondorff, G., & Goebbels, R. (1985). Diagnosis and grading of Malignancy in squamous epithelial lesions of the larynx with DNA cytophotometry. Cancer 56, 1600–1604.

Bocking, A., Chatelain, R., Biesterfeld, S., Noll, E., Biesterfeld, D., Wohltmann, D., & Goecke, C. (1989). DNA malignancy grading in breast cancer. Analytical and Quantitative Cytology and Histology 11(2), 73–80.

Bodmer, W.F., Bailey, C.J., Bodmer, J., Bussey, H.J.R., Ellis, A., Gorman, P., Lucibello, F.C., Murday, V.A., Rider, S.H., Scambler, P., Sheer, D., Solomon, E., & Spurr, N.K. (1987). Localization of the gene for familial adenomatous polyposis on chromosome 5. Nature 328, 614–616.

Bravard, A., Luccioni, C., Muleris, M., Lefrancois, D., & Dutrillaux, B. (1991). Relationships between UMPK and PGD activities and deletions of chromosome 1p in colorectal cancers. Canc. Genet. Cytogenet. 56, 45–50.

Bruderlein, S., van der Bosch, K., Schlag, P., & Schwab, M. (1990). Cytogenetics and DNA amplification in colorectal cancers. Genes, Chrom. & Canc. 2, 63–70.

Burt, R.W., Bishop, D.T., Cannon, L.A., Dowdle, M.A., Lee, R.G., & Skolnick, M.H. (1985). Dominant inheritance of adenomatous colonic polyps and colorectal cancer. New Engl. J. Med. 312, 1540–1544.

Cannon-Albright, L., Skolnick, M.H., Bishop, D.T., Lee, R.G., & Burt, R.W. (1988). Common inheritance of susceptibility to colonic adenomatous polyps and associated colorectal cancers. New Engl. J. Med. 319, 533–537.

Carpenter, R., Gibbs, N., Matthews, J., & Cooke, T. (1987). Importance of cellular DNA content in premalignant breast disease and pre-invasive carcinoma of the female breast. Brit. J. Surg. 74, 905–906.

Carter, B.S., Ewing, C.M., Ward, W.S., Treiger, B.F., Aalders, T.W., Schalken, J.A., Epstein, J.I., & Isaacs, W.B. (1990). Allelic loss of chromosomes 16q and 10q in human prostate cancer. Proc. Natl. Acad. Sci. USA 87, 8751–8755.

Chen, L.C., Dollbaum, C., & Smith, H.S. (1989). Loss of heterozygosity in human ductal breast tumors indicates a recessive mutation on chromosome 1q. Proc. Natl. Acad. Sci. USA 86, 7202–7204.

Clark, G.M., Dressler, L.G., Owens, M.A., Pounds, G., Oldaker, T., & McGuire, W.L. (1989). Prediction of relapse or survival in patients with node-negative breast cancer by DNA flow cytometry. New Engl. J. Med. 320, 627–633.

Coles, C., Thompson, A.M., Elder, P.A., Cohen, B.B., Mackenzie, I.M., Cranston, G., Chetty, U., Mackay, J., Macdonald, M., Nakamura, Y., Hoyheim, B., & Steel, C.M. (1990). Evidence implicating at least two genes on chromosome 17p in breast carcinogenesis. The Lancet 336, 761–763.

Coulson, P.B., Thornthwaite, J.T., Woolley, T.W., Sugarbaker, E.V. & Seckinger, D. (1984). Prognostic indicators including DNA histogram type, receptor content, and staging related to human breast cancer patient survival. Can. Res. 44, 4187–4196.

Crissman, J.D., Zarbo, R.J., Ma, C.K., & Visscher, D.W. (1989). Histopathologic parameters and DNA analysis in colorectal adenocarcinomas. In: Pathology Annual, (Rosen, P.P., Fechner, R.E., eds.), Vol. 24, pp. 103–147. Appleton and Lange, East Norwalk, CT.

Croce, C. (1987). Role of chromosome translocation in human neoplasia. Cell 46, 155–156.

Deschner, E.E. & Lipkin, M. (1975). Proliferative patterns in colonic mucosa in familial polyposis. Cancer 35, 413–418.

Desigan, G., Wang, M., Alberti-Flor, J., Dunn, G.D., Halter, S., & Vaughan, S. (1985). De novo carcinoma of the rectum. A case report. Am. J. Gastroenterol. 80, 553–556.

Devilee, P., van Vliet, M., Bardoel, A., Kievits, T., Kuipers-Dijkshoorn, N., Pearson, P.L., & Cornelisse, C.J. (1991). Frequent somatic imbalance of marker alleles for chromosome 1 in human primary breast carcinoma. Cancer Res. 51, 1020.

Devilee, P., van den Broek, M., Kuipers-Dijkshorn, N., Kolluri, R., Khan, P.M., Pearson, P.L., & Cornelisse, C.J. (1989). At least four different chromosomal regions are involved in loss of heterozygosity in human breast carcinoma. Genomics 5, 554–560.

Devilee, P., Thierry, R.F., Kievits, T., Kolluri, R., Hopman, A.H.N., Willard, H.F., Pearson, P.L., & Cornelisse, C.J. (1988). Detection of chromosome aneuploidy in interphase nuclei from human primary breast tumors using chromosome-specific repetitive DNA probes. Cancer Res. 48, 5825–5830.

Dressler, L.G., Seamer, L.C., Owens, M.A., Clark, G.M., & McGuire, W.L. (1988). DNA flow cytometry and prognostic factors in 1331 frozen breast cancer specimens. Cancer 61, 420–427.

Dutrillaux, B., Gerbault-Sereau, M., & Zafrani, B. (1990). A comparison of 30 paradiploid cases with few chromosome changes. Canc. Genet. Cytogenet. 49, 203–217.

Dutrillaux, B., Gerbault-Sereau, M., Remvikos, Y., Zafrani, B., & Prieur, M. (1991). Breast cancer genetic evolution, I. Data from cytogenetics and DNA content. Breast Canc. Res. Treat. 19(3), 245–255.

Emdin, S.O., Stenling, R., & Roos, G. (1987). Prognostic value of DNA content in colorectal carcinoma, A flow cytometric study with some methodologic aspects. Cancer 60, 1282–1287.

Ensley, J.F., Maciorowski, Z., Hassan, M., Pietraszkiewicz, H., Heilbrun, L., Kish, J.A., Tapazoglou, E., Jacobs, J.R., & Al-Sarraf, M. (1989). Cellular DNA content parameters in untreated and recurrent squamous cell cancers of the head and neck. Cytom. 10(3), 334–338.

Fallenius, A.G., Auer, G.U. & Carstensen, J.M. (1988). Prognostic significance of DNA measurements in 409 consecutive breast cancer patients. Cancer 62, 331–341.

Farber, E. & Rubin, H. (1991). Cellular adaptation in the origin and development of cancer. Canc. Res. 51, 2751–2761.

Fearon, E.R. & Vogelstein, B. (1990). A genetic model for colorectal tumorigenesis. Cell 61, 759–767.

Fearon, E.R., Cho, K.R., Nigro, J.M., Kern, S.E., Simons, J.W., Ruppert, J.M., Hamilton, S.R., Preisinger, A.C., Thomas, G., Kinzler, K.W., & Vogelstein, B. (1990). Identification of a chromosome 18q gene that is altered in colorectal cancers. Science 247, 49–56.

Fenoglio, C.M. & Lane, N. (1974). The anatomical precursor of colorectal carcinoma. Cancer 34, 819–823.

Ferti-Passantonopoulou, A., Panani, A.D., & Raptis, S. (1991). Preferential involvement of 11q23-24 and 11p15 in breast cancer. Canc. Genet. Cytogenet. 51, 183–188.

Filipe, M.I. (1984). Mucins of Normal, Premalignant and Malignant Colonic Mucosa. In: Markers of Colonic Cell Differentiation (Wolman, S.R., & Mastromarino, A.J. eds), pp. 237–251, Vol. 29, Progress in Cancer Research and Therapy, Raven Press, New York.

Fisher, B., Gunduz, N., Costantino, J., Fisher, E.R., & Redmond, C. (1991). DNA flow cytometric analysis of primary operable breast cancer. Cancer 68, 1465–1475.

Fisher, E.R. & Paulson, J.D. (1978). Karyotypic abnormalities in precursor lesions of human cancer of the breast. Am. J. Card. Path. 69(3), 284–288.

Fourth International Workshop on Chromosomes in Leukemia, 1982 (1984). Clinical significance of chromosomal abnormalities in acute non-lymphoblastic leukemia. Canc. Genet. Cytogenet. 11, 332–350.

Fu, Y.S.. Temmin, L., Olaizola, Y.M., & Reagan, J.W. (1980). Nuclear DNA characteristics of microinvasive squamous carcinoma of the uterine cervix. In: Progress in Surgical Pathology (Fenoglio, C.M., & Wolff, M.W., eds.). New York., NY: Masson Publishing USA Inc; 1, pp. 233–244.

Giaretti, W., Danova, M., Geido, E., Mazzini, G., Sciallero, S., Aste, H., Scivetti, P., Riccardi, A., Marsano, B., Merlo, F., & d'Amore, E.S.G. (1991). Flow cytometric DNA index in the prognosis of colorectal cancer. Cancer 67, 1921–1927.

Gebhart, E., Bruderlein, S., Augustus, M., et al. (1986). Cytogenetic studies on human breast carcinomas. Breast Canc. Res. Treat. 8, 125–138.

Geleick, D., Muller, H., Matter, A., Torhorst, J., & Regenass, U. (1990). Cytogenetics of breast cancer. Canc. Genet. Cytogenet. 46, 217–229.

Gendler, S.J., Cohen, E.P., Craston, A., Duhig, T., Johnstone, G., & Barnes, D. (1990). The locus of the polymorphic epithelial mucin (PEM). tumour antigen on chromosome 1q21 shows a high frequency of alteration in primary human breast tumours. Int. J. Canc. 45, 431–435.

Genuardi, M., Tshira, H., Anderson, D.E., & Saunders, G.F. (1989). Distal deletion of chromosome 1p in ductal carcinoma of the breast. Am. J. Human Genet. 45, 73–82.

Gerbault-Seureau, M., Vielh, P., Zafrani, B., Salmon, R., & Dutrillaux, B. (1987). Cytogenetic study of twelve human near-diploid breast cancers with chromosomal changes. Ann. Genet. 30, 138–145.

Giaretti, W., Danove, M., Geido, E., Mazzini, G., Sciallero, S., Aste, H., Scivetti, P., Riccardi, A., Marsano, B., Merlo, F., & d'Amore, E.S.G. (1991). Flow cytometric DNA index in the prognosis of colorectal cancer. Cancer 67, 1921–1927.

Goh, H.S., & Jass, J.R. (1986). DNA content and the adenoma-carcinoma sequence in the colorectum. J. Clin. Pathol. 39, 387–392.

Hainsworth, P.J., Raphael, K.L., Stillwell, R.G., Bennett, R.C., & Garson, O.M. (1991). Cytogenetic features of twenty-six primary breast cancers. Canc. Genet. Cytogenet. 52, 205–218.

Harlow, S.P., Eriksen, B.L., Poggensee, L., Chmiel, J.S., Scarpelli, D.G., Mural, T., & Bauer, K.D. (1991). Prognostic implications of proliferative activity and DNA Aneuploidy in Astler-Coller Dukes stage C colonic adenocarcinomas. Can. Res. 51, 2403–2409.

Heim, S., Mertens, F., Jin, Y., Mandahl, N., Johansson, B., Biorklund, A., Wennerberg, J., Jonsson, N., & Mitelman, F. (1989). Diverse chromosome abnormalities in squamous cell carcinomas of the skin. Canc. Genet. Cytogenet. 39, 69–76.

Heim, S. & Mitelman, F. (1989). Cytogenetically unrelated clones in hematological neoplasms. Leukemia 3, 6–8.

Heppner, G.H. (1984). Tumor heterogeneity. Can. Res. 44, 2259–2265.

Herrera, L., Kakati, S., Gibas, L., Pietrzak, E., & Sandberg, A.A. (1986). Gardner syndrome in a man with an interstitial deletion of 5q. Am. J. Med. Genet. 25, 473–476.

Herrington, C.S. & McGee, J. (1990). Interphase Cytogenetics. Biochem. Res. 15, 467–474.

Hirai, K., Campbell, G., & Defendi, V. (1974). Changes of regulation of host DNA synthesis and viral DNA integration in SV40-infected cells, In: Control of Proliferation in Animal Cells, pp. 151–166, Cold Spring Harbor Laboratory, Cold Spring Harbor, NY.

Hofstadter, F., Delgado, R., Jakse, G., & Judmaier, W. (1986). Urothelial dysplasia and carcinoma *in situ* of the bladder. Cancer 57, 356–361.

Hunt R. & Cherian, M. (1990). Endoscopic diagnosis of small flat carcinoma of the colon. Dis. Colon Rectum 33, 143–147.

James, C.D., Mikkelson, T., Cavenee, W.K., & Collins, V.P. (1990). Molecular genetic aspects of glial tumor evolution. Canc. Surv. 9, 632–644.

Joensuu, H., Toikkanen, S., & Klemi, P.J. (1990). DNA index and S-phase fraction and their combination as prognostic factors in operable ductal breast carcinoma. Cancer 66, 331–340.

Kallioniemi, O.P., Kallioniemi, A., Kurisu, W., Thor, A., Chen, L.C., Smith, H.S., Waldman, F., Pinkel, D., & Gray, J.W. (1992). C-ERBB-2 oncogene amplification in breast cancer analyzed by fluorescence *in situ* hybridization. PNAS 89, 5321–5325.

Kerbel, R.S., Waghorne, C., Korczak, B., Lagarde, A., & Breitman, M.L. (1988). Clonal dominance of primary tumors by metastatic cells; genetic analysis and biological implications. Canc. Surv. 7, 594–629.

Kim, Y.J., Ngoi, S.S., Godwin, T.A., DeCosse, J.J., & Staiano-Coico, L. (1991). Ploidy in invasive colorectal cancer. Cancer 68, 638–641.

Kobayashi., H., Kaneko, Y., Maseki, N., & Sakurai, M. (1990). Karyotypically unrelated clones in acute leukemias and myelodysplastic syndromes. Canc. Genet. Cytogenet. 47, 171–178.

Kokal, W., Sheibani, K., Terz, J., & Harada, J.R. (1986). Tumor DNA content in the prognosis of colorectal carcinoma. J. Am. Med. Assoc. 255, 3123–3127.

Koss, L.G., Czerniak, B., Herz, F., & Wersto, R.P. (1989). Flow cytometric measurements of DNA and other cell components in human tumors: a critical appraisal. Hum. Pathol. 20, 528–548.

Kute, T.E., Muss, H.B., Cooper, M.R., Case, L.D., Buss, D., Stanley, V., Gregory, B., Galleshaw, J., & Booher, K. (1990). The use of flow cytometry for the prognosis of stage II adjuvant treated breast cancer patients. Cancer 66, 1810–1816.

Lafage, M., Nguyen, C., Szepetowski, P., Pebusque, M.J., Simonetti, J., Courtois, G., Gaudray, P., deLapeyriere, O., Jordan, B., & Birnbaum, D. (1990). The 11q13 amplicon of a mammary carcinoma cell line. Genes Chrom. & Canc. 2, 171–181.

Larsson, C., Bystrom, C., Skoog, L., Rotstein, S., & Nordenskjold, M. (1990). Genomic alterations in human breast carcinomas. Genes Chrom. & Canc. 2, 191–197.

Laurent-Puig, P., Olschwang, S., Delattre, O., Validire, P., Melot, T., Mosseri, V., Salmon, R.J., & Thomas, G. (1991). Association of Ki-*ras* mutation with differentiation and tumor-formation pathways in colorectal carcinoma. Int. J. Canc. 49, 220–223.

Lehman, J.M. & Defendi, V. (1970). Changes in deoxyribonucleic acid synthesis regulation in Chinese hamster cells infected with simian virus 40. J. Virol. 6, 738–749.

Leibovitz A. (1989). Tissue culture manual. Third Int. Workshop on Chrom. in Solid Tumors, Workshop Lab. Man., 1–28.

Leister, I., Weith, A., Bruderlein, S., Cziepluch, C., Kangwanpong, D., Schlag, P., & Schwab, M. (1990). Human colorectal cancer: High frequency of deletions at chromosome 1p35. Canc. Res. 50, 7232–7235.

Leppert, M., Dobbs, M., Scambler, P., O'Connell, P., Nakamua, D., Stauffer, D., Woodward, S., Burt, R., Hughes, J., Gardner, E., Lathrop, M., Wasmuth, J., Lalouel, J.M., & White, R. (1987). The gene for familial polyposis coli maps to the long arm of chromosome 5. Science 238, 1411–1413.

Levack, P.A., Mullen, P., Anderson, T.J., Miller, W.R., & Forrest, A.P.M. (1987). DNA analysis of breast tumour fine needle aspirates using flow cytometry. Brit. J. Cancer 56, 643–646.

Levin, B. & Reichmann, A. (1986). Chromosomes and large bowel tumors. Canc. Genet. Cytogenet. 19, 159–162.

Longy, M., Saura, R., Schouler, L., Mauhin, C., Goussot, J.F., Grison, O., & Couzigou, P. (1990). Chromosomal analysis of colonic adenomatous polyps. Canc. Genet. Cytogenet. 49, 249–257.

Look, A.T., Hayes, A.F., Nitshke, R., McWilliams, N.B., & Green, A.A. (1984). Cellular DNA content as a predictor of response to chemotherapy in infants with unresectable neuroblastoma. New Engl. J. Med. 311, 231–235.

Lynch, H.T., Lanspa, S., Smyrk, T., Boman, B., Watson, P., & Lynch, J. (1991). Hereditary nonpolyposis colorectal cancer (Lynch Syndromes I & II). Genetics, Pathology, Natural History, and Cancer Control, Part I. Canc. Genet. Cytogenet. 53, 143–160.

Lynch, H.T., Smyrk, T.C., Lanspa, S.J., Marcus, J.N., Kriegler, M., & Lynch, J.F. (1988). Flat adenoma in a colon cancer–prone kindred. J. Natl. Cancer. Inst. 80, 278–282.

McKinley, M.F., Budman, D.R., & Kahn, E. (1987). High grade dysplasia in Crohn's colitis characterized by flow cytometry. J. Clin. Gastroenterol. 9, 452–455.

Merlo, G.R., Venesio, T., Bernardi, A., Canale, L., Gaglia, P., Lauro, D., Cappa, A.P.M., Callahan, R., & Liscia, D.S. (1992). Loss of heterozygosity on chromosome 17p13 in breast carcinomas identifies tumors with high proliferation index. Am. J. Path. 140(1), 215–223.

Micale, M.A., Mohamed, A.N., Sakr, W., Powell, I.J., Wolman, S.R. (1992). Cytogenetics of Primary Prostatic Adenocarcinoma—Clonality and Chromosome Instability. Canc. Genet. and Cytogenet., 61, 165–173.

Mitchell, E.L.D. & Santibanez-Koref, M.F. (1990). 1p13 is the most frequently involved band in structural chromosomal rearrangements in human breast cancer. Genes, Chrom. and Canc. 2, 278–289.

Mitelman, F. (1988). Catalog of Chromosome Aberrations in Cancer, 3rd ed., Alan R. Liss, Inc. New York, 4th ed. Wiley-Liss, Inc., New York, 1991.

Morson, B.C. (1976). Genesis of colorectal cancer. Clin. Gastroenterol. 5, 505–525.

Muleris, M., Salmon, R.J., & Dutrillaux, B. (1988). Existence of two distinct processes of chromosomal evolution in near-diploid colorectal tumors. Canc. Genet. Cytogenet. 32, 43–50.

Muleris, M., Delattre, O., Olschwang, S., Dutrillaux, A.M., Remvikos, Y., Salmon, R.J., Thomas, G., & Dutrillaux, B. (1990a). Cytogenetic and molecular approaches of polyploidization in colorectal adenocarcinomas. Canc. Genet. and Cytogenet. 44, 107–118.

Muleris, M., Dutrillaux, A.M., Salmon, R.J, & Dutrillaux, B. (1990b). Sex chromosomes in a series of 79 colorectal cancers: Replication pattern, numerical, and structural changes. Genes, Chrom. & Canc. 1, 221–227.

Muleris, M., Salmon, F.J., Zafrani, B., Girodet, J., & Dutrillaux, B. (1985). Consistent deficiencies of chromosome 18 and of the short arm of chromosome 17 in eleven cases of human large bowel cancer: A possible recessive determinism. Ann. Genet. 28, 206–213.

Muleris, M., Salmon, R.J., Dutrillaux, A., Vielh, P., Zafrani, B., Girodet, J., & Dutrillaux, B. (1987). Characteristic chromosomal imbalances in 18 near-diploid colorectal tumors. Canc. Genet. Cytogenet. 29, 289–301.

Muss, H.B., Kute, T.E., Case, L.D., Smith, R., Booher, C., Long, R., Kammire, L., Gregory, B., & Brockschmidt, J.K. (1989). The relation of flow cytometry to clinical and biologic characteristics in women with node negative primary breast cancer. Cancer 64, 1894–1900.

Muto, T., Kamiya, J., Sawada, T., Konishi, F., Sugihara, K., Kubota, Y., Adachi, M., Agawa, S., Saito, Y., Morioka, Y., & Tanprayoon, T. (1985). Small "flat adenoma" of the large bowel with special reference to its clinico-pathologic features. Dis. Colon Rectum 28, 847–851.

Muto, T., Bussey, H.J.R., & Morson, B.C. (1975). The evolution of cancer of the colon and rectum. Cancer 36, 2251–2270.

Nishisho, I., Nakamura, Y., Miyoshi, Y., Miki, Y., Ando, H., Horii, A., Koyama, K., Utsunomiya, J., Baba, S., Hedge, P., Markham, A., Krush, A.J., Petersen, G., Hamilton, S.R., Nilbert, M.C., Levy, D.B., Bryan, T.M., Preisinger, A.C., Smith, K.H., Su, L.K., Kinzler, K.W., & Vogelstein, B. (1991). Mutations of chromosome 5q21 genes in FAP and colorectal cancer patients. Science 253, 665–669.

Nowell, P.C. (1976). The clonal evolution of tumor cell populations. Science 194, 23–28.

Ochi, H., Takeuchi, J., Douglass, H., & Sandberg, A.A. (1983). Possible specific chromosome changes in large bowel cancer. Canc. Genet. Cytogenet. 10, 121–122.

Page, D.L. & Dupont, W.D. (1990). Anatomic markers of human premalignancy and risk of breast cancer. Cancer 66, 1326–1335.

Pandis, N., Heim, S., Bardi, G., Idvall, I., Mandahl, N., & Mitelman, F. (1993). Chromosome analysis of 20 breast carcinomas, cytogenetic multiclonality and karyotypic-pathologic correlations. Genes, Chromosomes & Cancer 6: 51–57.

Passantonopoulou, A.F., Panani, A.D., & Raptis, S. (1991). Preferential involvement of 11q23-24 and 11p15 in breast cancer. Canc. Genet. Cytogenet. 51, 183–188.

Peters, J.M., Visscher, D.W., & Crissman, J.D. (1991). DNA content in primary and metastatic prostate cancer. Cytometry 12(suppl.), 58 [abstr.].

Petersen, S.E., Madsen, A.L., & Bak, M. (1991). Chromosome number distribution and cellular DNA content in colorectal adenomas from polyposis and nonpolyposis patients. Canc. Genet. Cytogenet. 53, 219–228.

Pinkel, D., Landegent, J., Collins, C., Fuscoe, J., Segraves, R., Lucas, J., & Gray, J. (1988). Fluorescence *in situ* hybridization with human chromosome-specific libraries: Detection of trisomy 21 and translocations of chromosome 4. Proc. Natl. Acad. Sci. USA 85, 9138–9142.

Potluri, V.R., Helson, L., Elsworth, R.M., Reid, T., & Gilbert, F. (1986). Chromosomal abnormalities in human retinoblastoma, a review. Cancer 58, 663–671.

Remvikos, Y., Gerbault-Seurrbau, M., Vielh, P.H., Zafrani, B., Magorlanant, H., & Dutrillaux (1988a). Relevance of DNA ploidy as a measure of genetic deviation: a comparison of flow cytometry and cytogenetics in 25 cases of human breast cancer. Cytometry 9(6), 612–618.

Remvikos, Y., Muleris, M., Vielh, P., Salmon, R.J., & Dutrillaux, B. (1988b). DNA content and genetic evolution of human colorectal adenocarcinoma. A study by flow cytometry and cytogenetic analysis. Int. J. Canc. 42, 539–543.

Remvikos, Y., Laurent-Puig, P., Hammel, P., Tominaga, O., Validire, P., Salmon, R.J., Dutrillaux, B., & Thomas, G. (1991). Human colorectal cancer: From clinicopathological to molecular prognosis. Cytometry 12(suppl), 28 [abstr].

Rey, J.A., Bello, M.F., de Campos, J.M., Kusak, M.E., & Moreno, S. (1987). On trisomy of chromosome 7 in human gliomas. Canc. Genet. Cytogenet. 29,323–326.

Rodgers, C.S., Hill, S.M., & Hulten, M. (1984). Cytogenetic Analysis in Human Breast Carcinoma. Canc. Genet. Cytogenet. 13, 95–119.

Rognum, T.O., Lynd, E., Meling, G.I., & Langmark, F. (1991). Near diploid large bowel carcinomas have better five-year survival than aneuploid ones. Cancer 68, 1077–1081.

Rosai, J. (1991). Borderline epithelial lesions of the breast. Amer. J. Surg. Path. 15, 209–221.

Russo, A., Bazan, V., Plaja, S., Leonardi, P., & Bazan (1991). Patterns of DNA-ploidy in operable colorectal carcinoma: A prospective study of 100 cases. J. Surg. Onc. 48, 4–10.

Saint-Ruf, C., Gerbault-Sereau, M., Viegas-Pequignot, E., Zafrani, B., Cassingena, R., & Dutrillaux, B. (1990). Proto-oncogene amplification and homogeneously staining regions in human breast carcinomas. Genes, Chrom. Canc. 2, 18–26.

Sandberg, A.A. (1990). The Chromosomes in Human Cancer and Leukemia (2nd Ed.), Elsevier Science Publishing Co., New York.

Sato, T., Akiyama, F., Sakamoto, G., Kasumi, F., & Nakamura, Y. (1991). Accumulation of genetic alterations and progression of primary breast cancer. Can. Res. 51, 5794–5799.

Schmidt-Ullrich, R., Lin, P.S., Mikkelsen, R.B., & Monroe, M.M.T. (1986). Proliferative rates of cloned malignant mammary epithelial cells as a measure of clonal heterogeneity in human breast carcinomas. J. Natl. Canc. Inst. 77, 1001–1011.

Schutte, B., Reynders, M.M.J., Wiggers, T., Arends, J.W., Volovics, L., Bosman, F.T., & Blijham, G.H. (1987). Retrospective analysis of the prognostic significance of DNA content and proliferative activity in large bowel carcinoma. Can. Res. 47, 5494–5496.

Sciallero, S., Bruno, S., Divine, A., Geido, E., Aste, H., & Giaretti, H. (1988). Flow cytometric DNA ploidy in colorectal adenomas and family history of colorectal cancer. Cancer 61, 114–120.

Shapiro, J.R., Yung, W.K.A., & Shapiro, W.R. (1981). Isolation karyotype and clonal growth of heterogeneous subpopulations of human malignant gliomas. Canc. Res. 41, 2349–2359.

Slovak, M.L., Ho, J., & Simpson, J.F. Cytogenetic Studies of 52 Human Breast Carcinomas, in press.

Solis, V., Pritchard, J., & Cowell, J.K. (1988). Cytogenetic changes in Wilms' tumors. Canc. Genet. Cytogenet. 34, 223–234.

Stål, O., Wingren, S., Carstensen, J., Rutqvist, L.E., Skoog, L., Klintenberg, C., & Nordenskjold, B. (1988). Prognostic value of DNA ploidy and S-phase fraction in relation to estrogen receptor content and clinicopathological variables in primary breast cancer. Eur. J. Canc. 25, 301–309.

Tribukait, B. (1991). DNA flow cytometry in carcinoma of the prostate for diagnosis, prognosis and study of tumor biology. Acta Onc. 30, 187–192.

Tribukait, B., Granberg-Ohman, I., & Wijkstrom, H. (1986). Flow cytometric DNA and cytogenetic studies in human tumors: A comparison and discussion of the differences in modal values obtained by the two methods. Cytometry 7, 194–199.

Tubiana, M. & Koscielny, S. (1988). Cell kinetics, growth rate and the natural history of breast cancer: The Heuson memorial lecture. Eur. J. Canc. Clin. Oncol. 24, 9–14.

Uccelli, R., Calugi, A., Forte, D., Mauro, F., Polonio-Balbi, P., Vecchione, A., Vizzone, A., & De Vita, R. (1986). Flow cytometrically determined DNA content of breast carcinoma and benign lesions; correlations with histopathological parameters. Tumori 72, 171–177.

Viola, M.V., Chen, J., Fromowitz, F., Finkel, J., Compton, C., & Perucho, M. (1991). Ras mutations in precancerous cells in ulcerative colitis. In: Dysplasia and Cancer (Robert Riddell, ed.), pp. 135–159. Elsevier Press, New York.

Waldman, F.M., Balasz, M., Mayall, B.H., Pinkel, D., & Gray, J.W. (1992). Karyotypic heterogeneity and its relationship to labeling index in interphase breast tumor cells. Canc. Res., in press.

Winchester, D.J., Duda, R.B., August, C.Z., Goldschmidt, R.A., Wruck, D.M., Rademaker, A.W., Winchester, D.P., & Merkel, D.E. (1990). The importance of DNA flow cytometry in node-negative breast cancer. Arch. Surg. 125, 886–889.

Witzig, T.E., Loprinzi, C.L., Gonchoroff, N.J., Reiman, H.M., Cha, S.S., Wieand, H.S., Katzmann, J.A., Paulsen, J.K., & Moertel, C.G. (1991). DNA ploidy and cell kinetic measurements as predictors of recurrence and survival in stages B2 and C colorectal adenocarcinoma. Cancer 68, 879–888.

Wolber, R.A. & Owen, D.A. (1991). Flat adenoma of the colon. Hum. Pathol. 22, 70–74.

Wolley, R.C., Schreiber, K., Koss, L.G., Karas, M., & Sherman, A. (1982). DNA distribution in human colon carcinomas and its relationship to clinical behavior. J. Natl. Canc. Inst. 69, 15–22.

Wolman, S.R., Macoska, J.A., Micale, M.A., & Sakr, W.A. (1992). An approach to definition of genetic alterations in prostate cancer. Diagnostic Molecular Pathology, A. J. Surg. Path., Part B, 1, 192–199.

Wolman, S.R., Smith, H.S., Stampfer, M., & Hackett, A.J. (1985). Growth of diploid cells from breast cancers. Canc. Genet. Cytogenet. 16, 49–64.

Wolman, S.R., Camuto, P.M., & Feinberg, A.P. (1990). Genetic markers in colonic neoplasia. In: Familial Adenomatous Polyposis, pp. 371–382. Alan R. Liss, Inc., New York.

Wolman, S.R., Camuto, D.M., Golimbu, M., & Schinella, R. (1988). Cytogenetic, flow cytometric, and ultrastructural studies of twenty-nine nonfamilial human renal carcinomas. Canc. Res. 48, 2890–2897.

Wolman, S.R., & Heppner, G.H. (1992). Editorial: Genetic Heterogeneity in Breast Cancer. J. Natl. Canc. Inst. 84, 469–470.

Wolman, S.R., Waldman, F.M., & Balazs, M. (1993). Complementarity of interphase and metaphase chromosome analysis in human renal tumors. Genes, Chrom., and Canc. 6, 17–23.

Wurster-Hill, D.H., Pettengill, O.S., Noll, W.W., Gibson, S.H., & Brinck-Johnsen, T. (1990). Hypodiploid, Pseudodiploid, and Normal Karyotypes Prevail in Cytogenetic Studies of Medullary Carcinomas of the Thyroid and Metastatic Tissues. Canc. Genet. Cytogenet. 47, 227–241.

Zhang, R., Wiley, J., Howard, S.P., Meisner, L.F., & Gould, M.N. (1989). Rare clonal karyotypic variants in primary cultures of human breast carcinoma cells. Canc. Res. 49, 444–449.

ALTERED EXPRESSION OF TRANSFORMING GROWTH FACTOR-α AND TRANSFORMING GROWTH FACTOR-β AUTOCRINE LOOPS IN CANCER CELLS

M.G. Brattain, K.M. Mulder, S.P. Wu, G. Howell,
L. Sun, J.K.V. Willson, and B.L. Ziober

Advances in Molecular and Cell Biology
Volume 7, pages 35–59

ISBN: 1-55938-624-X

I. INTRODUCTION

Polypeptide growth factors control proliferation, differentiation, and organization of normal cells through interactions with specific membrane receptors. This interaction triggers an intracellular signalling cascade which leads to the activation and expression of specific subsets of target genes. Compelling evidence indicates that malignancy arises from a stepwise progression of genetic events which, among other effects, leads to abnormal regulation of growth factor expression, or of components of the associated signalling pathway that leads to the culmination of normal growth factor action (Fearon and Vogelstein, 1990; Aaronson, 1991). Thus, the ever expanding list of molecules with transforming activity resulting from abnormal expression or mutation includes growth factors, their receptors, signalling components such as tyrosine kinases and G proteins, and nuclear proteins involved in cell cycle control or transcriptional control. The accumulation of these alterations, as well as alterations in other classes of molecules, leads first to transformation and, as more changes occur, to progression of malignancy.

Highly progressed malignant cells are generally associated with a high degree of autonomy with respect to growth factor requirements and with respect to exogenous controls on growth. Excellent examples of progression and the loss of response to exogenous growth controls are the development of estrogen independence in breast carcinoma and androgen independence in prostatic carcinoma. Highly progressed cell lines of human melanoma and colonic carcinoma are capable of growing in medium devoid of exogenous growth factors (Boyd et al., 1988; Herlyn, 1990) while more differentiated, less progressed forms show some degree of dependence on exogenous growth factors (Herlyn, 1990; Mulder and Brattain, 1989).

The control of the cell cycle in normal cells is dependent upon the balance of autocrine negative and positive factors (Pardee, 1989). As indicated above, growth factor independence is the result of the substitution of abnormal, constitutive stimulatory activity and/or the breakdown of the response to controls on the cell cycle.

This review will focus on the role of major positive and negative autocrine loops in maintaining growth regulation in colon cancer and the mechanisms underlying the breakdowns in these autocrine loops whihc lead to growth factor independence in this system. While the details in this review and the specific autocrine loops described deal with colon cancer, the general principles of growth factor dependence and the role of the balance of positive and negative controls in maintaining growth factor dependence should be applicable to other systems. Again, while the molecular basis for the loss of growth factor dependence and progression to independence presented in this review pertains to colon cancer, the process should be instructive as to the types of events which may be occurring in other systems.

II. BIOLOGICAL PROPERTIES OF THE HUMAN COLON CARCINOMA MODEL SYSTEM

Development of a Human Colon Carcinoma Cell Line Bank for Studies of Differentiation and Growth Regulation. A normal colon has a well defined pattern of differentiation extending from the stem cells of the crypt to the functional, differentiated cells of the surface epithelium. To some degree, colon carcinomas reflect the various states of differentiation and heterogeneity of cell types found in the normal colon. We have developed a bank of more than 25 human colon carcinoma cell lines which are reflective of the heterogeneity of malignant cells encountered *in vivo* (Brattain et al., 1984). This bank of cell lines has been classified with respect to its biological properties and has diversity in terms of the differentiation and growth regulation of the available target cells (Brattain et al., 1984; Mulder and Brattain, 1989a).

Advances in tissue culture methodology have led to the establishment of banks of some histopathological types of tumors (Gallie et al., 1982; Smith et al., 1982; Brattain et al., 1984; Herlyn, 1992). As our concepts of tumor heterogeneity solidified in the 1970's it became apparent that an individual cell line, whether it be of animal or human origin, could not provide an adequate model for the biological, biochemical or pharmacological properties of a given histopathological type of tumor. These banks of cell lines were established in an attempt to obtain a broad spectrum of properties through a large number of cell lines which would be reflective of the diversity encountered *in vivo*. We do not regard the individual colon carcinoma cell lines as models for individual human tumors, but rather as models for the individual types of malignant cells which can be found in colon carcinoma. The tumors from which the lines were established were most likely composed of varying amounts of the subpopulations of malignant cells. Moreover, the cell line bank does not necessarily contain all of the types of malignant cells found in colon cancer. However, our methodology allowed for the successful development of cell lines from 75% of the specimens attempted, suggesting the establishment of a wider variety of cell types than obtained with previously utilized methods for colon carcinoma which showed success rates of 10–15% or less (Breborowicz et al., 1975; Leibovitz et al., 1976; Brattain et al., 1981a,b).

Classification of Biological Subclasses of Human Colon Carcinoma Cell Lines. The colon carcinoma cell lines could be subclassified into 3 broad groups on the basis of differences in growth properties *in vitro* and as xenografts. These groups included aggressive, undifferentiated cell lines (designated group I) on one extreme and unaggressive, well-differentiated cell lines (designated Group III) on the other (Brattain et al., 1984; Chantret et al., 1988; Mulder and Brattain, 1989). Large inocula of group III cells (10^7) are required for xenograft formation and xenografts always show differentiated characteristics. These cells also retain differentiated characteristics in tissue culture. These features include growth in a strict monolayer with basolateral polarity, microvilli formation, tight junction formation and the development of functional transport domes (Chantret et al., 1988). In contrast, Group I cells are poorly-differentiated both *in vitro* and *in vivo*. In tissue culture, they grow as multi-layered cells without basolateral polarity, do not have microvilli, do not form tight junctions or transport domes (Chantret et al., 1988). However, as with the group III cells they do express villin, indicating their epithelial and intestinal origin (Chantret et al., 1988). Group I cells form undifferentiated tumors in 100% of athymic mice injected with 10^6 cells. Gram sized tumors are formed within 25 days of inoculation.

A third subclass of cell lines with intermediate properties between the 2 extremes was designated group II cells. Our studies have been concentrated on the group I and group III cells since comparison of the extreme biological phenotypes was most likely to lead to the identification of key growth regulatory differences. The retention of functional differentiated characteristics or lack of differentiated characteristics in tissue culture is critical if the control of growth and differentiation are being studied. This is especially true of colon carcinoma because there are no normal colon cells which can be progressively grown in tissue culture as controls for these types of studies.

III. GROWTH REGULATORY PHENOTYPES IN THE COLON CARCINOMA BANK

Evidence for Growth Factor Independence in Group I Cells. One might expect that stem cells responsible for cell renewal would be under autonomous control of cell proliferation and produce large amounts and/or be highly responsive to stimulatory factors. On the other hand, as cells differentiate in normal colon they lose their proliferative capacity. This could be due to a combination of events involving reduced production and/or sensitivity to stimulatory factors along with increased production of differentiation-inducing inhibitory types of activities. If the states of differentiation of colon carcinoma cell lines reflect normal differentiation (Brattain et al., 1984), it would be expected that undifferentiated, aggressive group I cells would show an autonomous type of growth behavior whereas differentiated, unaggressive group III cells would be more dependent upon exogenous factors for proliferation.

Group I cells were grown in McCoy's serum free medium without any supplementation of growth factors such as epidermal growth factor (EGF), insulin (In) and transferrin (Tr) (Boyd et al., 1988). Growth in medium without growth factor supplementation was highly dependent on the concentration of cell inocula suggesting autocrine factors produced by group I cells were important in permitting growth. Conditioned medium from the group I cells grown without factor supplementation obviated the inoculum dependence further supporting this hypothesis.

Group III cells were unable to survive in serum free medium without growth factor supplementation.

Adaptation of the Cell Line Bank to Continuous Maintenance and Growth in Serum Free Chemically Defined Medium. The growth of group I cells in factor free medium could be augmented by Tr and further increased by In. EGF, sodium selenite, triiodothyronine and hydrocortisone are agents which are frequently employed to enhance serum free growth of a variety of histopathological types of malignant cells. None of these had any effect on serum free growth of group I human colon carcinoma cells (Boyd et al., 1988).

In contrast to group I cells, when group II cells were grown on plastic, they could not be developed into progressively growing cell lines unless the medium contained Tr. Like group I cells, In could augment the growth of group II lines but in contrast to group I, group II cells were also able to respond to EGF with increased growth in the presence of Tr supplemented medium. Group III cells required the combination of all 3 factor supplements to obtain serum free growth although some growth could be maintained in a medium supplemented with combinations of Tr + In or Tr + EGF (Wan et al., 1988).

Growth Factor Effects on Mitogenesis. At quiescence, transformed AKR-MCA fibroblasts can be stimulated to enter the S phase of the cell cycle by nutrient replenishment alone (Chakrabarty et al., 1984). In contrast, their untransformed counterpart cells require the addition of defined polypeptide growth factors in order to initiate DNA synthesis. This suggested that these transformed cells were able to produce autocrine growth factors and were, therefore, more independent of exogenous factors than their untransformed counterparts. In order to determine whether the growth controls in poorly-differentiated human colon carcinoma cells were altered relative to those in well-differentiated cells, we compared the mitogenic responses to nutrients and/or growth factors in cell lines representative of the two groups (Mulder and Brattain, 1989b).

Cells were rendered quiescent by growth factor and nutrient starvation until a basal level of ^3H incorporation into DNA was obtained (5–7 days depending upon the cell line). Cells were released from quiescence by the addition of nutrients or nutrients plus growth factors (Tr, In, EGF). A mitogenic response was elicited in the poorly-differentiated RKO and HCT 116 cells by addition of nutrients alone. Addition of growth factors, as well as nutrients, resulted in no additional stimulation of thymidine incorporation. Under both conditions, DNA synthesis was stimulated 10–20 fold above baseline levels in the group I cells.

In contrast, all of the well-differentiated group III cells (FET, CBS, and GEO) displayed a greater mitogenic response to the combination of nutrients and the growth factors than to nutrients alone. Thymidine incorporation was stimulated to levels 1.4–3 times greater than basal levels in response to nutrients alone. However, addition of nutrients and growth factors to the group III cells resulted in a 6–10 fold stimulation of DNA synthesis above basal levels. Thus, the polypeptide growth factor combination was 2 to 7 times more mitogenic than were nutrients alone in the well-differentiated cells.

Therefore, 2 growth regulatory phenotypes were recognized which reflected the biological subclasses previously recognized in the cell line bank. Both cell proliferation and DNA synthesis were growth factor independent in poorly-differentiated group I cells, whereas the well-differentiated group III cells were growth factor dependent. Next we utilized quiescence and the release from quiescence to compare cell cycle differences between the 2 phenotypes with respect to the expression of c-myc and TGF-α.

Expression of c-myc in Growth Factor Dependent and Independent Cells. The induction of a specific family of genes, known as "competence" genes, is a prerequisite to reentry of quiescent cells into the cell cycle (Stiles, 1983). Among these genes is the protooncogene *c-myc*, which is up-regulated before stimulation of DNA synthesis by a variety of growth stimuli. This protooncogene may be an intracellular mediator of mitogenic signals and appears to play a key role in determining cellular responsiveness to polypeptide growth factors (Cole, 1985; Sorrentino et al., 1986; Stern et al., 1987). It was of interest to determine whether the up-regulation of this competence gene would be related to the differential effects of nutrients and growth factors on thymidine incorporation in the well-differentiated cells. Kinetic analyses of the effects of nutrients or nutrients + growth factors on the expression of *c-myc* in a well-differentiated cell line designated CBS were performed. The results illustrated different kinetics for induction of *c-myc* in response to nutrients or growth factors. Nutrients alone resulted in peak *c-myc* levels (approximately 2-fold above baseline) at approximately 90 minutes following addition. In contrast, *c-myc* expression reached maximal levels (approximately 5-fold above baseline) at 4 hours following addition of nutrients plus growth factors. As observed with stimulation of DNA synthesis, nutrients plus the growth factors resulted in a greater increase in *c-myc* RNA than did nutrients alone in the well-differentiated CBS cells. The increase in *c-myc* expression under both conditions was transient and had returned to baseline levels between 8–24 hours (Mulder and Brattain, 1989b).

Since the poorly-differentiated cells responded equally well to either nutrients or to nutrients and growth factors, it was of interest to determine whether the kinetic pattern for *c-myc* expression in these growth factor independent cells would differ from that in the well-differentiated cells. Northern analysis of *c-myc* expression in poorly-differentiated HCT 116 cells, at various times following addition of nutrients and/or growth factors was performed (Mulder and Brattain, 1989b). The results indicated that regulation of *c-myc* expression by nutrients and/or growth factors in these cells differs substantially from that in the well-differentiated cells. Growth factor independent HCT 116 cells displayed an initial reduction in *c-myc* expression in response to either nutrients or nutrients plus growth factors. Expression levels of *c-myc* were restored to baseline values within 4–8 hours in both cases. Consequently, HCT 116 cells were characterized by a lack of the ordered control of *c-myc* expression seen in untransformed cells and in well-differentiated colon carcinoma cells. The inability of the HCT 116 cells to show reductions in *c-myc* levels during the course of mitogenic stimulation by nutrients indicated constitutive stimulation of this competence gene (Mulder and Brattain, 1989b). It seemed likely that this might be due to endogenous TGF-α since the expression of *c-myc* is responsive to EGF in untransformed cells or well-differentiated colon carcinoma cells. The presence of growth factors did not affect the final level of *c-myc* expression.

Cell Cycle Expression of TGF-α mRNA as a Function of Growth Regulatory Phenotype. Similarly to *c-myc*, we also showed that TGF-α was rapidly up-regulated in quiescent, well-differentiated cells in response to stimulation with growth factors (Mulder et al., 1990a,b; Mulder, 1991). This up-regulation was suppressed by simultaneous addition of TGF-β to the growth factor combination at the time of release from quiescence (Mulder et al., 1990a). Well-differentiated, but not poorly-differentiated colon carcinoma cell lines respond to TGF-β with inhibition (Hoosein et al., 1989). In contrast to the well-differentiated cells, TGF-α mRNA levels were not affected by growth factor stimulation in poorly-differentiated HCT 116 cells. HCT 116 cells showed a high level of constitutive expression of TGF-α which was not affected by cell cycle time or exogenous growth factors (Mulder et al., 1990a,b; Mulder, 1991). Furthermore, TGF-β was unable to repress these levels of TGF-α mRNA.

We thought that the constitutive expression of *c-myc* and TGF-α in HCT 116 cells might be due to autocrine activity of TGF-α since this polypeptide had been shown to be autoinducible (Coffey et al., 1987) and because as mentioned above, both TGF-α and EGF stimulate *c-myc* expression. However, we were still bothered by the fact that even if TGF-α were responsible for constitutive stimulation, acting as an autocrine factor it would presumably still be required to bind to cell surface EGF receptors which would then be down regulated and desensitized until the next cycle of stimulation. Therefore, one would still expect a cyclical up-regulation and down-regulation of *c-myc* and TGF-α as seen in GEO cells unless there were some other additional defect in control. Therefore, we examined the role of stimulatory autocrine activity in growth factor dependent and independent cells.

IV. STIMULATORY AUTOCRINE LOOPS IN COLON CARCINOMA CELLS

GEO Cells have an External Autocrine TGF-α Loop. TGF-α is a polypeptide mitogen which is homologous to EGF and acts through the same receptor. Production of TGF-α protein expression and its mRNA has been noted for a variety of solid human tumors and cell lines (Derynck et al., 1987; Smith et al., 1987; Arteaga et al., 1988; Bates et al., 1988; Mulder and Brattain, 1989a). Expression of the EGFr and its mRNA in the same systems have led to the implication that TGF-α acts as an autocrine growth factor in most human solid tumors. Hence, the EGFr and/or TGF-α would be possible targets for inhibiting cancer growth. More recently, as mentioned above, a number of studies have now shown expression of TGF-α and EGFr in normal cells and tissues (Coffey et al., 1987; Mydlo et al., 1989; Bates et al., 1990) and evidence for its function in an extracellular loop in normal cells has been presented (Bates et al., 1990) .

The availability of an EGF receptor (EGFr) blocking antibody permitted us to directly test the hypothesis that TGF-α was an autocrine factor in colon carcinoma cells. The response of HCT 116 cells and GEO cells to EGF and recombinant TGF-α were compared. HCT 116 cells (group I) were unresponsive to either factor whereas proliferation of GEO cells (group III) increased in a dose dependent manner equally well to both factors. Northern analysis showed HCT 116 cells have approximately 4-fold more TGF-α mRNA than GEO cells and secreted larger amounts of TGF-α protein. HCT 116 cells also had 3-fold higher intracellular levels of TGF-α than GEO cells. Comparison of EGFr levels by Scatchard analysis showed that HCT 116 cells had approximately 3-fold higher numbers of receptors than GEO cells without any significant differences in binding affinity. These differences in receptor number were consistent with different levels of mRNA for the EGFr in the two cell lines. Thus, both cell lines had all the necessary components for a TGF-α autocrine loop, but because HCT 116 cells were not responsive to exogenous EGF and because of the constitutive, non-controlled expression of c-myc and TGF-α itself we thought it more likely that these poorly-differentiated, growth factor independent cells would present an autocrine TGF-α loop.

Antibody to EGFr should block binding of autocrine TGF-α back to the cell in the absence of exogenous competing EGF. Cells that are grown in absence of exogenous EGF should be inhibited by blocking antibody if TGF-α is acting in an external autocrine loop. Ab528 is a blocking, non-activating antibody which is commercially available. Surprisingly, this antibody was capable of inhibiting proliferation of GEO cells in a dose dependent manner, but had essentially no effect on HCT 116 cells. Similarly, a commercially available TGF-α neutralizing antibody had no effect on HCT 116 cells, but was able to inhibit GEO cell growth. Anchorage independent growth of GEO cells but not HCT 116 cells was also inhibited by Ab528. Titration of the EGFr with Ab528 indicated that available receptors were present in both cell lines. Consequently, we were forced to conclude that GEO cells, while retaining the ability to respond to exogenous EGF, also had an operational external TGF-α autocrine loop. HCT 116 cells either did not have a classical external loop or they did not require autocrine TGF-α for optimum growth.

Internal Autocrine Loops. Information was beginning to evolve about the potential for internal autocrine loops in other systems at this time. Several growth factors have been implicated as having intracellular auto-activation mechanisms in recent years. These include PDGF, interleukin-3 (IL-3) and granulocyte macrophage colony stimulating factor (GM-CSF). The most extensively studied of these, PDGF, will be described below.

PDGF is a heterodimer of A and B chains which can be expressed in all of its permutations. Homodimers of either type can serve as potent mitogens for both normal and neoplastic mesenchymal cells. A viral transforming gene, v-*sis* was transduced from the PDGF gene. Several lines of evidence indicate that the-transforming activity of v-*sis* required interaction of the v-*sis* product with PDGF receptor β intracellularly in the secretory compartment. Antibodies neutralizing the v-*sis* product in conditioned medium did not fully inhibit growth, exposure to exogenous v-*sis* product did not transform cells, and some v-*sis* transformed lines did not secrete v-*sis* product (Bejcek et al., 1989; Browder et al., 1989). Removal of the signal peptide from the v-*sis* gene by *in vitro* mutagenesis prevented v-*sis* protein from reaching the secretory compartment and prevented transformation (Hannink and Donoghue, 1986).

Other groups suggested that although ligand and receptor might bind in secretory compartments, eventual cell surface localization of the complex was necessary for transformation (Hannink and Donoghue, 1988; Fleming et al., 1989). These experiments were based on dissociation of receptor-ligand complexes with suramin, an agent which has since been shown to be able to penetrate secretory compartments (Huang and Huang, 1988). Protamine, an agent which inhibits only extracellular binding was found to have no effect on autocrine growth (Huang and Huang, 1988; Lokeshwar et al., 1989).

Resolution of the controversy came from experiments (Bejcek et al., 1989) in which the v-*sis* gene was modified to express a carboxyl terminal sequence of Ser-Glu-Lys-Asp-Glu-Leu (SEKDEL) which had been shown to cause the retention of soluble proteins in the endoplasmic reticulum (ER) and Golgi apparatus (Munro and Pelham, 1987; Pelham, 1988). The SEKDEL modified v-*sis* product was fully transforming and could not be reversed by neutralizing antibody added to cultures or by protamine. No secretion or cell surface localization of SEKDEL v-*sis* could be detected. In contrast, SEKDEL v-*sis* was located in ER and Golgi after subcellular fractionation. Localization of SEKDEL v-*sis* in these fractions was 3-fold higher than in cells expressing a SEKDAS v-*sis*-mutant. SEKDAS v-*sis* also fully transforming, but can be secreted in a fashion analogous to unmodified v-*sis*.

The results with SEKDEL mutants provide convincing evidence that some autocrine mechanisms are intracellular rather than extracellular. Both autocrine mechanisms can confer a growth advantage, but as discussed below, potential implications of internal loops on growth control and approaches for therapeutic intervention in cancer are quite different from those of external loops.

Internal TGF-α Autocrine Loop in Growth Factor Independent Cells. The results described above regarding the lack of EGF/TGF-α response provides a molecular basis for growth factor independence in HCT 116 cells. An internal TGF-α loop would explain the lack of response to Ab528 as well as the uncontrolled, constitutive expression of *c-myc* and TGF-α since both of these genes are controlled by ligand binding to EGFr. Therefore, if we were going to inhibit HCT 116 growth by inhibiting TGF-α action we needed to develop a means of inhibiting intracellular TGF-α expression. Consequently, we decided to construct TGF-α anti-sense expression vectors. A full-length anti-sense vector driven by the CMV promoter was constructed and expressed in HCT 116 cells. The vector was effective in reducing TGF-α expression. Expression of TGF-α anti-sense mRNA resulted in reduced endogenous TGF-α expression of 3-fold. Protein expression was reduced 6-fold. Transfected HCT 116 cells showed an absolute requirement for exogenous EGF to grow. The addition of exogenous EGF to transfected HCT 116 cells rescued them from the effects of the anti-sense vector thus indicating that inhibition was specifically due to inhibition of intracellular expression of endogenous TGF-α. This is the first example of a "native" as opposed to genetically engineered internal autocrine loop.

Potential Effects of Internal Autocrine Loops on Growth Regulatory Mechanisms. As indicated above, normal growth factor response leads to an ordered rise and decline in expression of specific genes required for movement through the cell cycle as well as a similar controlled rise and decline of expression of the genes needed to initiate the next cycle of division (Pardee, 1989). The formation of an internal loop could have the effect of removing the cell from being subject to those controls by allowing for constitutive stimulation without interference from cell surface and extracellular controls (such as the desensitization of growth factor response through down regulation of receptors). Continuous intracellular ligand-receptor interactions would then have the effect of allowing for unregulated constitutive stimulation of key cell regulatory components without the normal ordered rise and decline of expression in response to external signals. Consequently, external autocrine loops may be normal self regulatory mechanisms whereas the expression of internal loops may be an aberrant type of control resulting in many of the biological properties associated with transformation and differentiation which in effect provides a molecular basis for growth factor independence.

In Vivo Effects of Modulation of the TGF-α Autocrine Loops. It was important to determine whether repression of TGF-α expression had an effect on the tumorigenic potential of the highly aggressive HCT 116 cells. Therefore, several clones of TGF-α anti-sense transfected cells were compared with neomycin transfected HCT 116 cells for their ability to form tumors in athymic nude mice at inocula of 10^6 or 5×10^6 cells (Ziober et al., in press). At 5×10^6 cells all clones formed tumors in 6 of 6 animals. However, the time required for tumor formation was delayed from ~7 days for control cells to ~15 days for cells transfected with TGF-α anti-sense. At the lower inoculum anti-sense expressing clones varied in their tumorigenicity from 1 of 6 animals to 6 of 6 animals. In those animals which did form tumors, growth was again delayed relative to the Neo control animals which formed tumors in 6 of 6 animals. To summarize, *in vivo* experiments have shown that repression of TGF-α expression leads to reduced tumorigenicity, and when tumors are formed, delayed growth and increased differentiation of these tumors.

Tumors resulting from cells transfected with TGF-α anti-sense DNA were analyzed for expression of sense and anti-sense TGF-α mRNA. In contrast to the cultured cells used to inoculate the tumors, xenografts of the anti-sense transfected cells showed much larger amounts of endogenous TGF-α sense mRNA than anti-sense mRNA. This suggested that selection of TGF-α anti-sense transfected cells expressing extraordinary levels of endogenous TGF-α was occurring during xenograft formation. When these cells were injected back into athymic mice, they showed the same tumorigenic properties as the control cells, thus providing additional evidence that selection of a subpopulation of cells occurred. Cells from xenografts were removed and cultured. These cells also showed a high sense: anti-sense mRNA ratio.

If TGF-α contributes to the tumorigenic properties of colon carcinoma cells, it would be expected that increasing its expression would enhance tumorigenicity in poorly tumorigenic cells. GEO cells are poorly tumorigenic and express low amounts of TGF-α. Consequently, GEO cells were transfected with the sense construct of the vector used for TGF-α anti-sense transfection. Transfectants expressed the expected TGF-α mRNA of ~1Kb, produced 5-fold higher levels of TGF-α protein and lost the ability to respond to exogenous EGF or TGF-α. TGF-α transfected cells were tested for tumorigenicity at 10^6 and 5×10^6 cells in athymic mice. At 10^6 cells control GEO cells are non-tumorigenic, but progressively growing tumors were formed in all animals inoculated with TGF-α transfected cells and attained volumes ranging from 10^3 to 3×10^3 mm^3 within 30 days.

Gastrin is an Autocrine Growth Factor in Colon Cancer Cell Lines. Gastrin is a gastrointestinal peptide produced by neuroendocrine cells and antral G cells (Dorkray, 1978; Walsh, 1981). In addition to its stimulatory action on normal gut tissue, gastrin has been reported to promote the growth of colon tumors *in vivo* (Winsett et al., 1982), some colon carcinoma cell lines in tissue culture (Murakami and Masui, 1980; Sirinek et al., 1985; Kusyk et al., 1986) and primary human colon tumor tissues (Watson et al., 1989).

Taken as a whole, these studies provided strong evidence that gastrin functions as a stimulatory factor in colon carcinoma, but did not indicate that the hormone was functioning in an autocrine manner. However, some colon carcinomas do share differentiation lineage markers in common with neuroendocrine cells which are known to produce gastrin (Dockray, 1978; Walsh, 1981). Hence, we hypothesized that gastrin might be an autocrine factor in colon carcinoma.

We were not able to detect a significant response to exogenously added gastrin in our various assays for continuously maintained serum-free cells. Hence, we hypothesized that endogenous production of gastrin-like peptides might be interfering with response to exogenously added peptide. Therefore, we undertook similar experiments as those described above for TGF-α to determine if agents which block gastrin action would inhibit growth of colon carcinoma cells.

Proglumide and benzotript, two amino acid derivatives which are antagonists to gastrin receptor binding were found to inhibit growth of colon cancer cells by as much as 90% although IC_{50} values were in the mM range (Hoosein et al., 1988). Removal of the gastrin receptor antagonists resulted in resumption of normal growth rates indicating that the effects were non-toxic and fully reversible. The effects of receptor antagonists could be overcome by the addition of exogenous gastrin-17. Anti-gastrin anti-serum was also effective in preventing colon carcinoma proliferation (up to 85% inhibition) whereas normal IgG had no effect at the same antibody concentration. The effects of anti-gastrin anti-serum could also be reversed by the addition of exogenous gastrin-17 (Hoosein et al., 1990).

These results strongly suggested that gastrin-like autocrine activity was stimulating the growth of colon cancer cells, but as yet, we had not provided direct evidence of the secretion of a gastrin-like activity in the conditioned medium of the cells. Consequently, in a follow-up study we characterized group I and group III colon carcinoma cells for the expression of gastrin-like activity. It was found that all of the colon carcinoma cell lines tested secreted an immunoreactive gastrin-like polypeptide as assessed by an RIA (Hoosein et al., 1990). However, due to potential cross-reactivity with CCK, it was not clear that the autocrine activity was due to gastrin.

Analysis of Gastrin mRNA in Colon Carcinoma Cell Lines. Due to low levels, gastrin mRNA was difficult to assess at high stringency conditions in Northern blots. In order to increase the sensitivity of our detection methods, we first used the polymerase chain reaction (PCR) to verify whether or not authentic gastrin mRNA was present in these cell lines. The human genomic gastrin gene consists of 3 exons of 60, 216 and 192 nucleotides, respectively (Kariya et al., 1986). Intron 1 comprises 3041 nucleotides and intron 2 has 130 nucleotides. We designed two primers for the PCR reaction to span the second intron and first strand cDNA synthesis was generated from 1 μg of poly A^+-enriched RNA. On gel electrophoresis, we observed the expected 382 bp band in group I HCT 116 and group III GEO and FET cell lines. Identity to correctly-spliced gastrin mRNA was confirmed by Southern blotting using the ^{32}P-labeled 269 bp Hind III fragment of gastrin cDNA as probe. Baldwin et al. (1990) reported similar findings after PCR of poly A^+ RNA from the gastric Okajima and colonic HCT 116 and LIM1215 lines by PCR.

We have also used an RNase protection assay to study gastrin mRNA production in our cell lines. The 269 bp Hind III gastrin fragment, which spans the second intron, was cloned into the pGEM3z(-) vector (Promega) and high specific activity riboprobe with incorporated [^{32}P]-UTP was synthesized using SP6-RNA-polymerase (Promega transcription system). We again detected correctly sized gastrin mRNA transcript at low levels in our cell lines. The group I HCT 116 cell line produced approximately 25-fold more gastrin mRNA.

Interaction of Gastrin and TGF-α Autocrine Loops. Gastrin promoter activity has been shown to be induced by addition of exogenous EGF to GH4 pituitary cells and an EGF response element has been identified (Godley and Brand, 1989; Merchant et al., 1990). Our development of an RNase protection assay for gastrin and the availability of our group III cell lines which respond to exogenous EGF have allowed us to determine whether EGF/TGF-α can induce gastrin mRNA expression. EGF was shown to increase gastrin mRNA expression by 3-fold in FET cells. This is in good agreement with the 2–3 fold levels of gastrin induction shown in the promoter studies mentioned above.

EGF induction of gastrin mRNA raises the question of whether strong autocrine TGF-α loops such as that seen in HCT 116 contribute to the high level of gastrin mRNA observed in this cell line. The availability of HCT 116 TGF-α anti-sense transfected cells allowed us to test this hypothesis. When TGF-α expression was repressed in HCT 116 cells, gastrin expression was also repressed several-fold relative to control transfected HCT 116 cells. These results show interaction between the TGF-α and gastrin autocrine systems and indicate that studies at the transcriptional level would yield more information about the mechanisms involved in this interaction.

We located a TGF-α response element (TRE) in the TGF-α promoter and determined that the TGF-α expression was transcriptionally autostimulated using HCT 116 colon carcinoma cells which had been transfected with a TGF-α, anti-sense expression vector. Since the gastrin promoter is also controlled by a TRE, it was not surprising when we found that TGF-α repression also led to gastrin repression. Thus, interference of the TGF-α autocrine expression also carried with it the added benefit of interference with the gastrin autocrine loop. The interaction of the 2 autocrine loops is shown in Figure 1.

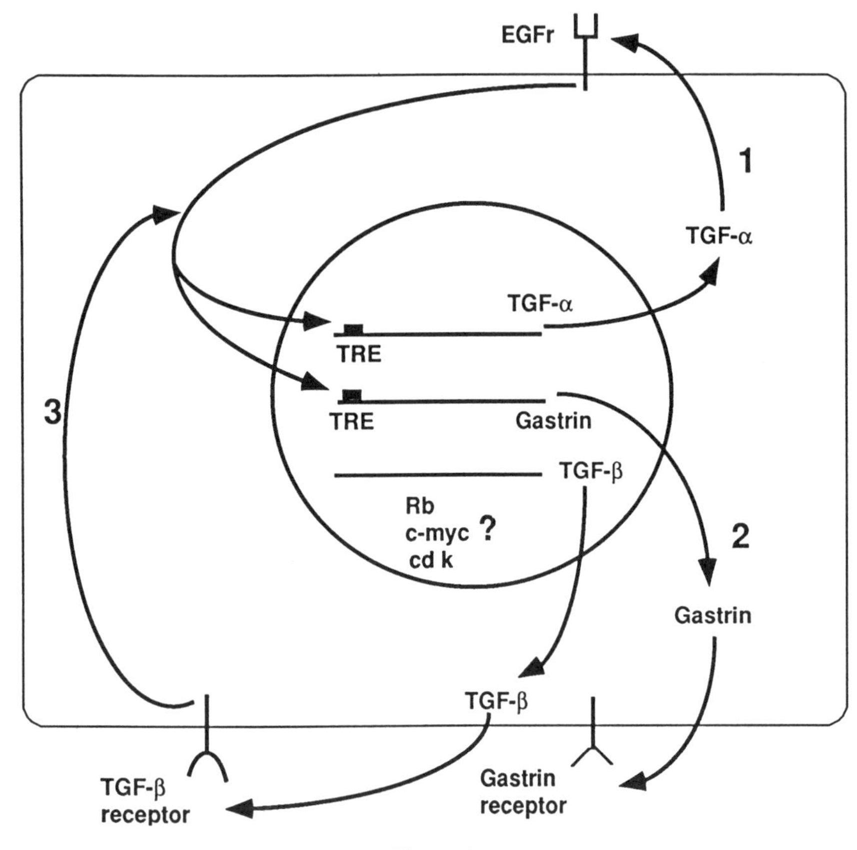

Figure 1.

V. TGF-β's ARE POTENT NEGATIVE REGULATORS OF CANCER CELLS

TGF-β Family. TGF-β's are a large family of structurally related peptide growth factors with diverse biological activities (Roberts and Sporn, 1990; Massague, 1990). Three human forms designated TGF-β_1, β_2 and β_3, respectively have been identified. The TGF-β's are potent inhibitors of cell proliferation in a variety of cell types including many types of cancer cells. Inhibition of colon cancer cells by exogenous TGF-β treatment is also associated with the acquisition of a more differentiated phenotype which includes the induction of extracellular matrix proteins and carcinoembryonic antigen and the repression of c-myc (Hoosein et al., 1987, Chakrabarty et al., 1988, 1989; Mulder and Brattain, 1988; Mulder et al., 1988a,b). Although the 3 forms of TGF-β show some differences in potency (Graycar et al., 1989) and are differentially regulated (Bascom et al., 1989), they have similar biological effects.

TGF-β Receptors and Signal Transduction. TGF-β elicits its responses through high affinity binding proteins which have been classified by their molecular masses in receptor crosslinking experiments (Massague, 1990). These binding proteins have been designated type I, II and III, respectively. Current evidence indicates that type, I and II are involved in signal transduction whereas the type III protein may function as a storage site for TGF-β.

The type I receptor is a 53 kDa glycoprotein, detectable as a 65 kDa affinity-labeled complex with TGF-β monomer (12 kDa) in crosslinking experiments (Massague, 1990). This receptor subtype binds TGF-β_1 and β_2 with high affinity and specificity. Evidence implicating this receptor type in signal transduction includes: (1) the presence of this receptor type in all cell types which respond to TGF-β, (2) the expression in certain hematopoietic cells of only the type I receptor, which binds the TGF-β_1 and β_2 isoforms with affinities that parallel the order of potencies for response, (3) the inability of chemically mutagenized mink lung epithelial cells lacking the type I receptor to respond to TGF-β with growth inhibition, fibronectin induction, and cellular flattening (Boyd and Massague, 1989) and (4) the non-complementarity of mutants with defects in signalling, crossed with mutants lacking the type I receptor (a recessive, non-complimentary trait), in somatic cell hybrid experiments (Laiho et al., 1991).

Additional evidence, however, suggests that expression of both type I and type II receptors is required for TGF-β responsiveness (Laiho et al., 1990, 1991). The 73 kDa type II glycoprotein is detectable as an 85 kDa affinity labeled complex which displays high affinity binding to TGF-β isoforms. Although cell surface expression of this receptor type is not always detectable in cells which respond to TGF-β, the high frequency of isolation of receptor mutants with defects simultaneously in the two receptor types provided evidence that both the type I and II receptors are required for biological responsiveness to TGF-β. Genetic complementation studies supported this hypothesis and indicated that the type I and II receptors may interact in mediating TGF-β responses. The type II receptor has recently been cloned (Lin et al., 1992).

The type III or betaglycan receptor has also been cloned (LopezCassilas et al., 1991; Wang et al., 1991) and consists of a 110–130 kDa (molecular mass with TGF-β monomer bound) core protein containing the receptor binding site attached to approximately 210 kDa of glycan chains. The expression of this proteoglycan is often absent from cells which are TGF-β responsive. It does not appear to transduce cell signals, but may function as a storage site or clearance form for TGF-β.

Some indirect evidence indicating involvement of G proteins in TGF-β response has been reported (Howe et al., 1990) but for the most part signalling mechanisms involved in TGF-β response are not well characterized. However, recently direct evidence implicating $p21^{ras}$ involvement in TGF-β response has been reported (Mulder and Morris, 1992). TGF-β_1 and β_2 elicited a rapid increase in GTP on $p21^{ras}$ in TGF-β sensitive lung and intestinal epithelial cells. The activation of p21 occurred at growth inhibitory concentrations of the TGF-β isoforms, whereas no activation was seen in TGF-β resistant variants of the epithelial cells.

Exogenous TGF-β Causes a G_1/S Block of the Cell Cycle. Several lines of evidence indicate that TGF-β's inhibitory action is based on a G_1/S block of the cell cycle. Treatment of mouse keratinocytes with TGF-β results in down-regulation of *c-myc* (Pietenpol et al., 1990). Treatment with *c-myc* anti-sense oligonucleotides reduced c-myc expression and inhibited cell proliferation as effectively as TGF-β (Pietenpol et al., 1990). Exogenous treatment with TGF-β isoforms was shown to be effective in the inhibition of proliferation of growth factor dependent, but not growth factor independent cell lines (Hoosein et al., 1989). The mechanistic basis for this was investigated and inhibition of DNA synthesis was shown to be associated with a TGF-β induced block of growth factor stimulated *c-myc* and TGF-α mRNA's (Mulder et al., 1990b). Addition of TGF-β prior to the peak of growth factor stimulated upregulation of *c-myc* was necessary to achieve reduction of mitogenesis (Mulder et al., 1990a).

Cell cycle-dependent late G_1 events appear to be targets through which TGF-β mediated growth inhibition might be regulated. Recently, Howe et al. (1991) reported that CCL64 mink lung epithelial cells were reversibly blocked in late G_1 at the G_1/S-phase boundary by exogenous TGF-β. In addition, exogenous TGF-β_1 anti-proliferative activity is associated with a decrease in the phosphorylation of the $PB4^{cdc2}$ kinase. $p34^{cdc2}$ is a serine-threonine kinase whose activity is required both before DNA synthesis (G_1/S boundary) and before mitosis (G_2/M boundary).

Other cell cycle related molecules such as the retinoblastoma gene product (RB) have also been implicated as mediating the anti-proliferative activity of exogenous TGF-β_1. RB has been described as a tumor suppressor gene with a growth inhibitory function (Marshall, 1991). Loss of the RB gene as a result of mutation or deletion results in malignant transformation. In normal cells, RB is expressed throughout the cell cycle but its activity appears to be controlled by phosphorylation. Phosphorylated forms which predominate during the S phase do not have growth suppressive activity, whereas the active underphosphorylated forms predominate during G_0 and G_1 (Buchkovich et al., 1989; Chen et al., 1989; Decaprio et al., 1989; Furukawa et al., 1990; Ludlow et al., 1990). Exogenous treatment with TGF-β_1 prevents or counteracts the activity of a kinase in G_1 which leads to the. phosphorylation of RB, thus blocking entry into S phase (Laiho et al., 1990). Thus, TGF-β_1 and RB appear to function in a common growth-inhibitory pathway in which TGF-β_1 acts to retain RB in the under-phosphorylated, growth-suppressive state. Recently, Kim et al. (1991) showed that TGF-β_1 gene expression was regulated by RB. In addition, several sequences homologous to the *c-fos* RB control element (RCE) have been found in the promoters for TGF-β_1, β_2 and β_3. Therefore, not only may TGF-β control RB function, but TGF-β expression may also be regulated by RB.

TGF-β Autocrine Activity. Direct evidence that TGF-β_1 and β_2 are negative autocrine activities in some types of cancer cells has come from studies with TGF-β neutralizing antibodies. Treatment with antibodies to remove TGF-β activity from tissue culture medium resulted in stimulation of proliferation, increased colony formation in semisolid medium and enhanced mitogenesis (Arteaga et al., 1990; Hafez et al., 1990). These results indicated that autocrine negative activity of TGF-β was operative in control of cell growth. Nevertheless, there are some distinct limitations in using neutralizing antibodies to demonstrate autocrine TGF-β inhibitory activity in transformed cells. One of these is the low level of induction of proliferation or enhancement of anchorage-independent growth seen in these studies as a result of neutralization (Arteaga et al., 1990; Hafez et al., 1990). This may be due to relatively weak effects of autocrine TGF-β on proliferation or to the loss of neutralizing antibody activity during the course of an experiment. A second and perhaps more important limitation is the difficulty of extending neutralizing antibody experiments from tissue culture to animals to determine autocrine effects of TGF-β on the biological behavior of tumors. Therefore, we developed an expression vector for expression of TGF-β_1 anti-sense RNA in order to obtain constitutive repression of TGF-β in cells expressing autocrine activity.

Constitutive Repression of Autocrine TGF-β Leads to Enhanced Tumorigenicity Without Alterations in Growth Rate. It was apparent from our work and the literature that TGF-β isoforms were expressed by colon carcinoma cells as well as other cell types. Given the role of exogenous TGF-β_1 in counter-acting growth factor induced release from quiescence we hypothesized that autocrine TGF-β isoforms might have the same effect and act as a cellular negative regulator of DNA synthesis. Subsequently, we developed a TGF-β_1 anti-sense expression vector to investigate this possibility. Constitutive repression of autocrine TGF-β_1 by this vector in a poorly tumorigenic, well-differentiated growth factor dependent Group III cell line resulted in the acquisition of the ability to form colonies with high efficiency in anchorage independent growth assays and to form tumors with 100% efficiency in athymic mice at cell inocula of control cells which would not form tumors (Wu et al., 1992). Significantly, the repression of autocrine TGF-β_1 did not alter growth rates, but did reduce the lag time of transfected cells prior to their achievement of exponential growth. These results showed that autocrine TGF-β_1 repressed tumorigenicity in FET cells. The lag time prior to attainment of exponential growth is generally associated with the time required for the inoculated cells to condition the tissue culture medium with sufficient concentrations of positive factors necessary for optimal growth. Therefore, the reduction in lag time by the transfected cells suggested a more growth factor independent phenotype, an attribute associated with progression to a high level of tumorigenicity in this model system.

Autocrine TGF-β Maintains Quiescence and Growth Factor Dependence for Release from Quiescence. The hypothesis that autocrine TGF-β repressed growth factor independence was investigated in two Group III differentiated cell lines, designated FET and CBS. Unlike FET cells, CBS did not express high levels of any TGF-β isoform in exponentially growing cells. Thus, they did not exhibit autocrine TGF-β activity in exponential phase. However, when the cells were made quiescent after nutrient and growth factor deprivation, negative autocrine TGF-β_1 and β_2 activity was demonstrated as nutrient stimulation of DNA synthesis by treatment with TGF-β neutralizing antibodies resulting in ^{3}H-thymidine incorporation levels of 2–3 fold over those of nutrient stimulated control cells. These results showed that TGF-β_1 and β_2 were autocrine growth factors in non-dividing quiescent cells, but not exponential phase cells. Moreover, the results suggested that the acquisition of the quiescent state could induce autocrine expression of TGF-β's.

TGF-β_1 and β_2 mRNA levels were determined by Northern analysis of cells harvested daily during the establishment of quiescence. Both isoforms were expressed at low levels in pre-quiescent cells, but were expressed at high levels in quiescence. Active and latent TGF-β_1 and β_2 protein levels were also determined. There were no detectable levels of active polypeptide in the conditioned media of pre-quiescent cells. The quiescent cells secreted 4–5 fold higher levels of both proteins and approximately 5% of the activity of both isoforms was present in the active form.

Quantitative RNAse protection assays were developed in order to determine whether the signal for increased TGF-β expression was cell contact or quiescence. TGF-β_1 was induced only by quiescence. TGF-β_2 was predominantly induced by quiescence, but was also induced by cell contact. The mechanisms underlying the increased expression of TGF-β_1 and β_2 in quiescent cells were investigated by nuclear run-on analysis and mRNA half-life determinations. Increased TGF-β_1 mRNA levels appeared to be exclusively due to an increase in stability whereas TGF-β_2 mRNA levels were due to increased transcription.

These studies showed that TGF-β_1 and β_2 are autocrine negative factors which can be situationally expressed by cells in response to their growth state. The control of this expression is complex, but in CBS cells is distinct for each isoform, thus providing potential flexibility for response to different stimuli. Autocrine TGF-β does not affect exponentially growing cells (Wu et al., 1992), but rather appears to function by maintaining a quiescent state and/or by blocking progression into the cell cycle.

Autocrine TGF-β Maintains the Quiescent State by Repressing TGF-α and EGFr Expression. Additional studies showed that autocrine TGF-β_1 has the function of maintaining cells in a quiescent, non-dividing state and appears to carry this out through the modulation of the expression of the components of the TGF-α autocrine loop. Repression of autocrine TGF-β_1 in FET cells by a TGF-β_1 antisense cDNA altered the growth regulatory phenotype of the cells with respect to the establishment of and release from quiescence. A high anti-sense expressing clone designated FET B and a pool of FET TGF-β_1 anti-sense transfected clones were more resistant than control cells to the establishment of quiescence by nutrient and growth factor depletion as they showed 2-fold higher levels of ^{3}H-thymidine incorporation until 6 days after the initiation of depletion. Quiescent FET cells showed little or no induction of DNA synthesis unless they were replenished with both nutrients and growth factors (Mulder and Brattain, 1989b). In contrast, nutrient replenishment alone was as effective in stimulating DNA synthesis by quiescent TGF-β_1 anti-sense transfected cells as growth factor and nutrient mediated stimulation of parental cells. Furthermore, TGF-β_1 anti-sense transfected cells showed a higher capacity for mitogenesis than control FET cells. These results indicated that one function of autocrine TGF-β_1 is the maintenance of cells in the quiescent state in the absence of exogenous stimulation, as well as the attenuation of the degree of response to exogenous signals. They also suggested that repression of autocrine TGF-β_1 leads to an imbalance between positive and negative controls of the cell cycle in the favor of positive factors and growth factor independence.

Early studies in this project indicated that exogenous TGF-β_1 abrogated TGF-α expression in FET cells released from quiescence (Mulder et al., 1990b). Therefore, we determined the effects of autocrine TGF-β_1 on the expression of the components of the TGF-α autocrine system. Surprisingly, repression of autocrine TGF-β_1 led to the reduction of both TGF-α and EGFr mRNA levels in exponential growth phase. This suggested autocrine TGF-β_1 controlled expression of these components. TGF-α promoter-CAT reporter assays indicated that the repression of TGF-α mRNA was transcriptionally regulated in TGF-β_1 anti-sense transfected cells. In addition, it was found that exogenous TGF-β_1 treatment of untransfected FET cells induced transcription of TGF-α in exponential phase cells.

In contrast to exponential phase cells, both TGF-α and EGFr were increased several fold in quiescent TGF-β_1 anti-sense transfected cells. Therefore, autocrine TGF-β_1 activity either directly or indirectly attenuates the expression of this autocrine stimulatory loop in quiescent cells. An increase in TGF-α autocrine loop activity concomitant with the loss of TGF-β_1 autocrine activity appears to lead to growth factor independence and relaxation of cell cycle control.

CBS cells represent a potentially more complex system than FET cells since they express autocrine TGF-β_2 as well as TGF-β_1 and express both isoforms as a function of quiescence and/or cell contact (see above). The results with the FET TGF-β_1 anti-sense transfected cells described above and in the published tumorigenicity studies indicated that autocrine TGF-β functioned in quiescent or non-dividing states rather than actively proliferating cells. The restriction of CBS autocrine activity to non-active growth states offered the opportunity to further test this conclusion by determining whether poorly tumorigenic CBS cells would progress toward a highly tumorigenic, growth factor independent phenotype if autocrine TGF-β activity was repressed. However, a potential technical issue involved the use of the TGF-β_1 anti-sense expression system to accomplish this.

Would the TGF-β_1 anti-sense mRNA lead to repression of TGF-β_2 expression as well as TGF-β_1? Transfection of CBS cells led to repression of both TGF-β_1 and β_2 autocrine activity. Moreover, this repression was due to direct action of anti-sense mRNA rather than the control of TGF-β_2 expression by TGF-β_1 as has been shown for some other model systems (Bascom et al., 1989). Repression of autocrine TGF-β_1 and β_2 expression also led to dramatic increases in CBS tumorigenicity. Therefore, we conclude that it is the autocrine TGF-β expressed at quiescence which is critical for repression of tumorigenicity.

The next issue was whether repression of autocrine TGF-β activity in CBS cells led to growth factor independence as observed for FET cells. TGF-β_1 anti-sense transfected CBS cells demonstrated the same growth factor independence exhibited by FET cells. This raised the question of TGF-α autocrine loop expression in these cells. Again, like FET cells, quiescent TGF-β_1 anti-sense transfected cells showed increased expression of TGF-α and the EGFr. Growth factor independence of the TGF-β anti-sense transfected cells could be reversed by antibodies to the EGFr. These results show that one mechanism by which autocrine TGF-β activity restricts the movement of cells through the cell cycle in the absence of exogenous growth factors is by repressing autocrine TGF-α expression.

VI. SUMMARY

In summary, we have described growth factor dependent and independent phenotypes of colon carcinoma cell lines. Growth factor dependence and unaggressive biological behavior as evidenced by reduced tumorigenicity was tightly coupled to the expression of a TGF-β autocrine loop. The TGF-β autocrine loop maintains cells in a quiescent or nondividing state in the absence of exogenous growth factor stimulation. One way in which it carries out this function out is by repressing TGF-α expression (represented by loop 3 in Figure 1).

TGF-α autoinduces its own transcription and thus tends to promote its own autocrine activity. Progressed, growth factor independent cells have a strong autocrine loop which is autoinductive (represented by loop 1 in Figure 1) through a TGF-α responsive element in the TGF-α promoter (TRE in Figure 1). Gastrin also contains a TRE in its promoter. Therefore, the autoinduction of TGF-α is also accompanied by gastrin induction (represented by loop 2). The presence of these and perhaps other positive loops in growth factor independent cells leads to a phenotype which constitutively expresses positive acting molecules such as TGF-α and c-myc throughout the cell cycle. This effectively removes the cell from any extracellular growth controls, thereby leading to a highly aggressive and tumorigenic phenotype.

REFERENCES

Aaronson, S.A. (1991). Growth factors and cancer. Science 254, 1146–1153.

Arteaga, C.L., Coffey, R.J., Dugger, T.C., McCutchen, C.M., Moses, H.L., & Lyons, R.M. (1990). Growth stimulation of human breast cancer cells with anti-transforming growth factor-β antibodies: evidence for negative autocrine regulation by transforming growth factor-β. Cell Growth and Diff. 1, 367–374.

Baldwin, G.S., Casey, A., Mantamidiotis, T., McBride, K., Sizeland, A.M., & Thunwood, C.M. (1990). PCR cloning and sequence of gastrin mRNA from carcinoma cell lines. Biochem. Biophys. Res. Comm. 170, 691–697.

Bascom, C.C., Wolfshohl, J.R., Coffey, R.J., Madison, K., Webb, N.R., Purchio, A.R., Derynck, R., & Moses, H.L. (1989). Complex regulation of transforming growth factor β_1, β_2 and β_3 mRNA expression in mouse fibroblasts and keratinocytes by transforming growth factor β_1 and β_2. Mol. Cell. Biol. 9, 5508–5515.

Bates, S.E., Valverius, E.M., Ennis, B.W., Bronzert, D.A., Sheridan, J.P., Stampfer, M.R., Mendelsohn, J., Lippman, M.E., & Dickson, R.B. (1990). Expression of the transforming growth factor alpha/epidermal growth factor receptor pathway in normal human breast epithelial cells. Endocrin. 126, 596–607.

Bejcek, B.E., Li, D.Y., & Deuel, T.F. (1989). Transformation by v-*sis* occurs by an internal autoactivation mechanism. Science 245, 1496–1498.

Boyd, D., Levine, A.E., Brattain, D.E. McKnight, M.K., & Brattain, M.G. (1988). A comparison of growth requirements of two human intratumoral colon carcinoma cell lines in monolayer and soft agarose. Canc. Res. 48, 2469–2474.

Boyd, F.T. & Massague, J. (1989). Transforming growth factor β inhibition of epithelial cell proliferation linked to the expression of a 53 kDa-membrane receptor. J. Biol. Chem. 254, 2272–2278.

Brattain, M.G., Brattain, D.E., Fine, W.D., Khaled, F.M., Marks, M.E., Arcolano, L.A., & Danbury, B.H. (1981). Initiation of cultures of human colon carcinoma with different biological properties on confluent fibroblasts. Onco. Dev. Biol. 2, 355–367.

Brattain, M.G., Fine, W.D., Khaled, F.M., Thompson, J., & Brattain, D.E. (1981). Heterogeneity of malignant cells from a human colonic carcinoma. Canc. Res. 41, 1751–1756.

Brattain, M.G., Levine, A.E., Chakrabarty, S., Yeoman, L.C., Willson, J.K.V., & Long, B.H. (1984). Heterogeneity of human colon carcinoma. Canc. Metastas. Rev. 3, 177–191.

Breborowicz, J., Easty, G.C., Birbach, M., Robertson, D., Nery, R., & Neville, A.M. (1975). The monolayer and organ culture of human colorectal carcinoma and associated normal colonic mucosa and their production of carcinoembryonic antigens. Br. J. Canc. 31, 557–569.

Browder, T.M., Dunbar, C., & Nienhuis, A. (1989). Private and public autocrine loops in neoplastic cells. Canc. Cells 1, 9–17.

Buchkovich, K., Duffy, L.A., & Harlow, E. (1989). The retinoblastoma protein is phosphorylated during specific phases of the cell cycle. Cell 58, 1097–1105.

Chantret, I., Barbat, E., Dussaulx, E., Brattain, M.G., & Zweibaum, A. (1988). Epithelial polarity, villin expression, and enterocyte differentiation of cultured human colon carcinoma cells: a survey of twenty-cell lines. Canc. Res. 48, 1936–1942.

Chakrabarty, S., Jan, Y., Brattain, M.G., Tobon, A., & Varani, J. (1989). Diverse cellular responses elicited from human colon carcinoma cells by TGF-β. Canc. Res. 49, 2112–2117.

Chakrabarty, S., McRae, L.J., Levine, A.E., & Brattain, M.G. (1984). Restoration of normal growth control and membrane antigen comparison in malignant cells by N,N-dimethyl-formamide. Canc. Res. 44, 2181–2185.

Chakrabarty, S., Tobon, A., Varani, J., & Brattain, M.G. (1988). Induction of carcinoembryonic antigen-secretion and modulation of protein secretion/expression and fibronectin/laminin expression in human colon carcinoma cells by TGF-β. Canc. Res. 48, 4059–4064.

Chen, P.-L., Schully, P., Shew, J.-Y., Wang, J.Y.J; Lee, W.-H. (1989). Phosphorylation of the retinoblastoma gene product is modulated during the cell cycle and cellular differentiation. Cell 58, 1193–1198.

Coffey, R.J., Goustin; A.S., Soderquist, A.M., Shipley, G.D., Wolfshohl, J., Carpenter, G., & Moses, H.L. (1987). TGF-α and β expression in human colon cancer lines: implications for an autocrine model. Canc. Res. 47, 4590–4594.

Cole, M.D. (1986). The myc oncogene: its role in transformation and differentiation. Ann. Rev. Genet. 20, 361–384.

DeCaprio, J.A., Ludlow, J.W., Lynch, D., Furukawa, Y., Griffin, J., Piwnica-Worms, H., Huang, C-M., & Livingston, D.M. (1989). The product of the retinoblastoma susceptibility-gene has properties of a cell-cycleregulatory element. Cell 58, 1085–1095.

Derynck, R., Goeddel, D.V.M., Ullrich, A., Gutterman, J.U., Williams, R.D., Bringman, T.S., & Berger, W.J. (1987). Synthesis of mRNAs for transforming growth factor-α and β and the epidermal growth factor receptor by human tumors. Canc. Res. 47, 707–712.

Dockray, G.J. (1978). Gastrin Overview. In: Gut Hormones (Bloom, S.R., ed.), pp. 129–139, Churchill Livingston, New York.

Fearon, E.R. & Vogelstein, B. (1990). A genetic model for colorectal tumorigenesis. Cell 61, 759–767.

Fleming, T.P., Matsui, T., Molloy, C.J., Robbins, K.C., & Aaronson, S.A. (1989). Autocrine mechanism for v-sis transformation requires cell surface localization of internally activated growth factor receptors. Proc. Natl. Acad. Sci. USA 86, 8063–8067.

Furukawa, Y., DeCaprio, J.A., Freedman, A., Kanakura, Y., Nakamura, M., Ernst, T.J., Livingston, D.M., & Griffin, J.D. (1990). Expression and state of phosphorylation of the retinoblastoma susceptibility gene product in cycling and noncycling human hematopoietic cells. Proc. Natl. Acad. Sci. USA 87, 2770–2774.

Gallie, B.L., Holmes, W., & Phillips, R.A. (1982). Reproducible growth in tissue culture of retinoblastoma specimens. Canc. Res. 42, 301–305.

Godley, J.M. & Brand, S.J. (1989). Regulation of the gastrin promoter by EGF and neuropeptides. Proc. Natl. Acad. Sci. USA 86, 3036–3040.

Graycar, J.L., Miller, D.A., Arrick, B.A., Lyons, R.M., Moses, H.L Derynck, R. (1989). Human transforming growth factor β_3: recombinant expression, purification, and biological activities in comparison with transforming growth factor β_1 and β_2. Mol. Endocrin. 3, 1977–1986.

Hafez, M.M., Infante, D., Winawer, S., & Friedman, E. (1990). Transforming growth factor-β_1 acts as an autocrine negative growth regulator in colon enterocytic differentiation but not in goblet cell maturation. Cell Growth and Diff. 1, 617–626.

Hannink, M. & Donoghue, D.J. (1986). Biosynthesis of the v-sis gene product: signal sequence cleavage, glycosylation, and proteolytic processing. Mol. Cell. Biol. 6, 1343–1348.

Hannink, M., Sauer, M.K., & Donoghue, D.J. (1986). Deletions in the C-terminal coding region of the v-sis gene: dimerization if required for transformation. Mol. Cell. Biol. 6, 1304–1314.

Herlyn, M. (1990). Human melanoma: development and progression. Canc. Metast. Rev. 9, 101–112.

Hoosein, N.M., Brattain, D.E., McKnight, M.K., Levine, A.E., & Brattain, M.G. (1987). Characterization of the inhibitory effects of transforming growth factor-β on a human colon carcinoma cell line. Canc. Res. 47, 2950–2954.

Hoosein, N.M., Kiener, P.A., Curry, R.C., & Brattain, M.G. (1990). Evidence for autocrine growth stimulation of human colon carcinoma cells by a gastrin-like polypeptide. Exp. Cell. Res. 186, 15–21.

Hoosein, N.M., Kiener, P.A., Curry, R.C., McGilbra, D.K., & Brattain, M.G. (1988). Anti-proliferative effects of gastrin receptor, antagonists and antibodies to gastrin on human colon carcinoma cell lines. Canc. Res. 48, 7179–7183.

Hoosein, N.M., McKnight, M.K., Levine, A.E., Mulder, K.M., Childress, K.E., Brattain, D.E., & Brattain, M.G. (1989). Differential sensitivity of subclasses of human colon carcinoma cell lines to the growth inhibitory effects of transforming growth factor-β_1. Exp. Cell Res. 181, 442–453.

Howe, P.H., Draetta, G., & Leoff, E.B. (1991). Transforming growth factor-β_1 inhibition of $p34^{cdc2}$ phosphorylation and histone H1 kinase activity is associated with G_1/S-phase growth arrest. Mol. Cell. Biol. 11, 1185–1194.

Huang, S.S. & Huang, J.S. (1988). Rapid turnover of the platelet-derived growth factor receptor in sis-transformed cells and reversal by suramin. Implications for the mechanism of autocrine transformation. J. Biol. Chem. 263, 12608–12618.

Huang, J.J., Yee, J.K., Shew, J.Y., Chen, P.L., Bookstein, R., Friedmann, T., Lee, E.Y., & Lee, W.H. (1988). Suppression of the neoplastic phenotypes by replacement of the RB gene in human cancer cells. Science 242, 1563–1565.

Kariya, Y., Kiato, K., Hayashizaki, Y., Himeno, S., Tarui, S., & Matsubara, K. (1986). Gene 50, 342–350.

Kim, S.J., Lee, H.D., Robbins, P.D., Busam, K., Sporn, M.B., & Roberts, A.B. (1991). Regulation of transforming growth factor β_1 gene expression by the product of the retinoblastoma-susceptibility gene. Proc. Natl. Acad. Sci. USA 88, 3052–3056.

Kusyk, C.J., McNiel, N.O., & Johnson, L.R. (1986). Stimulation of growth of a colon cancer cell line by gastrin. Am. J. Physiol. 251, G597–G601.

Laiho, M., DeCaprio, J.A., Ludlow, W.J., Livingston, D.M., & Massague, J. (1990). Growth inhibition by TGF-β linked to suppression of retino blastoma protein phosphorylation. Cell 62, 175–185.

Laiho, M., Weis, F., Boyd, F., Ignotz, R., & Massague, J. (1991). Responsiveness to TGF-β restored by genetic complementation between cells defective in TGF-β receptors I and II. J. Biol. Chem. 266, 9108–9112.

Laiho, M., Wels, F., & Massague, J. (1990). Concomitant loss of TGF-β receptor type I and II in TGF-β resistant cell mutants implicates both receptor types in signal transduction. J. Biol. Chem. 265, 18518–18524.

Leibovitz, A., Stinson, J.C., McCombs, W.B., McCoy, C.E., Masur, K. Mabry, N.D. (1976). Classification of human colorectal adenocarcinoma cell lines. Canc. Res. 36, 4562–4569.

Lin, H.Y., Wang, X.F., Eaton, E., Weinberg, R.A., & Lodish, H.F. (1992). Expression cloning of the TGF-β type II receotir, a functional transmembrane serine/threonine kinase. Cell 68, 1–20.

Lokeshwar, V.B., Huang, S.S., & Huang, J.S. (1990). Intracellular turnover, novel secretion, and mitogenically active intracellular forms of v-sis gene product in simian sarcoma virus-transformed cells. Implications for intracellular loop autocrine transformations. J. Biol. Chem. 265, 1665–1675.

Lopez-Cassilas, F., Cheifetz, S., Doody, J., Andres, J.L., Kane, W.S., & Massague, J. (1991). Structure and expression of the membrane proteolycan, a component of the TGF-β receptor system. Cell 67, 785–797.

Ludlow, J.W., Shon, J., Pipas, J.M., Livingston, D.M., & DeCaprio, A. (1990). The retinoblastoma susceptibility gene product undergoes cell cycle-dependent phosphorylation binding to and release from SV 40 and large T-antigen. Cell 60, 387–396.

Massague, J. (1990). The transforming growth factor-β family. Ann. Rev. Cell. Biol. 6, 597–641.

Merchant, J.L., Demediuk, B., & Brand, S.J. (1991). A GC-rich element confers EGF responsiveness to transcription from the gastrin promoter. Mol. Cell. Biol. 11, 2686–2696.

Mulder, K.M. (1991). Differential regulation of c-myc and transforming growth factor-α messenger RNA expression in poorly-differentiated and well-differentiated colon carcinoma cells during the establishment of a quiescent state. Canc. Res. 51, 2256–2262.

Mulder, K.M. & Brattain, M.G. (1989a). Growth factor expression and response in human colon carcinoma cells. In: The Cell and Molecular Biology of Colon Cancer (Augenlicht, L., ed.), pp. 45–67, CRC Press, Boca Raton, FL.

Mulder, K.M. & Brattain, M.G. (1989b). Effects of growth stimulatory factors on mitogenicity and c-myc expression in poorly differentiated and well-differentiated human colon carcinoma cells. Mol. Endocrin. 3, 1215–1222.

Mulder, K.M. & Brattain, M.G. (1988). Alterations in c-myc expression in relation to maturational status of human colon carcinoma cells. Int. J. Canc. 42, 64–70.

Mulder, K.M., Humphrey, L.E., Choi, H.G., Childress-Fields, K.E., & Brattain, M.G. (1990). Evidence for c-myc in the signalling pathway for TGF-β in well-differentiated human colon carcinoma cells. J. Cell. Physiol. 145, 501–507.

Mulder, K.M., Levine, A.E., Hernandez, C., McKnight, M.K., Brattain, D.E. & Brattain, M.G. (1988). Modulation of c-myc by transforming growth factor-β in human colon carcinoma cells. Biochem. Biophys. Res. Comm. 2, 711–716.

Mulder, K.M. & Morris, S.L. (1992). Activation of $p21^{ras}$ by transforming growth factor-β in epithelial cells. J. Biol. Chem. 287, 5029–5031.

Mulder, K.M., Ramey, M.K., Hoosein, N.M., Levine, A.E., Hinshaw, X., Brattain, D.E., & Brattain, M.G. (1988). Characterization of transforming growth factor-β resistant subclones isolated from a transforming growth factor-β-sensitive human colon carcinoma cell line. Canc. Res. 48, 7120–7125.

Mulder, K.M., Zhong, Q., Choi, H.G., Humphrey, L.E., & Brattain, M.G. (1990). Inhibitory effects of transforming growth factor-β_1 on mitogenic response, transforming growth factor-α, and c-myc in quiescent, well-differentiated human colon carcinoma cells. J. Cell. Physiol. 145, 501–507.

Munro, S. & Pelham, H.R. (1987). A C-terminal signal prevents secretion of ER proteins. Cell 48, 899–907.

Murakami, H. & Masui, H. (1980). Hormonal control of human colon carcinoma cell growth in serum-free medium. Proc. Natl. Acad. Sci. USA 77, 3464–3468.

Mydlo, J.H. Michaeli, J., Cordon-Cardo, C., Goldenberg, A.S., Heston, W.D. & Fair, W.R. (1989). Expression-of-transforming growth factor alpha and epidermal growth factor receptor messenger RNA in neoplastic and nonneoplastic human kidney tissue. Canc. Res. 49, 3407–3411.

Pardee, A.B. (1989). G_1 events and regulation of cell proliferation. Science 246, 603–608.

Pelham, H.R. (1988). Evidence that luminal ER proteins are sorted from secreted proteins in a post-ER compartment. EMBO J. 7, 913–918.

Pientenpol, J.A., Holt, J.T., Stein, R.W., & Moses, H.L. (1990). Transforming growth factor-β_1 suppression of c-myc gene transcription: role in inhibition of keratinocyte proliferation. Proc. Natl. Acad. Sci. USA 87, 3758–3762.

Roberts, A.B. & Sporn, M.B. (1990). The transforming growth factor-betas. In: Peptide growth factors and their receptors (Sporn, M.B. & Roberts, A.B., eds.), pp. 419–472, Springer-Verlag, Heidelberg.

Sirinek, K.R., Levine, B.A., & Moyer, M.P. (1985). Pentagastrin stimulates in vitro growth by normal and malignant human colon epithelial cells. Am. J. Surg. 149, 35–39.

Smith, H.S., Lan, S., Ceriani, R., Hackett, A.J., & Stampfer, M.R. (1981). Clonal proliferation of cultured nonmalignant and malignant human breast epithelia. Canc. Res. 41, 4637–4643.

Smith, J.J., Derynck, R., & Korc, M. (1987). Production of transforming growth factor alpha in human pancreatic cancer cells: evidence for a superagnoist autocrine cycle. Proc. Natl. Acad. Sci. USA 84, 7567–7570.

Sorrentino, V., Drozdoff, V., McKinney, M.D., Zeitz, L., & Fleissner, E. (1986). Potentiation of growth factor activity by exogenous c-myc expression. Proc. Natl. Acad. Sci. USA 83, 8167–8171.

Stiles, C.D. 91983). The molecular biology of platelet derived growth factor. Cell 33, 653–655.

Walsh, J.G. (1981). Gastrin. In: Gut Hormones (Bloom, S.R., & Polak, J.M., eds.), pp. 163–170, Churchill Livingston, New York.

Wan, C.W., McKnight, M.K., Brattain, D.E., Brattain, M.G., & Yeoman, L.C. (1988). Different epidermal growth factor responses and receptor levels in human colon carcinoma cell lines. Canc. Lett. 43, 139–143.

Wang, X.F., Lin, H.Y., Eaton, N., Downward, J., Lodish, H.F., & Weinberg, R.A. (1991). Expression cloning and characterization o the TGF-β type III receptor. Cell 67, 797–805.

Watson, S.A., Durrant, L.G., Crosbie, J.D., & Morris, D.L. (1989). The in vitro growth response of primary human colorectal and gastrin cancer cells to gastrin. Int. J. Canc. 43, 692–696.

Winsett, O.E., Townsend, C.M., Glass, E.J., & Thompson, J.C. (1986). Gastrin stimulates growth of colon cancer. Surg. 99, 302–307.

Wu, S.P., Theodorescu, D., Kerbel, R., Willson, J.K.V., Mulder, K.M., Humphrey, L.E., & Brattain, M.G. (1992). TCF-β_1 is an autocrine negative growth regulator of human colon carcinoma FET cells in vivo as revealed by transfection of an anti-sense expression vector. J. Cell. Biol. 116, 186–197.

Ziober, B.L., Willson, J.K.V., Humphrey, L.E., Childress-Fields, K.E. Brattain, M.G. (1992). Evidence for an intracellular TGF-α autocrine loop in HCT 116 colon carcinoma cells. J. Biol. Chem., in press.

ALTERED SIGNAL TRANSDUCTION IN CARCINOGENESIS[1]

Catherine A. O'Brian, Nancy E. Ward, and Constantin G. Ioannides

Advances in Molecular and Cell Biology
Volume 7, pages 61–88

ISBN: 1-55938-624-X

I. INTRODUCTION

Recent advances in the understanding of carcinogenesis at the molecular level (Bishop, 1991; O'Brian, 1989) have led to the concept of carcinogenesis as a progressive disorder in signal transduction (Weinstein, 1990). A large body of evidence indicates that oncogene activation and tumor suppressor gene inactivation are critical genetic events in carcinogenesis (Bishop, 1991; Hunter, 1991). Numerous oncogene products and tumor suppressor gene products are key components of signal transduction pathways (Weinstein, 1990; Bishop, 1991; Hunter, 1991). For example, oncogene products include protein-tyrosine kinases (Hunter et al., 1990), growth factors (Aaronson, 1991), growth factor receptors (Ullrich and Schlessinger, 1990; Aaronson, 1991), G proteins, and transcription factors (Bishop, 1991). Studies in several organ systems indicate that carcinogenesis is a multistep process that entails initiation, promotion, and progression, which, are qualitatively distinct processes (Weinstein, 1991). Altered signal transduction is implicated in each of the three stages of the disease. Initiation is an event that is associated with DNA damage, and it is thought that mutagenic events causing oncogene activation and tumor suppressor gene inactivation account for this stage of carcinogenesis (Weinstein, 1990). The phorbol ester tumor promoter receptor is a critical target of tumor promoter action, and this receptor protein is protein kinase C (PKC), a protein-ser/thr kinase that is activated by phorbol ester tumor promoters and by the second messenger *sn*-1,2-diacylglycerol (O'Brian, 1989; Weinstein, 1990). Recent studies also provide evidence that signal transduction plays important roles in metastasis, which is itself a multistage process (Miller and Heppner, 1990; Radinsky, 1991).

In order to illustrate the usefulness of altered signal transduction as a conceptual framework for the molecular events associated with carcinogenesis, this review will focus on altered signal transduction in colon cancer and in the multidrug resistance (MDR) phenotype of tumor cells, which is at least in part responsible for the intrinsic drug resistance commonly observed in colon cancer. The paper is organized into five sections, which are: (1) Altered expression of PKC activity in colon carcinogenesis; (2) Altered regulation of PKC activity by environmental factors that influence the development of colon cancer; (3) Oncogene activation and tumor suppressor gene inactivation: Critical events in human colon carcinogenesis; (4) Role of altered signal transduction in the intrinsic drug resistance of human colon cancer; and (5) Tumor antigen transport and the multidrug resistance (MDR) phenotype—a hypothesis.

II. ALTERED EXPRESSION OF PKC ACTIVITY IN COLON CARCINOGENESIS

Measurements of the level of PKC activity in normal, premalignant, and transformed human colonic epithelial tissues provide evidence that PKC plays a role in proliferative disorders of the intestinal mucosa (Guillem et al., 1987; Kopp et al., 1991; Sakanoue et al., 1991). A comparison of the PKC activity present in different segments of normal human intestinal mucosa reveals that the level of PKC activity is highest in the distal ileum, lowest in the rectum, and intermediate in the intervening segments (Kopp et al., 1991). Membrane-associated PKC is generally thought to be the activated form of the enzyme *in vivo*, and the percentage of PKC activity that is membrane-associated is approximately the same in each segment (Kopp et al., 1991). Because carcinogenesis is a rare event in the small intestine, these results indicate a positive correlation between reduced PKC activity levels and increased cancer risk in the intestinal mucosa of healthy individuals. This correlation is also observed when the PKC activity level in the colonic epithelium of healthy subjects is compared with the PKC activity level in the uninvolved colonic mucosa of colon cancer patients (Sakanoue et al., 1991). Increased cellular proliferation is observed in the uninvolved colonic mucosa of colon cancer patients and in the colonic mucosa of persons at high risk for colon cancer. A significant reduction in the level of PKC activity is observed in both cytosolic and particulate fractions of the uninvolved colonic mucosa of colon cancer patients when compared with normal controls, and the reduction in PKC activity is accompanied by an increased percentage of membrane-associated PKC activity (Sakanoue et al., 1991). Furthermore, colonic adenomas appear to represent a premalignant and hyperproliferative state of the intestinal mucosa, and the level of PKC activity is significantly reduced in human colon adenomas when compared with either adjacent normal- appearing mucosa or colonic mucosa of control subjects (Kopp et al., 1991). Taken together, the levels of PKC activity observed in various intestinal mucosa specimens that are associated with increased cancer risk provide evidence that reduction in the level of PKC activity in the colonic epithelium may be an early event in colon carcinogenesis.

In human colon carcinomas, the level of PKC activity is significantly reduced with respect to adjacent normal mucosa in both cytosolic and particulate fractions (Guillem et al., 1987; Kopp et al., 1991). Taken together with the studies discussed above, this suggests that a progressive downregulation of PKC activity occurs during colon carcinogenesis. PKC isozymes are downregulated at different rates in cultured mammalian cells, and the relative rates of downregulation among the isozymes depend on the cell type studied (Cooper et al., 1989; Huang et al., 1989). Therefore, if PKC expression is downregulated during human colon carcinogenesis, it appears likely that the PKC isozyme profiles of normal human intestinal mucosa, colonic adenomas, and colon carcinomas would differ. It is not yet known whether PKC expression is reduced in human colonic adenomas and carcinomas. Reduced PKC expression could allow the use of PKC as a marker for the detection of human colon cancer, particularly if specific PKC isozymes could be used to distinguish normal and transformed tissues. Furthermore, the potential usefulness of PKC isozyme expression as an intermediate marker in chemoprevention trials is suggested by the reduced level of PKC activity in premalignant lesions of the human intestinal mucosa and in the uninvolved colonic mucosa of colon cancer patients.

Chronic activation of PKC by phorbol-ester tumor promoters generally results in downregulation of the enzyme in mammalian cells (O'Brian and Ward, 1989). The level of the endogenous PKC activator *sn*-1,2-DAG is elevated in a variety of oncogene-transformed cultured fibroblasts (Wolfman et al., 1987) including ras-transformed lines (Fleischman et al., 1986; Wolfman et al., 1987), and PKC is expressed at a reduced level in the oncogene-transformed cells (Wolfman et al., 1987; Weyman et al., 1988). It is therefore thought that *sn*-1,2-DAG mediates a partial downregulation of PKC in the fibroblasts (Wolfman et al., 1987; Weyman et al., 1988). Since 30–40% of human colon cancers display mutations in codon 12 of c-K-*ras* (Vogelstein et al., 1986), it would seem plausible that the reduced level of PKC activity observed in human colon carcinomas (Guillem et al., 1987; Kopp et al., 1991) and premalignant adenomas (Kopp et al., 1991) could be a consequence of a partial downregulation of PKC in these tissues by chronically elevated *sn*-1,2-DAG. In fact, the increased percentage of membrane-associated PKC activity in the uninvolved colonic mucosa of colon cancer patients (Kopp et al., 1991; Sakanoue et al., 1991) and in colon carcinomas (Kopp et al., 1991) suggests the presence of elevated levels of endogenous PKC activators in these tissues. However, in contrast with oncogene-transformed cultured fibroblasts, human colon carcinomas have significantly reduced levels of *sn*-1,2-DAG compared to adjacent normal mucosa specimens, whether or not the tumors express mutations in codon 12 of c-K-*ras* (Sauter et al., 1990; Phan et al., 1991). A significant reduction in the level of *sn*-1,2-DAG has also been observed in human premalignant colonic adenomas (Sauter et al., 1990), but the *sn*-1,2-DAG levels in the uninvolved colonic mucosa of colon cancer patients and in the normal mucosa of patients without colonic abnormalities are indistinguishable (Sauter et al., 1990). The reduction in *sn*-1,2-DAG appears to be an early event in the adenoma-carcinoma sequence of human colon carcinogenesis, since it is observed in both premalignant adenomas and carcinomas. Thus, *sn*-1,2-DAG levels in human colon adenomas and carcinomas cannot account for the reduced PKC activity levels observed in these tissues.

Overexpression of rat brain PKC-β_1 in human colon cancer HT29 cells inhibits the growth of the cells *in vitro* in the presence of TPA and markedly reduces their tumorigenicity in nude mice (Choi et al., 1990). Thus, it appears that at least one PKC isozyme can function as a tumor suppressor in human colon cancer cells (Choi et al., 1990). It will therefore be important to the understanding of the regulation of growth in normal and transformed colonic epithelial tissues to ascertain the basis for the reduced level of PKC activity observed in human colon adenomas and carcinomas. PKC is a family of at least 10 closely related isozymes (Bell and Burns, 1991). Clearly, the first step toward identifying the basis for altered PKC activity levels in human colon adenomas and carcinomas should be to determine PKC isozyme expression in the tissues through measurements of PKC isozyme immunoreactivity and message abundance. Prolonged exposure to phorbol-ester PKC activators downregulates PKC by increasing the rate of degradation of the enzyme (Young et al., 1987; Borner et al., 1988; Cooper et al., 1989; Huang et al., 1989) without affecting the abundance of its message (Young et al., 1987) or its rate of synthesis (Young et al., 1987; Borner et al., 1988) in several cell systems. If the expression levels of some PKC isozymes are reduced in the transformed colon tissues and their message levels are unchanged, this would suggest that the PKC isozymes are degraded at an accelerated rate in the transformed tissues. Such a finding would justify both the measurement of calpain expression in the tissues, because calpain is the protease most clearly implicated in the degradation of cellular PKC (Kikkawa et al., 1989; Pontremoli et al., 1990), and the measurement of the expression of the catalytic fragment of PKC, which is a fully active and lipid-independent catalytic domain of PKC that is generated from the activated enzyme by calpain (Kikkawa et al., 1989). These studies could provide direct evidence of accelerated PKC degradation in the transformed tissues, by determining the level of the enzyme primarily responsible for PKC degradation and the level of a major product of PKC degradation in the tissues. Furthermore, the expression of the catalytic fragments of PKC isozymes is of interest in itself, because the catalytic fragment may play an important role in signal transduction (Pontremoli et al., 1986; Farago and Nishizuka, 1990; Pontremoli et al., 1990). On the other hand, PKC isozyme expression may be unchanged in the transformed colonic epithelial tissues, and altered expression of phosphatases or endogenous PKC inhibitors may account for the altered levels of PKC activity observed in the transformed human colon mucosa.

The development of 1,2-dimethylhydrazine (DMH)-induced colon cancer in the rat provides an important experimental model of colon carcinogenesis, since the pathology of the DMH-induced tumors resembles that of human colon cancer (LaMont and O'Gorman, 1978). The importance of this experimental model to the understanding of human colon cancer is further indicated by the high frequency of point mutations in c-K-*ras* observed in both human colon tumors (Vogelstein et al., 1986) and DMH-induced rat colon tumors (Jacoby et al., 1989). Similarities between human colon carcinogenesis (Guillem et al., 1987; Kopp et al., 1991; Sakanoue et al., 1991) and DMH-induced rat colon carcinogenesis (Baum et al., 1990; Wali et al., 1991) have also been observed in studies of the level of PKC activity and its intracellular distribution in normal, premalignant, and transformed colonic epithelial tissues. Prior to the development of premalignant lesions or tumors in the colonic epithelium of DMH-treated rats, the percentage of membrane-associated PKC activity is increased, and the total level of PKC activity is reduced in the colonic epithelium of the carcinogen-treated rats (Baum et al., 1990). The altered level and intracellular distribution of PKC activity in the preneoplastic colonic mucosa of DMH-treated rats (Baum et al., 1990) parallel the alterations in PKC activity observed in the uninvolved colonic epithelium in human colon cancer (Sakanoue et al., 1991). Furthermore, as in the case of human colon cancer (Guillem et al., 1987; Kopp et al., 1991; Sakanoue et al., 1991), the level of PKC activity in the uninvolved colonic epithelium of tumor-bearing DMH-treated rats is reduced compared with control rats, and an even sharper decline in PKC activity is observed in the colon tumors themselves (Wali et al., 1991).

The alterations in colonic epithelial PKC activity observed in human and DMH-induced rat colon carcinogenesis (Guillem et al., 1987; Baum et al., 1990; Kopp et al., 1991; Sakanoue et al., 1991; Wali et al., 1991) lend support to the hypothesis that the progressive loss of PKC activity is an important event in both early and late stages of human colon carcinogenesis. In the experimental rat model, increased membrane association of PKC activity is observed in the preneoplastic colonic mucosa prior to any alteration in the total level of PKC activity (Baum et al., 1990). This suggests that PKC activation may be an early event in colon carcinogenesis that is followed by downregulation of the enzyme (Baum et al., 1990). Studies of PKC expression in the colonic epithelium of DMH-treated rats will be needed, however, to ascertain whether translocation and downregulation of PKC actually occur during rat colon carcinogenesis.

The striking similarities between human and DMH-induced rat colon carcinogenesis with respect to PKC activity indicate the relevance of the experimental rat model to the elucidation of the role of PKC in human colon carcinogenesis. However, while the level of the PKC activator *sn*-1,2-DAG is reduced in human colon tumors (Sauter et al., 1990; Phan et al., 1991), the *sn*-1,2-DAG level is actually increased in DMH-induced rat colon tumors (Wali et al., 1991). Thus, increased *sn*-1,2-DAG levels may contribute to the reduction in PKC activity observed in rat colon tumors by causing translocation and partial downregulation of the enzyme (Wali et al., 1991). It will be important to ascertain whether a different endogenous PKC activator promotes PKC downregulation in human colon cancer.

Altered Regulation of PKC Activity by Environmental Factors that Influence the Development of Colon Cancer. A number of dietary factors implicated in human colon carcinogenesis modulate PKC activity (O'Brian and Ward, 1989; O'Brian and Ward, 1991). These factors include agents that increase cancer risk, such as bile acids (Fitzer et al., 1987), and others that may protect against colon cancer incidence, such as vitamin D (Wali et al., 1990). Epidemiological studies indicate a positive correlation between high fat diets and increased colon cancer risk (Guillem et al., 1987b). High fat diets are associated with elevated levels of bile acids and dietary fat metabolites in the colonic lumen. Bile acids function as endogenous tumor promoters according to animal models of colon (Narisawa et al., 1974; Reddy et al., 1976), liver (Cameron et al., 1982), and stomach (Furihata et al., 1987) carcinogenesis and tissue culture models of cell transformation (Kaibara et al., 1984). For example, intrarectal instillation of deoxycholate in MNNG-treated rats causes an increase in the incidence of colon adenocarcinomas, whereas no tumors form in rats treated with deoxycholate alone (Reddy et al., 1976). Similarly, deoxycholate promotes the transformation of MNNG-treated

C3H10T½ mouse fibroblasts under conditions where deoxycholate alone fails to transform the cells (Kaibara et al., 1984). PKC-activating phorbol-esters are among the most potent tumor promoters known (Weinstein, 1988). Because the phorbol-ester tumor promoters are specific PKC activators (Weinstein, 1988; O'Brian and Ward, 1989), it is thought that the stimulation of PKC activity observed with bile acids (Craven et al., 1987; Fitzer et al., 1987) may be a major contributing factor in bile acid-mediated tumor promotion (Weinstein, 1988; O'Brian and Ward, 1989; O'Brian and Ward, 1991). Likewise, free fatty acids derived from dietary fat are environmental factors in the colonic lumen that have tumor-promoting activity *in vivo* (Diamond et al., 1980) and PKC stimulatory activity *in vitro* (McPhail et al., 1984). Thus, several environmental factors present in the colonic lumen appear to affect colon cancer risk at least in part by perturbing the regulation of signal transduction pathways in the colonic epithelium.

Unsaturated fatty acids comprise a major component of dietary fat that stimulates proliferation of colonic epithelial cells, according to studies involving the intra-colonic instillation of fatty acids in rats (Bull et al., 1984; Bull et al., 1988; Craven and DeRubertis, 1988). The proliferative response of the colonic epithelial cells to the fatty acids is associated with the translocation of PKC activity to the membrane fraction of the cells, providing evidence that PKC activation participates in the proliferative response (Craven and DeRubertis, 1988). In fact, PKC appears to play a major role in the proliferative response of colonic epithelial cells to fatty acids, because the protein kinase inhibitor H7 antagonizes both fatty acid-induced cell proliferation and PKC translocation (Craven and DeRubertis, 1988), and because the specific PKC activator TPA also translocates PKC activity and induces a proliferative response in the colonic epithelial cells (Craven and DeRubertis, 1987). Unsaturated fatty acids directly activate isolated colonic epithelial PKC and also stimulate production of the endogenous PKC activator *sn*-1,2-DAG in rat colonic epithelial cells (Craven and DeRubertis, 1988). It is not yet clear whether the fatty acids activate PKC in the colonic epithelial cells through direct effects on the enzyme, *sn*-1,2-DAG production, or both. However, autoxidation products of unsaturated fatty acids are more potent than the parental unoxidized fatty acids in both activation of purified PKC (O'Brian et al., 1988) and stimulation of colonic epithelial cell proliferation (Bull et al., 1984; Bull et al., 1988), suggesting that the direct activation of PKC by fatty acids may be a critical event in the proliferative response. In any event, it is clear that fatty acids are environmental factors in the colon that perturb PKC-mediated signal transduction pathways in colonic epithelial cells.

Measurements of human fecal diglycerides indicate that various long-chain diglycerides are present in the colonic lumen at estimated concentrations of 50–500 μM (Friedman et al., 1989). Diglycerides of the *sn*-1,2 configuration (*sn*-1,2-DAGs) activate PKC by the same mechanism as phorbol-esters (O'Brian and Ward, 1989; Bell and Burns, 1991). *sn*-1,2-DAG is one component of human fecal diglyceride, although the percentage of total fecal diglyceride that contains the *sn*-1,2 configuration is not known (Morotomi et al., 1990). Synthetic long-chain *sn*-1,2-DAGs induce mitogenesis in primary cultures of human colon adenoma cells and also, in some cases, in primary cultures of human colon carcinoma cells. In contrast, these diglycerides do not stimulate proliferation of primary cultures of normal human colonic epithelial cells (Friedman et al., 1989). It should be noted, however, that the highly cell-permeable short-chain DAG 1-oleoyl-2-acetyl-glycerol, which is a nonphysiological DAG that appears to act primarily by activating PKC, does stimulate normal colonic epithelial cell proliferation when intracolonically instilled in rats (Craven et al., 1987). The differential effects of the long-chain *sn*-1,2-DAG's on the growth of normal, premalignant and malignant human colonic epithelial cells *in vitro* appear to be a consequence of their differential effects on signal transduction. Thus, *sn*-1,2-DAGs stimulate the phosphorylation of a 63 kDa protein at tyrosine in cultured human colon carcinoma cells and in primary cultures of human colonic adenoma cells, but not in primary cultures of normal human colonic epithelial cells (Marian et al., 1989). The effects of the long-chain *sn*-1,2-DAG's on mitogenesis in the normal, premalignant and malignant colonic epithelial cells overlap with the effects of TPA, but there are clear distinctions between the responses to the diglyceride and the phorbol ester (Marian et al., 1989). Therefore, PKC activation appears to contribute to, but does not totally account for, the perturbations of signal transduction pathways in human colonic epithelial cells by long chain *sn*-1,2-DAGs. Taken together, these results suggest that long-chain *sn*-1,2-DAGs present in the colonic lumen may promote the selective outgrowth of premalignant and malignant cells in the colonic epithelium (Friedman et al., 1989).

In view of the strong evidence that long-chain *sn*-1,2-DAGs in the colonic lumen (Morotomi et al., 1990) can induce mitogenesis of premalignant and transformed colonic epithelial cells (Friedman et al., 1989; Marian et al., 1989), it is clear that studies on the regulation of the production of these dietary metabolites in the colonic lumen represent an important new area of investigation (Morotomi et al., 1990). Human fecal bacteria produce *sn*-1,2-DAG and other lipid metabolites from phosphatidylcholine, phosphatidylinositol and phosphatidylethanolamine (Morotomi et al., 1990). The production of *sn*-1,2-DAG by the bacterial enzymes *in vitro* is completely dependent on the presence of bile acids. This suggests that interactions between intestinal bacteria, dietary fat and bile acids in the colonic lumen may lead to the induction of mitogenesis in colonic epithelial cells and thereby increase the risk of cancer (Morotomi et al., 1990). The importance of intestinal bacteria in the production of PKC activators in the colonic lumen is also suggested by studies with a synthetic fecapentaene (Hoshina et al., 1991). Fecapentaenes are potent mutagens that are frequently produced from polyunsaturated ether phospholipids by intestinal bacteria in the colonic lumen. It was recently demonstrated that a synthetic fecapentaene is also a potent PKC activator that replaces the phospholipid cofactor requirement of the enzyme (Hoshina et al., 1991). Thus, fecapentaenes may increase colon cancer risk not only by their mutagenic actions but also by perturbing signal transduction in colonic epithelial cells through aberrant activation of PKC (Hoshina et al., 1991).

Bile acids appear to be endogenous tumor promoters, since they behave as classical tumor promoters in various animal models of multistage carcinogenesis (Narisawa et al., 1974; Reddy et al., 1976; Cameron et al., 1982; Furihata et al., 1987). This classification of bile acids as the mediators of the clonal expansion of initiated cells is supported by reports that intracolonic instillation of bile acids stimulates ornithine decarboxylase activity (Takano et al., 1981, 1984; Craven et al., 1987) and DNA synthesis (Takano et al., 1984; Craven et al., 1987) in the rat colonic epithelium. Recent studies provide evidence that the proliferative response of colonic epithelial cells to bile acids is mediated by PKC.

Studies with purified PKC indicate that under certain reaction conditions its enzyme activity can be stimulated by a wide variety of bile acids and bile acid analogs, including deoxycholate, chenodeoxycholate, cholate, and the glyco- and tauro-conjugated forms of these bile acids (Fitzer et al., 1987; Ward and O'Brian, 1988). Bile acids can replace the requirement for phosphatidylserine (PS) in the activation of purified PKC by TPA (Ward and O'Brian, 1988; O'Brian et al., 1991), and they can enhance the stimulation of PKC activity by Ca^{2+} plus PS and by TPA plus PS (Fitzer et al., 1987; Ward and O'Brian, 1988; O'Brian et al., 1991). For example, 100 μM deoxycholate fully replaces the requirement for PS in the activation of PKC-α by 100 nM TPA, under conditions where neither TPA nor deoxycholate alone affects the activity of the isozyme. In addition, 100 μM deoxycholate enhances TPA- plus PS-stimulated PKC-α activity by more than 1.5-fold, and 500 μM deoxycholate enhances the activity over 3-fold (O'Brian et al., 1991). Deoxycholate and related bile acids are generally present in the colonic lumen at concentrations that exceed 100 μM (McJunkin et al., 1981). Therefore, bile acids may directly stimulate the activity of colonic epithelial PKC *in vivo* (Fitzer et al., 1987; Ward and O'Brian, 1988; O'Brian et al., 1991).

Deoxycholate also enhances the activity of a phosphatidylinositol-specific phospholipase C (PLC) purified from rat liver (Takenawa and Nagai, 1981), and deoxycholate stimulates PLC activity in intact rat colonic epithelial cells (Craven et al., 1987). Incubation of isolated rat colonic crypts with deoxycholate significantly increases the cellular levels of products of PLC catalysis, including the second messengers *sn*-1,2-DAG and inositol (1,4,5)-trisphosphate, by approximately twofold and significantly reduces the cellular level of the PLC substrate phosphatidylinositol (4,5)-bisphosphate (Craven et al., 1987). Concomitant with PLC activation, deoxycholate induces PKC translocation to the membrane fraction of rat colonic epithelial cells (Craven et al., 1987; Craven and DeRubertis, 1987). Deoxycholate-induced PKC translocation appears to reflect PKC activation, because the translocation is associated with events that can also be induced by the specific PKC activator TPA, e.g., overexpression of the ornithine decarboxylase gene in murine fibroblasts (Guillem et al., 1987c) and increased reactive oxygen production in rat colonic epithelial cells (Craven et al., 1987). Furthermore, deoxycholate-induced DNA synthesis is suppressed by the protein kinase inhibitor H7 in rat colonic epithelial cells (Craven and DeRubertis, 1987), and the co-mitogenic activity of deoxycholate in Swiss 3T3 cells is abolished when PKC is downregulated in the cells by prolonged exposure to phorbol esters (Takeyama et al., 1985).

In general, the effect of PKC activation on cellular proliferation depends on the cell system studied. Some cells proliferate in response to PKC activation by TPA, whereas other types of cells become quiescent and in some cases differentiate upon exposure to TPA (Gescher, 1985). TPA and deoxycholate each stimulate the proliferation of rat colonic epithelial cells (Craven et al., 1987; Craven and DeRubertis, 1987), and the amount of membrane-associated PKC activity is higher in the proliferating fraction of colonic epithelial cells isolated from rat colonic mucosa than in the nonproliferating fraction of the cells (Craven and DeRubertis, 1987). Therefore, PKC activation appears to be a positive signal for cell growth in the colonic epithelium (Nishino et al., 1986; Craven et al., 1987; Craven and DeRubertis, 1987). Bile acids may stimulate PKC activity in colonic epithelial cells through direct effects on the enzyme (Fitzer et al., 1987; Ward and O'Brian, 1988; O'Brian et al., 1991), stimulation of *sn*-1,2-DAG production (Craven et al., 1987), and induction of reactive oxygen production (Craven et al., 1987), which may produce oxygenated lipids that potently stimulate PKC (O'Brian et al., 1988; O'Brian and Ward, 1991). The relative contributions of these potential mechanisms of bile acid-mediated PKC activation to the stimulation of PKC activity in colonic epithelial cells is not yet known. Glycyrrhetic acid is a structural analog of bile acids that antagonizes tumor promotion in mouse skin (Nishino et al., 1986), and our recent observation that glycyrrhetic acid inhibits purified PKC (O'Brian et al., 1990) suggests that the direct effects of bile acids and their analogs on PKC may play an important role in their regulation of the enzyme *in vivo*. In future studies, the determination of the primary mechanism of PKC activation by bile acids in mammalian cells should provide an important contribution to our understanding of bile acid-mediated tumor promotion in intestinal tissues.

The hormonal form of vitamin D, 1,25-dihydroxyvitamin D_3 ($1,25(OH)_2D_3$) regulates calcium absorption and cell proliferation and differentiation in the colonic epithelium (Norman et al., 1982; Niendorf et al., 1987; Wali et al., 1990). In contrast with bile acids, $1,25(OH)_2D_3$ is an antagonist of tumor promotion in mouse skin (Wood et al., 1983; Chida et al., 1985) and in the rat colon. $1,25(OH)_2D_3$ suppresses the tumor-promoting action of lithocholic acid in the colonic epithelium of N-methyl-N-nitrosourea-treated rats (Kawaura et al., 1989). $1,25(OH)_2D_3$ also antagonizes the tumor-promoting effects of a high fat diet in the colonic epithelium of dimethylhydrazine-treated rats, but the vitamin does not affect tumor incidence in dimethylhydrazine-treated rats that are maintained on a low-fat diet (Pence and Buddingh, 1988). Further evidence that dietary vitamin D protects against colon carcinogenesis was recently provided by a study that showed that dietary Ca^{2+} supplements reduce the number and size of colon tumors in dimethylhydrazine-treated rats only when adequate levels of vitamin D are present in their diet (Sitrin et al., 1991). Surprisingly, 1,25$(OH)_2D_3$ shares with bile acids the ability to rapidly stimulate phosphatidylinositol turnover in rat colonic epithelial cells, and the vitamin causes a significant drop in the cellular level of phosphatidylinositol (4,5)-bisphosphate and a significant increase in the cellular levels of *sn*-1,2-DAG and inositol (1,4,5)-trisphosphate (Wali et al., 1990). Furthermore, 1,25$(OH)_2D_3$ also has in common with bile acids the ability to translocate PKC activity to the membrane fraction of colonic epithelial cells (Wali et al., 1990). The translocation of PKC activity is associated with enhanced DNA synthesis (Wali et al., 1990), providing evidence that the translocated enzyme is in fact an activated form of PKC. Because 1,25$(OH)_2D_3$ does not affect the activity of isolated PKC, it appears that the induction of *sn*-1,2-DAG production accounts for the activation of PKC by the vitamin in rat colonic epithelial cells (Wali et al., 1990). Moreover, the 1,25$(OH)_2D_3$-induced increase in inositol (1,4,5)-trisphosphate production appears to have functional significance, because it is accompanied by an elevation in the intracellular Ca^{2+} concentration (Wali et al., 1990). The effects of 1,25$(OH)_2D_3$ on phosphatidylinositol turnover and PKC activation in rat colonic epithelial cells are very rapid and therefore appear to be independent of the receptor for this steroid (Wali et al., 1990).

It is striking that 1,25$(OH)_2D_3$ induces proliferation of rat colonic epithelial cells but actually antagonizes tumor promotion in the colonic epithelium. However, the effects of 1,25$(OH)_2D_3$ on cell growth and differentiation are complex. Although 1,25$(OH)_2D_3$ inhibits proliferation of several human colon cancer cell lines *in vitro* (Lointier et al., 1987; Niendorf et al., 1987; Brehier and Thomasset, 1988; Tanaka et al., 1990), the vitamin enhances the chemical transformation of Balb 3T3 fibroblasts (Kuroki et al., 1983) and Syrian hamster embryo cells (Jones et al., 1984) *in vitro*. Furthermore, several studies on the effects of 1,25$(OH)_2D_3$ on PKC activity in various cell systems indicate that the fundamental effects of the vitamin on PKC-mediated signal transduction pathways depend on the cell system studied (Sasaki et al., 1986; Martell et al., 1987; Mezzetti et al., 1987; Ways et al., 1987; Wali et al., 1990). Thus, the effects of 1,25$(OH)_2D_3$ on signal transduction are also complex. An understanding of the molecular mechanisms of 1,25$(OH)_2D_3$ action may disclose the basis for its protection against cancer in some systems and its promotion of cell transformation in others.

Future studies should include investigations devoted to a resolution of the paradox that PKC activation and phosphatidylinositol turnover are induced in the colonic epithelium by tumor-promoting bile acids (Craven et al., 1987; Craven and DeRubertis, 1987) and by the tumor promotion antagonist 1,25$(OH)_2D_3$ (Wali et al., 1990). Intracolonically instilled bile acids increase the percentage of membrane-associated PKC activity in colonic epithelial cells within ten minutes, and this effect persists quantitatively for at least 4 hours (Craven et al., 1987). In contrast, the stimulation of phosphatidylinositol (4,5)-diphosphate hydrolysis and membrane association of PKC activity by 1,25$(OH)_2D_3$ in isolated rat colonic epithelial cells is transient (Wali et al., 1990). The persistent membrane association of PKC activity in bile acid-treated colonic epithelial cells (Craven et al., 1987) may be a consequence of the direct stimulatory effects of bile acids on phospholipase C (Takenawa and Nagai, 1981) or PKC (Fitzer et al., 1987; Ward and O'Brian, 1988). Previous studies with TPA indicate that short-term exposure to the phorbol ester generally activates PKC in mammalian cells, whereas long-term exposure to the phorbol ester alters PKC isozyme expression through downregulation of the enzyme (Huang et al., 1989; O'Brian and Ward, 1989). Consequently, profound differences are often observed between the effects of short-term and prolonged exposure to TPA on cell phenotypes (Huang et al., 1989; O'Brian and Ward, 1989). By analogy, in view of the persistent activation of PKC by bile acids in colonic epithelial cells (Craven et al., 1987) and the increased susceptibility of activated PKC to downregulation (Huang et al., 1989), it is likely that long-term exposure to bile acids may alter PKC isozyme expression in the cells. Thus, differences between the effects of long-term exposure to bile acids and $1,25(OH)_2D_3$ on PKC isozyme expression might account, at least in part, for the opposing effects of these agents on tumor promotion in the colonic epithelium. It should be of particular interest to determine if identical PKC isozyme(s) are lost in transformed colonic epithelial tissues and in bile acid-exposed colonic epithelial cells. $1,25(OH)_2D_3$ inhibits the proliferation of several human colon cancer cell lines including HT29 *in vitro* (Lointier et al., 1987; Niendorf et al., 1987; Brehier and Thomasset, 1988; Tanaka et al., 1990). Since PKC-β_1 overexpression inhibits the growth of HT29 cells *in vitro* (Choi et al., 1990), it will be important to determine whether $1,25(OH)_2D_3$ upregulates PKC-β_1 expression or the expression of other PKC isozymes in cultured human colon cancer cells. In addition, retinoic acid antagonizes deoxycholate-induced PKC translocation and reactive oxygen production in rat colonic epithelial cells (Craven et al., 1987), and retinoic acid suppresses cell transformation (Bertram, 1988) and promotes differentiation (Weinstein, 1991b) in various cell systems. The effects of retinoic acid on PKC isozyme expression in human colon cancer cells under conditions where retinoic acid exerts antiproliferative effects are therefore worth exploring.

III. ONCOGENE ACTIVATION AND TUMOR SUPPRESSOR GENE INACTIVATION: CRITICAL EVENTS IN HUMAN COLON CARCINOGENESIS

Proto-oncogene products include a number of signal-transducing proteins such as protein kinases, growth factors, and growth factor receptors. Oncogenes are mutated forms of proto-oncogenes, and oncogene products exhibit aberrant signal-transducing behavior that can result in uncontrolled cell growth (Stanbridge, 1990; Weinberg, 1991). Tumor suppressor gene products include signal-transducing proteins, and tumor suppressor gene inactivation also results in altered signal transduction that leads to uncontrolled cell growth (Stanbridge, 1990; Weinberg,

1991). Recent genetic studies of human colorectal cancer indicate that the development of colon cancer is associated with the activation of certain oncogenes and the inactivation of specific tumor suppressor genes (Fearon and Vogelstein, 1990).

Ras gene products are a family of proteins that bind and hydrolyze GTP. *Ras* proteins are thought to serve as G-proteins in transmembrane signal transduction. Activated *ras* oncogenes typically have reduced intrinsic GTPase activities and appear to perturb cellular growth control mechanisms by altering signal transduction pathways (Barbacid, 1987). Activated *ras* oncogenes are observed in the tissue specimens of approximately 40% of human colorectal carcinomas (Forrester et al., 1987; Bos et al., 1987; Vogelstein et al., 1988). Most of the *ras* mutations observed are in Ki-*ras*, although activated N-*ras* oncogenes occur in a small percentage of human colorectal carcinoma tissue specimens (Bos et al., 1987; Forrester et al., 1987; Vogelstein et al., 1988). Furthermore, the predominant mutations in the activated Ki-*ras* oncogenes in human colon carcinomas occur at codon 12 (Forrester et al., 1987; Bos et al., 1987; Vogelstein et al., 1988). Likewise, the activated Ki-*ras* oncogene is observed in roughly 30% of colon carcinomas in dimethylhydrazine-treated rats (Jacoby et al., 1989). Based on clinical and histopathological evidence, it has been postulated that human colorectal carcinomas generally arise from preexisting benign colon adenomas (Sugarbaker et al., 1985). In support of this model for human colon carcinogenesis, an examination of several human colon adenocarcinomas revealed that in most cases identical activated *ras* oncogenes were present in the adenomatous and malignant fractions of each tissue specimen (Bos et al., 1987). As colon adenomas increase in size and dysplasia and acquire a villous morphology, their propensity for malignant conversion increases (Sugarbaker et al., 1985). Activated *ras* oncogenes are present at similar frequencies in human colon carcinomas and large adenomas, but they occur in only about 10% of small adenomas (Vogelstein et al., 1988). Thus, it appears that *ras* oncogene activation generally precedes malignant conversion in human colon carcinogenesis and may contribute to the development of large colon adenomas with an increased potential for malignant conversion from small adenomas (Vogelstein et al., 1988; Fearon and Vogelstein, 1990).

The proto-oncogene product $pp60^{c\text{-}src}$ is a membrane-associated protein-tyrosine kinase (PTK) (Hunter and Cooper, 1985), and its participation in transmembrane signal transduction is evident from its enzyme activity. Elevated $pp60^{c\text{-}src}$ PTK-activity levels are consistently observed in tissue specimens of human colon carcinoma (Bolen et al., 1987). In addition, cultured human colon cancer cells have higher levels of $pp60^{c\text{-}src}$ PTK-activity than cultured normal human colonic epithelial cells. Increased $pp60^{c\text{-}src}$ PTK-activity has been observed in 15/15 human colon carcinoma tissue specimens and in 21/21 human colon cancer cell lines (Bolen et al., 1987). The increased $pp60^{c\text{-}src}$ PTK-activity levels in human colon carcinomas appear to reflect alterations in the intrinsic activity of $pp60^{c\text{-}src}$, because $pp60^{c\text{-}src}$ expression is not altered in the colon carcinoma tissue specimens. Therefore, the increased $pp60^{c\text{-}src}$ PTK-activity in human colon cancer cells may be a consequence of mutations or post-translational modifications of pp60$^{c\text{-}src}$ (Bolen et al., 1987). Increased levels of pp60$^{c\text{-}src}$ PTK activity are also present in tissue specimens of human colon adenomas, and the increase in the activity appears to be proportional to the propensity of the adenomas to give rise to carcinomas (Cartwright et al., 1990). Thus, as in the case of *ras*, activation of the proto-oncogene *c-src* may be an early event in human colon carcinogenesis that precedes malignant conversion and contributes to the development of large dysplastic adenomas from small adenomas. In addition to *src* and *ras*, activation of the oncogenes *neu*, *myc*, and *myb* has been implicated in human colon carcinogenesis, based on the amplification of these genes observed in human colon carcinoma tissue specimens and cell lines (Alitalo et al., 1984; D'Emilia et al., 1989; Finley et al., 1989).

Although oncogene activation is believed to play a critical role in human colon carcinogenesis, the predominant genetic changes that occur during the development of this disease appear to reflect tumor suppressor gene inactivation (Fearon and Vogelstein, 1990). Specific allelic deletions associated with various stages of human colon carcinogenesis provide evidence that inactivation of tumor suppressor genes is an important contributing factor in the development of the disease (Fearon and Vogelstern, 1990; Stanbridge, 1990; Weinberg, 1991). Allelic deletions in chromosome 5q, which contains the familial adenomatous polyposis (FAP) gene (Bodmer et al., 1987), are observed in about 30% of human colon carcinoma and adenoma tissue specimens (excluding adenoma specimens from FAP patients) (Solomon et al., 1987; Vogelstein et al., 1988) allelic loss in chromosome 17p occurs in over 75% of human colon carcinomas but is rare in colon adenomas (Vogelstein et al., 1988), and allelic loss in chromosome 18q occurs in about 70%, 50%, and 10% of human colon carcinomas, advanced adenomas, and small adenomas, respectively (Vogelstein et al., 1988). Based on these allelic losses, candidate tumor suppressor genes involved in the progression of human colon carcinogenesis have been identified as p53 (chromosome 17p) (Baker et al., 1989) and DCC ("deleted in colorectal carcinomas") (chromosome 18q) (Fearon et al., 1990). Inactivation of these candidate tumor suppressor genes would be expected to alter signal transduction in the colonic epithelium, because p53 is a DNA binding protein implicated in the regulation of transcription (Stanbridge, 1990; Weinberg, 1991), and the DCC product bears homology with neural cell adhesion molecules and appears to be a signal-transducing transmembrane protein (Weinberg, 1991) that may regulate cell-surface interactions in the colonic epithelium (Fearon et al., 1990). Furthermore, studies with cultured human colon cancer cells suggest that certain PKC isozymes may be tumor suppressors in the colonic epithelium (Choi et al., 1990), although allelic loss of PKC-encoding genes has not been reported in any tissue. Taken together, genetic studies of oncogene activation and tumor suppressor gene inactivation in human colon carcinogenesis provide strong evidence that the accumulation of genetic changes and not their order dictates the course of the disease (Fearon and Vogelstein, 1990). These results lend support to the concept of carcinogenesis as a progressive disorder in signal transduction (Weinstein, 1988).

IV. ROLE OF ALTERED SIGNAL TRANSDUCTION IN THE INTRINSIC DRUG RESISTANCE OF HUMAN COLON CANCER

Multidrug resistant (MDR) tumor cells are resistant to the cytotoxic effects of structurally and pharmacologically diverse anticancer drugs including anthracyclines such as Adriamycin and vinca alkaloids such as vincristine (Moscow and Cowan, 1988; Endicott and Ling, 1989). The primary mechanism of resistance in MDR tumor cells is a reduced capacity to accumulate the cytotoxic drugs, and the predominant biochemical feature of MDR tumor cell populations is overexpression of the transmembrane energy-dependent drug efflux pump P-glycoprotein. Thus, the defect in drug accumulation in the MDR tumor cells appears to result from the drug transport activity of P-glycoprotein (Moscow and Cowan, 1988; Endicott and Ling, 1989).

Chronic *in vitro* exposure of drug-sensitive cultured tumor cells to anticancer drugs that are affected by MDR can result in the acquisition of an MDR phenotype by the tumor cell population (Moscow and Cowan, 1988; Tsuruo, 1988; Endicott and Ling, 1989). In addition, certain human tumors, including renal, gastric and colon cancers, typically display intrinsic resistance against cytotoxic anticancer drugs *in vivo* (Tsuruo, 1988). This intrinsic resistance is associated with elevated expression of the gene encoding P-glycoprotein (*mdr1*) in several different types of human tumors including colorectal carcinomas (Goldstein et al., 1989). Thus, P-glycoprotein may play a crucial role in the intrinsic drug resistance of human colorectal cancer.

Recent studies provide evidence for a regulatory role for PKC in P-glycoprotein-mediated MDR phenotypes (O'Brian and Ward, 1989). Elevated PKC activity levels are commonly observed in drug-selected MDR tumor cells *in vitro* (Palayoor et al., 1987; Aquino et al., 1988; Fine et al., 1988; O'Brian et al., 1989; Chambers et al., 1990; O'Brian et al., 1991), and phorbol ester-induced PKC activation often protects drug-sensitive and MDR tumor cells from the cytotoxic effects of drugs affected by MDR *in vitro* (Ferguson and Cheng, 1987; Fine et al., 1988; O'Brian et al., 1991; Yu et al., 1991). Furthermore, PKC catalyzes the phosphorylation of purified P-glycoprotein (Chambers et al., 1990), although the functional significance of this phosphorylation event remains unclear.

The presence of PKC-activating bile acids (Craven et al., 1987; Fitzer et al., 1987; Ward and O'Brian, 1988; O'Brian et al., 1991) and dietary fat metabolites (Craven and DeRubertis, 1988; O'Brian et al., 1988; Friedman et al., 1989; Morotomi et al., 1990) in the colonic lumen suggests that these endogenous and environmental PKC activators may enhance the intrinsic drug resistance of human colon cancer *in vivo* (Dong et al., 1991). In support of a role for bile acids as endogenous factors that enhance intrinsic drug resistance phenotypes of colon tumors *in vivo*, we have found that deoxcholate transiently protects drug-sensitive and Adriamycin-selected MDR cultured murine fibrosarcoma cells *in vitro* against cytotoxic drugs affected by MDR but not against unrelated cytotoxic drugs. Similar protective effects were observed *in vitro* with phorbol-ester tumor promoters, and the protein kinase inhibitor H7 antagonized deoxycholate-induced protection, providing evidence that deoxycholate protected the cells through activation of PKC (O'Brian et al., 1991). In addition, *in vitro* exposure of cultured human colon cancer KM12L4a cells to phorbol esters and to synthetic DAGs reduces the intracellular accumulation of drugs affected by MDR but not that of unrelated cytotoxic drugs, thus providing *in vitro* evidence that implicates dietary fat metabolites in the intrinsic drug resistance of human colon cancer (Dong et al., 1991). The phorbol esters also protect the cultured colon cancer cells against the cytotoxic effects of Adriamycin, vincristine and vinblastine but not against 5-fluorouracil; the absence of protection by synthetic DAGs in this *in vitro* system is accounted for by their rapid hydrolysis upon incubation with the cultured cells (Dong et al., 1991). Since phorbol esters and *sn*-1,2-DAGs activate PKC by the same mechanism (O'Brian and Ward, 1989; Bell and Burns, 1991), the *in vitro* results with KM12L4a colon cancer cells provide evidence that the constantly replenished supplies of dietary DAGs may protect colon cancer cells *in vivo* against cytotoxic drugs affected by MDR (Dong et al., 1991). Thus, altered signal transduction may enhance the intrinsic drug resistance of human colon carcinomas *in vivo*.

V. TUMOR ANTIGEN TRANSPORT AND THE MULTIDRUG RESISTANCE (MDR) PHENOTYPE: A HYPOTHESIS

Recent studies concerning antigen expression on tumor cells suggest the possibility of a link between expression of the *mdr1* gene product P-glycoprotein and the susceptibility of tumor cells to killing by cytotoxic T lymphocytes. Decreased MHC Class I expression on tumor cells correlates with enhanced progression and metastasis of certain human and experimental animal tumors (Gooding, 1982; Tanaka et al., 1988; Cardo et al., 1991; Maudsley and Pound, 1991). The relationship between MHC Class I antigen expression and tumorigenesis has been inferred from studies of cells transformed with various DNA viruses. Mouse cells transformed with simian virus 40 (SV40) are immunogenic and rarely form tumors in immunocompetent syngeneic hosts, and cytotoxic T lymphocytes (CTL) restricted by syngeneic MHC Class I molecules are responsible for clearing tumor cells from these animals. Immunoselected variants of DNA virus-transformed tumor cells that fail to express MHC Class I are highly tumorigenic, even in immunocompetent syngeneic animals (Gooding, 1982). Downregulation of MHC Class I expression has been proposed as a mechanism of evasion of CTL surveillance by tumors (Cox et al., 1990; Jefferies and Burgert, 1990; Restifo et al., 1991). A key element in the development of specific cell-mediated immunosurveillance is the ability of T cells to develop a response to tumor Ag expressed in association with either MHC Class I chains or MHC Class II molecules, in those tumors which constitutively express these surface molecules.

For clarity of presentation, we note that the notion of tumor Ag recognition by T cells reflects recognition of tumor-derived peptides (Boon and van Pel, 1989) that are nonapeptides in the case of CTL (Falk et al., 1990; Rotzschke et al., 1990) and somewhat longer (14–18 residues) for $CD4^+$ cells (Rudensky et al., 1991; Sadegh-Nasseri and Germain, 1991) and that lack serological definition, in contrast with, for example, p 97 and CEA tumor associated Ag (Kahn et al., 1991). These peptides may be drawn from large pools of products of intracellularly digested endogenous tumor proteins, which are continuously recycled as products of cellular turnover (Van Bleek and Nathenson, 1990; Jardetsky et al., 1991). In this regard, T cell recognition of peptides derived from mutated *ras* oncogene has recently been demonstrated in experimental tumors (Peace et al., 1991). Endogenous peptides appear to associate with MHC Class I chains (Van Bleek and Nathenson, 1990), whereas exogenous (endocytosed) peptides associate preferentially with MHC Class II molecules (Harding and Unanue, 1989; Harding and Unanue, 1990; Harding et al., 1991). However, binding of antigenic peptides by histocompatibility antigens is a function of the assembly, intracellular trafficking and endocytosis of the complexes. For Class I molecules, assembly appears to be the most important step for peptide binding, because the presence of β-2 microglobulin and a tightly bound peptide confer on these molecules their conformational integrity (Townsend et al., 1989; Rock et al., 1991; Rock et al., 1991b). It is of particular interest that *in vivo* binding of peptides to MHC Class I chains is restricted topologically to the endoplasmic reticulum. However, the peptides are generated from proteins located in different cellular organelles and in the case of MHC Class I associated Ag, the corresponding peptides are generated by proteolytic digestion in the cytosol. Therefore, such peptides lacking N-terminal signal sequences are prevented from diffusing through intracellular membranes. This observation has led to the prediction that active transport systems (Townsend and Bodmer, 1989) are located on the membranes of the endoplasmic reticulum, and that these systems can function as "pumps," delivering peptides from the cytoplasm to the endoplasmic reticulum. Further assembled MHC Class I chains-peptide complexes of stable conformation can reach the plasma membrane through the Golgi apparatus. Therefore, association with peptides soon after biosynthesis is a necessary condition for transport of MHC Class I chains to the cell surface. These predictions have been strengthened by results with mutant cell lines such as RMA-S which fail to present Ag to CTL (Townsend et al., 1989; Hosken and Bevan, 1990). This deficiency is believed to be due to a defect in Ag transport, because these cells present exogenously delivered peptides at higher rates than normal cells, regardless of the lower levels of expression on the cell membrane of both Class I MHC chain and β_2 microglobulin, suggesting the presence of increased proportions of MHC Class I chains with free binding grooves in the mutant lines (Townsend et al., 1989).

It should be pointed out that MHC Class I molecules on normal cells do not spontaneously bind and exchange peptides. The rate of exchange is in general very low, as indicated by the small percentage of purified Class I molecules that can bind peptides (0.3%) (Chen and Parham, 1989). Therefore, one of the most critical elements in ensuring normal MHC Class I expression on tumors and accessory cells and the induction of a specific cellular immune response to the tumors rests in the normal function of peptide transporters located on the endoplasmic reticulum membrane of the tumor cells.

The nature and function of these transporters were until recently unknown. However, on the basis of results with mutant cell lines, it was predicted that genes for these transporters would map in the MHC. Several groups (Deverson et al., 1990; Monaco et al., 1990; Spies et al., 1990; Trowsdale et al., 1990) working intensely during the last decade have just reported that several genes located in the MHC Class II region which are designated HAM1 and HAM2 for histocompatibility antigen modifiers in the mouse (Monaco et al., 1990), *mtp.1* and *mtp.2* for (MHC-linked transporter proteins) located within the cim (Class I modification) locus within the rat MHC Class I region (RTIA) (Deverson et al., 1990), and PSF (Y3) and RING4 in humans may be candidates for this function (Spies et al., 1990; Trowsdale et al., 1990). The human RING4 gene is apparently homologous with the murine HAM1 (Parham, 1990). The human homologue for HAM2 is yet to be identified. The RING4 gene also maps in the MHC Class II locus on the short arm of chromosome 6. The RING4 gene can be positioned by two Not I sites separated by 9 kb, and lies at the CpG island approximately between the DNA and 25 kb centromeric from the HLA-DOB gene (Trowsdale et al., 1990).

Analysis of the amino acid sequences deduced from the cDNA sequences of the HAM, *mtp* and RING genes revealed proteins that contain multiple blocks of hydrophobic amino acids, and most surprisingly these proteins belong to the ABC superfamily of transporters (Deverson et al., 1990; Monaco et al., 1990; Parham, 1990; Spies et al., 1990; Trowsdale et al., 1990). This family includes ATP-dependent-transmembrane transporter proteins that comprise both eukaryotic and procaryotic members. Most important for this discussion is a comparison of the putative peptide transporters with the multi-drug resistance (*mdr*) gene family. The C-terminal domain of the putative transporter proteins contains a consensus nucleotide-binding site (ATP binding cassette, ABC). In contrast with the *mdr* gene products and other mammalian transporters which have very strong hydrophilic-termini that are believed to be cytoplasmic, the putative peptide transporters have very strong hydrophobic domains according to predicted hydropathy plots (Deverson et al., 1990; Monaco et al., 1990; Spies et al., 1990; Trowsdale et al., 1990). An additional structural difference between the putative peptide transporters and other ABC transporters including *mdr* gene products that may have functional implications is that the former, (e.g., RING4) consist of only one hydrophobic and one ATP binding domain. This suggests that putative peptide transporters may function as either homo-/or heterodimers with another as yet unknown protein (Parham, 1990; Spies et al., 1990; Trowsdale et al., 1990; Parham, 1991; Spies and DeMars, 1991).

By extrapolation, one can hypothesize the potential functions of the putative peptide transporters from evidence available for other ABC transporters. If the peptide transporters are functional (alone or in association with the as yet unknown component), they should be able to accommodate structural differences in the transported material in a fashion analogous with the *mdr1* gene product P-glycoprotein. Therefore, they are expected to evolve adaptively, and show sequence variability at the putative peptide binding sites. It is not yet known whether such structural variability is achieved by the recently discovered RING4, HAM or *mtp* proteins alone or in association with another protein, i.e., either in a similar fashion with the *mdr* gene family or in association with Ag binding domains of MHC Class II molecules or TCR (Parham, 1990). Direct evidence in support of the putative role played by peptide-transducing ABC-transporters is awaiting direct demonstration of their transporter function. In this regard, Spies and DeMar have recently shown that transfection with the proposed transporter gene (PSF, Y3) in a mutant B cell line LCL.134 that has a non-functional Y3 gene and consequently diminished expression of MHC Class I reversed its phenotype and resulted in normal levels of cell surface expression of human MHC Class I (HLA-A2 and HLA-B5) antigens (Spies and DeMars, 1991). In contrast, when the Y3 gene was transfected in another mutant cell line, LCL 721.174, which has a very large homozygous deletion in the MHC including PSF and other closely linked genes, it failed to up-regulate MHC Class I molecules. This suggests that another gene product in addition to PSF (RING4) may be needed for generation of a functional peptide pump (Spies and DeMars, 1991).

These experiments demonstrate a very important point. Namely, the expression of a functional putative peptide-transporter gene is a necessary condition (although not sufficient in itself) for Class I MHC expression on the cell surface (Parham, 1991). Although the second member of the functional peptide pump needs to be identified (the most likely candidates are HAM2, *mtp*, and RING4-like genes, which are homologous with PSF, HAM1, *mtp1* and RING4), and the direct demonstration of peptide transport across membranes is also needed, the newly discovered class of ABC-transporters may have important implications for our understanding of the pathways of altered signal transduction in carcinogenesis that may be governed by the products of these *mdr*-related genes. Increased resistance to chemotherapeutic drugs is often associated with increased expression of the *mdr1* gene product P-glycoprotein (Endicott and Ling, 1989). On the other hand, tumor invasion and metastasis most often correlate with decreased levels of MHC Class I Ags on tumor cells. Since peptide transporters related to *mdr* gene products have only recently been demonstrated to be critical for MHC Class I expression on mutant human tumor cell lines, *this may raise the question of whether the impaired expression of MHC Class I chains reflects in part altered signal transduction, where the signal in this case would be an endogenous peptide.* Both MHC and the transporter genes can be transcriptionally upregulated by γ-interferon in many human tumors (Maudsley and Pound, 1991; Spies and DeMars, 1991). In this process, TNF-α synergizes with IFN-γ in amplifying the MHC Class I expression on various tumors. Although these observations may be interpreted as supporting the hypothesis of altered signal transduction (peptide-transporter-MHC Class I) in tumors, the evidence in favor of this hypothesis is still circumstantial.

Both human and experimental tumors resistant to chemotherapy have shown in certain instances innate drug resistance *in vivo* (Tsuruo, 1988). MDR has been shown to be frequently associated *in vitro* with elevated levels of P-glycoprotein, the product of the mdr1 gene, reflecting the overexpression of this gene (Moscow and Cowan, 1988; Tsuruo, 1988; Endicott and Ling, 1989). It is still not known whether enhanced *mdr1* expression correlates with enhanced peptide-transporter gene expression, and whether the commonly used chemotherapeutic compounds have any effect on transcription or expression of the putative peptide-transporters. Although at the moment there is no evidence regarding this, it would be tempting to speculate that if up-regulation of peptide-transporters is observed following chemotherapy, this may provide a rationale for hypotheses attempting to address the nature of specific cellular immune responses to autologous tumors. Such responses from T cells infiltrating tumors resistant to chemotherapy or even malignant ascites have recently been observed (Ioannides et al., 1991; Ioannides et al., 1991b). In contrast, when LAK cells, which recognize tumors by mechanisms different than T cells, were used to assess the relationship between drug-resistance and susceptibility to lysis, an inverse relationship between *mdr1* gene expression and susceptibility to LAK cells was found (Kimmig et al., 1990). Both LAK cells and NK cells can lyse tumors in a non-MHC restricted fashion. Moreover, tumor lysability by NK cells has correlated more often with decreased MHC Class I expression (Ljunggren et al., 1989), suggesting that if functional relationships between *mdr1*, peptide transporters, MHC Class I and tumor susceptibility to lysis by immune cells will be investigated, the studies should focus on T cells. It should be noted that Mokyr and Dray have reported that low doses of chemotherapeutic drugs potentiated antitumor immunity in experimental tumor models (Mokyr and Dray, 1987). Although high doses of the drugs either potentiated to a lesser extent or failed to augment the antitumor immunity, these findings show that chemotherapy can interfere with specific immune responses to autologous tumors. Several mechanisms may account for these and more recent observations reporting augmentation of T cell-mediated specific immunity to experimental tumors by Adriamycin (Gautam et al., 1991). Clearly, the effects of chemotherapeutic drugs on peptides-transporters need to be investigated. Direct evidence that the peptide transporters discovered this year and altered signal transduction play roles in specific immunity against tumors awaits exciting studies and discoveries in the future.

ACKNOWLEDGMENTS

We thank Ms. Patherine Greenwood for her excellent assistance in the preparation of the manuscript. Supported by NIH Grants CA57293 and CA52460, Robert A. Welch Foundation Grant G-1141, and an award from the Sid W. Richardson Foundation.

REFERENCES

Aaronson, S.A. (1991). Growth factors and cancer. Science 254, 1146–1153.

Alitalo, K., Winqvist, R., Lin, C.C., de la Chapelle, A., Schwab, M., & Bishop, J.M. (1984). Aberrant expression of an amplified *c-myb* oncogene in two cell lines from a colon carcinoma. Proc. Natl. Acad. Sci. 81, 4534–4538.

Aquino, A., Hartman, K.D., Knode, M.C., Grant, S., Huang, K-P., Niu, C-H., & Glazer, R.I. (1988). Role of protein kinase C in phosphorylation of vinculin in Adriamycin-resistant HL-60 leukemia cells. Canc. Res. 48, 3324–3329.

Baker, S.J., Fearon, E.R., Nigro, J.M., Hamilton, S.R., Preisinger, A.C., Jessup, J.M., van Tuinen, P., Ledbetter, D.H., Barker, D.F., Nakamura, Y., White, R., & Vogelstein, B. (1989). Chromosome 17 deletions and p53 gene mutations in colorectal carcinomas. Science 244, 217–221.

Barbacid M. (1987). *Ras* genes. Ann. Rev. Biochem. 56, 779–827.

Baum, C.L., Wali, R.K., Sitrin, M.D., Bolt, M.J.G., & Brasitus, T.A. (1990). 1,2-dimethylhydrazine-induced alterations in protein kinase C activity in the rat preneoplastic colon. Canc. Res. 50, 3915–3920.

Bell, R.M. & Burns, D.J. (1991). Lipid activation of protein kinase C. J. Biol. Chem. 266, 4661–4664.

Bertram, J.S. (1988). The role of retinoids as inhibitors of tumor promotion. Progr. in Canc. Res. and Therapy 34, 223–236.

Bishop, J.M. (1991). Molecular themes in oncogenesis. Cell 64, 235–248.

Bodmer, W.F., Bailey, C.J., Bodmer, J., Bussey, H.J.R., Ellis, A., Gorman, P., Lucibello, F.C., Murday, V.A., Rider, S.H., Scambler, P., Sheer, D., Solomon, E., & Spurr, N.K. (1987). Localization of the gene for familial adenomatous polyposis on chromosome 5. Nature 328, 614–616.

Bolen, J.B., Veillette, A., Schwartz, A.M., DeSeau, V., & Rosen, N. (1987). Activation of $pp60^{c\text{-}src}$ protein kinase activity in human colon carcinoma. Proc. Natl. Acad. Sci. 84, 2251–2255.

Boon, T. & Van Pel, A. (1989). T cell-recognized antigenic peptides derived from the cellular genome are not protein degradation products but can be generated directly by transcription and translation of short subgenic regions, a hypothesis. Immunogenetics 29, 75–79.

Borner, C., Eppenberger, U., Wyss, R., & Fabbro, D. (1988). Continuous synthesis of two protein kinase C-related proteins after downregulation by phorbol-esters. Proc. Natl. Acad. Sci. 85, 2110–2114.

Bos, J.L., Fearon, E.R., Hamilton, S.R., de Vries, M.V., van Boom, J.H., van der Eb, A.J., & Vogelstein, B. (1987). Prevalence of ras gene mutations in human colorectal cancers. Nature 327, 293–297.

Brehier, A. & Thomasset, M. (1988). Human colon cell line HT29: characterization of 1,25-dihydroxy-vitamin D_3 receptor and induction of differentiation by the hormone. J. Steroid. Biochem. 29, 265–270.

Bull, A.W., Nigro, N.D., Golembieski, W.A., Crissman, J.D., & Marnett, L.J. (1984). *In vivo* stimulation of DNA synthesis and induction of ornithine decarboxylase in rat colon by fatty acid hydroperoxides, autoxidation products of unsaturated fatty acids. Canc. Res. 44, 4924–4928.

Bull, A.W., Nigro, N.D., & Marnett, L.J. (1988). Structural requirements for stimulation of colonic cell proliferation by oxidized fatty acids. Canc. Res. 48, 1771–1776.

Cameron, R.G., Imaida, K., Tsuda, H., & Ito N. (1982). Promotive effects of steroids and bile acids on hepatocarcinogenesis initiated by diethylnitrosamine. Canc. Res. 42, 2426–2428.

Cardo, C.C., Fuks, Z., Drobnjak, M., Moreno, C., Eisenbach, L., & Feldman, M. (1991). Expression of HLA-A,B,C antigens on primary and metastatic tumor cell populations of human carcinomas. Canc. Res. 51, 6372–6380.

Cartwright, C.A., Meisler, A.I., & Eckhart, W. (1990). Activation of the $pp60^{c\text{-}src}$ protein kinase is an early event in colonic carcinogenesis. Proc. Natl. Acad. Sci. 87, 558–562.

Chambers, T.C., McAvoy, E.M., Jacobs, J.W., & Eilon, G. Protein kinase C phosphorylates P-glycoprotein in multidrug resistant human KB carcinoma cells. (1990). J. Biol. Chem. 265, 7679–7686.

Chen, B.P. & Parham, P. (1989). Direct binding of influenza peptides to class I HLA molecules. Nature 337, 743–745.

Chida, K., Hashiba, H., Fukushima, M., Suda, T., & Kuroki, T. (1985). Inhibition of tumor promotion in mouse skin by 1,25-dihydroxyvitamin D_3. Canc. Res. 45, 5426–5430.

Choi, P.M., Tchou-Wong, K.M., & Weinstein, I.B. (1990). Overexpression of protein kinase C in HT29 colon cancer cells causes growth inhibition and tumor suppression. Mol. Cell. Biol. 10, 4650–4657.

Cooper, D.R., Watson, J.E., Acevedo-Duncan, M., Pollet, R.J., Standaert, M.L., & Farese, R.V. (1989). Retention of specific protein kinase C isozymes following chronic phorbol ester treatment in BC3H-1 myocytes. Biochem. Biophys. Res. Commun. 161, 327–334.

Cox, J.H., Yewdell, J.W., Eisenlohr, L.C., Johnson, P.R., & Bennink, J.R. (1990). Antigen presentation requires transport from the endoplasmic reticulum. Science 247, 715–718.

Craven, P.A. & DeRubertis, F.R. (1987). Subcellular distribution of protein kinase C in rat colonic epithelial cells with different proliferative activities. Canc. Res. 47, 3434–3438.

Craven, P.A. & DeRubertis, F.R. (1988). Role of activation of protein kinase C in the stimulation of colonic epithelial proliferation by unsaturated fatty acids. Gastroenterology 95, 676–685.

Craven, P.A., Pfanstiel, J., & DeRubertis, F.R. (1987). Role of activation of protein kinase C in the stimulation of colonic epithelial proliferation and reactive oxygen formation by bile acids. J. Clin. Investig. 79, 532–541.

D'Emilia, J., Bulovas, K., D'Erole, K., Wolf, B., Steele, G., & Summerhayes, I.C. (1989). Expression of the c-erbB-2 gene product (P185) at different stages of neoplastic progression in the colon. Oncogene 4, 1233–1239.

Deverson, E.V., Gow, I.R., Coadwell, W.J., Monaco, J.J., Butcher, G.W., & Howard, J.C. (1990). MHC class II region encoding proteins related to the multidrug resistance family of transmembrane transporters. Nature 348, 738–741.

Diamond, L., O'Brien, T.G., & Baird, W.M. (1980). Tumor promoters and the mechanism of tumor promotion. Adv. Canc. Res. 32, 1–74.

Dong, Z., Ward, N.E., Fan, D., Gupta, K.P., & O'Brian, C.A. (1991). *In vitro* model for intrinsic drug resistance, Effects of protein kinase C activators on the chemosensitivity of cultured human colon cancer cells. Molec. Pharmacol. 39, 563–569.

Endicott, J.A. & Ling, V. (1989). The biochemistry of P-glycoprotein-mediated multidrug resistance. Annu. Rev. Biochem. 58, 137–171.

Falk, K., Rotzschke, O., & Rammensee, H.G. (1990). Cellular peptide composition governed by major histocompatibility complex class I molecules. Nature 348, 248–251.

Farago, A. & Nishizuka, Y. (1990). Protein kinase C in transmembrane signalling. FEBS Letts. 268, 350–354.

Fearon, E.R., Cho, K.R., Nigro, J.M., Kern, S.E., Simons, J.W., Ruppert, J.M., Hamilton, S.R., Preisinger, A.C., Thomas, G., Kinzler, K.W., & Vogelstein, B. (1990). Identification of a chromosome 18q gene that is altered in colorectal cancers. Science 247, 49–56.

Fearon, E.R. & Vogelstein, B. (1990). A genetic model for colorectal tumorigenesis. Cell 61, 759–767.

Ferguson, P.J. & Cheng, Y. (1987). Transient protection of cultured human cells against antitumor agents by 12-0-tetradecanoylphorbol-13-acetate. Canc. Res. 47, 433–441.

Fine, R.L., Patel, J., & Chabner, B.A. (1988). Phorbol esters induce multidrug resistance in human breast cancer cells. Proc. Natl. Acad. Sci. USA 85, 582–586.

Finley, G.G., Schulz, N.T., Hill, S.A., Geiser, J.R., Pipas, J.M., Meisler, A.I. (1989). Expression of the myc gene family in different stages of human colorectal cancer. Oncogene 4, 963–971.

Fitzer, C.J., O'Brian, C.A., Guillem, J.G., & Weinstein, I.B. (1987). The regulation of protein kinase C by chenodeoxycholate, deoxycholate, and several structurally related bile acids. Carcinogenesis 8, 217–220.

Fleischman, L.F., Chahwala, S.B., & Cantley, L. (1986). Ras-transformed cells: Altered levels of phosphatidylinositol-4,5-bisphosphate and catabolites. Science 231, 407–410.

Forrester, K., Almoguera, C., Han, K., Grizzle, W.E., & Perucho, M. (1987). Detection of high incidence of K-*ras* oncogenes during human colon tumorigenesis. Nature 327, 298–303.

Friedman, E., Isaksson, P., Rafter, J., Marian, B., Winawer, S., & Newmark, H. (1989). Fecal diglycerides as selective endogenous mitogens for premalignant and malignant human colonic epithelial cells. Canc. Res. 49, 544–548.

Furihata, C., Takezawa, R., Matsushima, T., & Tatematsu, M. (1987). Potential tumor-promoting activity of bile acids in rat glandular stomach. Gann 78, 32– 39.

Gautam, S.C., Chikkala, N.F., Ganapathi, R., & Hamilton, T.A. (1991). Combination therapy with adriamycin and interleukin 2 augments immunity against murine renal cell carcinoma. Canc. Res. 51, 6133–6137.

Gescher A. (1985). Antiproliferative properties of phorbol ester tumor promoters. Biochem. Pharmacol. 34, 2587–2592.

Goldstein, L.J., Galski, H., Fojo, A., Willingham, M., Lai, S-L., Gazdar, A., Pirker, R., Green, A., Crist, W., Brodeur, G.M., Lieber, M., Cossman, J., Gottesman, M.M., & Pastan, I. (1989). Expression of a multidrug resistance gene in human cancers. J. Natl. Canc. Inst. 81, 116–124.

Gooding LR. (1982). Characterization of a progressive tumor from C3H fibroblasts transformed *in vitro* with SV40 virus. Immunoresistance *in vivo* correlates with phenotypic loss of H-$2K^k$. J. Immunol. 129, 1306–1312.

Guillem, J.G., Matsui, M.S., & O'Brian, C.A. (1987b). Nutrition in the prevention of neoplastic disease in the elderly. Clin. Geriatr. Medic. 3, 373–387.

Guillem, J.G., O'Brian, C.A., Fitzer, C.J., Forde, K.A., LoGerfo, P., Treat, M., & Weinstein, I.B. (1987). Altered levels of protein kinase C and Ca^{2+}-dependent protein kinases in human colon carcinomas. Canc. Res. 47, 2036–2039.

Guillem, J.G., O'Brian, C.A., Fitzer, C.J., Johnson, M.D., Forde, K.A., LoGerfo, P., & Weinstein, I.B. (1987c). Studies on protein kinase C and colon carcinogenesis. Arch. Surg. 122, 1475–1478.

Harding, C.V. & Unanue, E.R. (1989). Antigen processing and intracellular Ia. Possible roles of endocytosis and protein synthesis in Ia function. J. Immunol. 142, 12–19.

Harding, C.V. & Unanue, E.R. (1990). Quantitation of antigen-presenting cell MHC class II/peptide complexes necessary for T-cell stimulation. Nature 346, 574–576.

Harding, C.V., Collins, D.S., Slot, J.W., Geuze, H.J., & Unanue, E.R. (1991). Liposome encapsulated antigens are processed in lysosomes, recycled, and presented to T cells. Cell 64, 393–401.

Hoshina, S., Ueffing, M., Morotomi, M., & Weinstein, I.B. (1991). Effects of a fecapentaene on protein kinase C. Biochem. Biophys. Res. Commun. 176, 505–510.

Hosken, N.A. & Bevan, M.J. (1990). Defective presentation of endogenous antigen by a cell line expressing class I molecules. Science 248, 367–370.

Huang, F.L., Yoshida, Y., Cunha-Melo, J.R., Beaven, M.A., & Huang, K.P. (1989). Differential downregulation of protein kinase C isozymes. J. Biol. Chem. 264, 4238–4243.

Hunter, T. (1991). Cooperation between oncogenes. Cell 64, 249–270.

Hunter, T. & Cooper, J.A. (1985). Protein-tyrosine kinases. Annu. Rev. Biochem. 54, 897–930.

Hunter, T., Lindberg, R.A., & Middlemas, D.S. (1990). Novel receptor protein-tyrosine kinases. In: The Biology and Medicine of Signal Transduction (Nishizuka, Y., ed.), pp. 260–265, Raven Press, New York.

Ioannides, C.G., Freedman, R.S., Platsoucas, C.D., Rashed, S., & Kim, Y.P. (1991). Cytotoxic T cell clones isolated from ovarian tumor-infiltrating lymphocytes recognize multiple antigenic epitopes on autologous tumor cells. J. Immunol. 146, 1700–1707.

Ioannides, C.G., Platsoucas, C.D., Rashed, S., Wharton, J.T., Edwards, C.L., & Freedman, R.S. (1991b). Tumor cytolysis by lymphocytes infiltrating ovarian malignant ascites. Canc. Res. 51, 4257–4265.

Jacoby, R.F., Westbrook, C., Llor, X., & Brasitus, T.A. (1989). Gastroenterology 96, A234.

Jardetzky, T.S., Lane, W.S., Robinson, R.A., Madden, D.R. & Wiley, D.C. (1991). Identification of self peptides bound to purified HLA–B27. Nature 353, 326–329.

Jefferies, W.A. & Burgert, H. (1990). E3/19K from adenovirus 2 is an immunosubversive protein that binds to a structural motif regulating the intracellular transport of major histocompatibility complex class I proteins. J. Exp. Med. 172, 1653–1664.

Jones, C.A., Callaham, M.F., & Huberman, E. (1984). Enhancement of chemical-carcinogen-induced cell transformation in hamster embryo cells by 1α,25- dihydroxycholecalciferol, the biologically active metabolite of vitamin D_3. Carcinogenesis 5, 1155–1159.

Kahn, M., Sugawara, H., McGowan, P., Okuno, K., Nagoya, S., Hellstrom, K.E., Hellstrom, I., & Greenberg, P. (1991). CD4 + T cell clones specific for the human p97 melanoma associated antigen can eradicate pulmonary metastases from a murine tumor expressing the p97 antigen. J. Immunol. 146, 3235–3241.

Kaibara, N., Yurugi, E., & Koga, S. (1984). Promoting effect of bile acids on the chemical transformation of C3H/10T½ fibroblasts *in vitro*. Canc. Res. 44, 5482–5485.

Kawaura, A., Tamda, N., Sawada, K., Oda, M., & Shimoyama, T. (1989). Supplemental administration of 1-hydroxyvitamin D_3 inhibits promotion by intrarectal instillation of lithocholic acid in N-methyl-N-nitrosourea-induced colonic tumorigenesis in rats. Carcinogenesis 10, 647–649.

Kikkawa, U., Kishimoto, A., & Nishizuka, Y. (1989). The protein kinase C family, heterogeneity and its implications. Ann. Rev. Biochem. 58, 31–44.

Kimmig, A., Gekeler, V., Neumann, M., Frese, G., Handgretinger, R., Kardos, G., Diddens, H., & Niethammer, D. (1990). Susceptibility of multidrug-resistant human leukemia cell lines to human interleukin 2-activated killer cells. Canc. Res. 50, 6793–6799.

Kopp, R., Noelke, B., Sauter, G., Schildberg, F.W., Paumgartner, G., & Pfeiffer, A. (1991). Altered protein kinase C activity in biopsies of human colonic adenomas and carcinomas. Canc. Res. 51, 205–210.

Kuroki, T., Sasaki, K., Chida, K., Abe, E., & Suda, T. (1983). 1α,25-dihydroxyvitamin D_3 markedly enhances chemically induced transformation in Balb 3T3 cells. Gann 74, 611–614.

LaMont, J.T. & O'Gorman, T.A. (1978). Experimental colon cancer. Gastroenterology 75, 1157–1169.

Ljunggren, H.G., Paabo, S., Cochet, M., Kling, G., Kourilsky, P., & Karre, K. (1989). Molecular analysis of H-2-deficient lymphoma lines: distinct defects in biosynthesis and association of MHC class 1/β-2 microglobulin observed in cells with increased sensitivity to NK cell lysis. J. Immunol. 142, 2991–2997.

Lointier, P., Wargovich, M.J., Saez, S., Levin, B., Wildrick, D.M., & Boman, B.M. (1987). The role of vitamin D_3 in the proliferation of a human colon cancer cell line *in vitro*. Anticanc. Res. 7, 817–822.

Marian, B., Winawer, S., & Friedman, E. (1989). Tyrosine phosphorylation of a Mr 63,000 protein induced by an endogenous mitogen in human colon carcinoma cells, but not in normal colonocytes. Canc. Res. 49, 4231–4236.

Martell, R.E., Simpson, R.U., & Taylor, J.M. (1987). 1,25-dihydroxyvitamin D_3 ($1,25(OH)_2D_3$) acts through protein kinase C to induce HL-60 differentiation. Fed. Proc. 46, 603 (Abstr).

Maudsley, D.J. & Pound, J.D. (1991). Modulation of MHC antigen expression by viruses and oncogenes. Immunol. Today 12, 429–431.

McJunkin, B., Fromm, H., Sarva, R.P., & Amin, P. (1981). Factors in the mechanism of diarrhea in bile acid malabsorption: Fecal pH—a key determinant. Gastroenterology 80, 1454–1464.

Mezzetti, O., Bagnara, G.P., Monti, M.C., Casolo, L.P., Bonsi, L., & Brunelli, M.A. (1987). Phorbol esters, but not the hormonal form of vitamin D, induces change in protein kinase C during differentiation of human histiocytic lymphoma cell line (U937). Life Sci. 40, 2111–2117.

McPhail, L.C., Clayton, C.C., & Snyderman, R. (1984). A potential second messenger role for unsaturated fatty acids: Activation of Ca^{2+}-dependent protein kinase. Science 224, 622–625.

Miller, F.R. & Heppner, G.H. (1990). Cellular interactions in metastasis. Canc. Metastas. Rev. 9, 21–34.

Mokyr, M.B. & Dray, S. (1987). Interplay between the toxic effects of anticancer drugs and host antitumor immunity in cancer therapy. Canc. Invest. 5, 31–38.

Monaco, J.J., Cho, S., & Attaya, M. (1990). Transport protein genes in the murine MHC: possible implications for antigen processing. Science 250, 1723–1726.

Morotomi, M., Guillem, J.G., LoGerfo, P., & Weinstein, I.B. (1990). Production of diacylglycerol, an activator of protein kinase C, by human intestinal microflora. Canc. Res. 50, 3595–3599.

Moscow, J.A. & Cowan, K.H. (1988). Multidrug resistance. J. Natl. Canc. Inst. 80, 14–20.

Narisawa, T., Magadia, N.E., Weisburger, J.H., & Wynder, E.L. (1974). Promoting effect of bile acids on colon carcinogenesis after intrarectal instillation of N-methyl-N′-nitro-N-nitrosoguanidine in rats. J. Natl. Canc. Inst. 53, 1093–1095.

Niendorf, A., Arps, H., & Dietel, M. (1987). Effect of 1,25-dihydroxyvitamin D_3 on human cancer cells *in vitro*. J. Steroid Biochem. 27, 825–828.

Nishino, H., Yoshioka, K., Iwashima, A., Takizawa, H., Konishi, S., Okamoto, H., Okabe, H., Shibata, S., Fujiki, H., & Sugimura, T. (1986). Glycyrrhetic acid inhibits tumor-promoting activity of teleocidin and 12-0-tetradecanoylphorbol-13-acetate in two-stage mouse skin carcinogenesis. Gann 77, 33–38.

Norman, A.W., Roth, J., & Orci, L. (1982). The vitamin D endocrine system: steroid metabolism, hormone receptors and biological response (calcium binding proteins). Endocr. Rev. 3, 331–366.

O'Brian, C.A. (1989). Transmembrane signal transduction. Oncol. Overviews, Editorial. vii–x.

O'Brian, C.A. & Ward, N.E. (1989). Biology of the protein kinase C family. Canc. Metast. Revs. 8, 199–214.

O'Brian, C.A., Fan, D., Ward, N.E., Dong, Z., Iwamoto, L., Gupta, K.P., Earnest, L.E., & Fidler, I.J. (1991). Transient enhancement of multidrug resistance by the bile acid deoxycholate in murine fibrosarcoma cells *in vitro*. Biochem. Pharmacol. 41, 797–806.

O'Brian, C.A., Ward, N.E., & Vogel, V.G. (1990). Inhibition of protein kinase C by the 12-0-tetradecanoylphorbol-13-acetate antagonist glycyrrhetic acid. Canc. Letts. 49, 9–12.

O'Brian, C.A. & Ward, N.E. (1991). Relevance of the tumor promoter receptor protein kinase C in colon carcinogenesis. In: Dysplasia and Cancer in Colitis (Riddell, R., ed.), pp. 135–139. Elsevier Press, New York.

O'Brian, C.A., Fan, D., Ward, N.E., Seid, C., & Fidler, I.J. (1989). Level of protein kinase C activity correlates directly with resistance to Adriamycin in murine fibrosarcoma cells. FEBS Letts. 246, 78–82.

O'Brian, C.A., Ward, N.E., Weinstein, I.B., Bull, A.W., & Marnett, L.J. (1988). Activation of rat brain protein kinase C by lipid oxidation products. Biochem. Biophys. Res. Commun. 155, 1374–1380.

Palayoor, S.T., Stein, J.M., & Hait, W.N. (1987). Inhibition of protein kinase C by antineoplastic agents: Implications for drug resistance. Biochem. Biophys. Res. Commun. 148, 718–725.

Parham P. (1990). Transporters of delight. Nature 348, 674–675.

Parham P. (1991). Half of a peptide pump. Nature 351, 271–272.

Peace, D.J., Chen, W., Nelson, H., & Cheever, M.A. (1991). T cell recognition of transforming proteins encoded by mutated ras proto-oncogenes. J. Immunol 146, 2059–2065.

Pence, B.C. & Buddingh, F. (1988). Inhibition of dietary fat-promoted colon carcinogenesis in rats by supplemental calcium or vitamin D. Carcinogenesis 9, 187–190.

Phan, S.C., Morotomi, M., Guillem, J.G., LoGerfo, P., & Weinstein, I.B. (1991). Decreased levels of 1,2-*sn*-diacylglycerol in human colon tumors. Canc. Res. 51, 1571–1573.

Pontremoli, S., Melloni, E., Michetti, M., Sacco, O., Salamino, F., Sparatore, B., & Horecker, B.L. (1986). Biochemical responses in activated human neutrophils mediated by protein kinase C & a Ca^{2+}-requiring proteinase. J. Biol. Chem. 261, 8309–8313.

Pontremoli, S., Michetti, M., Melloni, E., Sparatore, B., Salamino, F., & Horecker, B.L. (1990). Identification of the proteolytically activated form of protein kinase C in stimulated human neutrophils. Proc. Natl. Acad. Sci. 87, 3705–3707.

Radinsky R. (1991). Growth factors and their receptors in metastasis. Sem. Canc. Biol. 2, 169–177.

Reddy, B.S., Narasawa, T., Weisburger, J.H., & Wynder, E.L. (1976). Promoting effect of sodium deoxycholate on colon adenocarcinomas in germfree rats. J. Natl. Canc. Inst. 56, 441–442.

Restifo, N.P., Esquivel, F., Asher, A.L., Stotter, H., Barth, R.J., Bennink, J.R., Mule, J.J., Yewdell, J.W., & Rosenberg, S.A. (1991). Defective presentation of endogenous antigens by a murine sarcoma. J. Immunol. 147, 1453–1459.

Rock, K.L., Gamble, S., Rothstein, L., Gramm, C., & Benacerraf, B. (1991). Dissociation of beta-2-microglobulin leads to the accumulation of a substantial pool of inactive class I MHC heavy chains on the cell surface. Cell 65, 611–620.

Rock, K.L., Gamble, S., Rothstein, L., & Benacerraf, B. (1991b). Reassociation with beta 2-microglobulin is necessary for Db class I major histocompatibility complex binding of an exogenous influenza peptide. Proc. Natl. Acad. Sci. U.S.A. 88, 301–304.

Rotzschke, O., Falk, K., Deres, K., Schild, H., Norda, M., Metzger, J., Jung, G., & Rammensee, H.G. (1990). Isolation and analysis of naturally processed viral peptides as recognized by cytotoxic T-cells. Nature 348, 252–254.

Rudensky, A.Y., Preston-Hurlburt, P., Hong, S.C., Barlow, A., & Janeway, C.A. (1991). Sequence analysis of peptides bound to MHC class II molecules. Nature 353, 622–627.

Sadegh-Nasseri, S. & Germain, R.N. (1991). A role for peptide in determining MHC class II structure. Nature 353, 167–170.

Sakanoue, Y., Hatada, T., Kusunoki, M., Yanagi, H., Yamamura, T., & Utsunomiya, J. (1991). Protein kinase C activity as marker for colorectal cancer. Int. J. Canc. 48, 803–806.

Sasaki, K., Chida, K., Hashiba, H., Kamata, N., Abe, E., Suda, T., & Kuroki, T. (1986). Enhancement by 1 alpha, 25-dihydroxyvitamin D_3 of chemically induced transformation of BALB 3T3 cells without induction of ornithine decarboxylase or activation of protein kinase C. Canc. Res. 46, 604–610.

Sauter, G., Nerlich, A., Spengler, U., Kopp, R., & Pfeiffer, A. (1990). Low diacylglycerol values in colonic adenomas and colorectal cancer. Gut 31, 1041–1045.

Sitrin, M.D., Halline, A.G., Abrahams, C., & Brasitus, T.A. (1991). Dietary calcium and vitamin D modulate 1,2-dimethylhydrazine-induced colonic carcinogenesis in the rat. Canc. Res. 51, 5608–5613.

Solomon, E., Voss, R., Hall, V., Bodmer, W.F., Jass, J.R., Jeffreys, A.J., Lucibello, F.C, Patel, I., & Rider, S.H. (1987). Chromosome 5 allele loss in human colorectal carcinomas. Nature 328, 616–619.

Spies, T., Bresnahan, M., Bahram, S., Arnold, D., Blanck, G., Mellins, E., Pious, D., & DeMars, R. (1990). A gene in the human major histocompatibility complex class II region controlling the class I antigen presentation pathway. Nature 348, 744–747.

Spies, T., & DeMars, R. (1991). Restored expression of major histocompatibility class I molecules by gene transfer of a putative peptide transporter. Nature 351, 323–324.

Stanbridge, E.J. (1990). Identifying tumor suppressor genes in human colorectal cancer. Science 247, 12–13.

Sugarbaker, J.P., Gunderson, L.L., & Wittes, R.E. (1985). Colorectal cancer. In: Cancer: Principles and Practices of Oncology (DeVita, V.T., Hellman, S., and Rosenberg, S.A., eds), pp. 800–803, JB Lippincott, Philadelphia.

Takano, S., Akagi, M., & Bryan, G.T. (1984). Stimulation of ornithine decarboxylase activity and DNA synthesis by phorbol esters or bile acids in rat colon. Gann 75, 29–35.

Takano, S., Matsushima, M., Erturk, E., & Bryan, G.T. (1981). Early induction of rat colonic epithelial ornithine and S-adenosyl-L-methionine decarboxylase activities by N-methyl-N′-nitro-N-nitrosoguanidine or bile salts. Canc. Res. 41, 624–628.

Tanaka, K., Yoshioka, T., Bieberich, C., & Jay, G. (1988). Role of the major histocompatibility complex class I antigens in tumor growth and metastasis. Ann. Rev. Immunol. 6, 359–380.

Tanaka, Y., Bush, K.K., Eguchi, T., Ikekawa, N., Taguchi, T., Kobayashi, Y., & Higgins, P.J. (1990). Effects of 1,25-dihydroxyvitamin D_3 and its analogs on butyrate-induced differentiation of HT-29 human colonic carcinoma cells and on the reversal of the differentiated phenotype. Arch. Biochem. Biophys. 276, 415–423.

Takenawa, T. & Nagai, Y. (1981). Purification of phosphatidylinositol-specific phospholipase C from rat liver. J. Biol. Chem. 256, 6769–6775.

Takeyama, Y., Kaibuchi, K., Ohyanagi, H., Saitoh, Y., & Takai, Y. (1985). Enhancement of growth factor-induced DNA synthesis by colon tumor-promoting bile acids in Swiss 3T3 cells. FEBS Letts. 193, 153–158.

Townsend, A. & Bodmer, H. (1989). Antigen recognition by Class I-restricted *T* lymphocytes. Ann. Rev. Immunol. 7, 601–624.

Townsend, A., Ohlen, C., Bastin, J., Ljunggren, H.G., Foster, L., & Karre, K. (1989). Association of class I major histocompatibility heavy and light chains induced by viral peptides. Nature 340, 443.

Trowsdale, J., Hanson, I., Mockridge, I., Beck, S., Townsend, A., & Kelly, A. (1990). Sequences encoded in the class II region of the MHC related to the "ABC" superfamily of transporters. Nature 348, 741–744.

Tsuruo, T. (1988). Mechanisms of multidrug resistance and implications for therapy. Jpn. J. Canc. Res. 79, 285–296.

Ullrich, A. & Schlessinger, J. (1990). Signal transduction by receptors with tyrosine kinase activity. Cell 61, 203–212.

Vogelstein, B., Fearon, E.R., Hamilton, S.R., Kern, S.E., Preisinger, A.C., Leppert, M., Nakamura, Y., White, R., Smits, A.M.M., & Bos, J.L. (1988). Genetic alterations during colorectal-tumor development. New Engl. J. Med. 319, 525–532.

Vogelstein, B., Fearon, E., Hamilton, S., Kern, S., Preisinger, A., Leppert, M., Nakamura, Y., White, R., Smits, A., & Bos, J. (1986). Genetic alterations during colorectal tumor development. N. Engl. J. Med. 319, 525–532.

Van Bleek, G.M. & Nathenson, S.G. (1990). Isolation of an endogenously processed immunodominant viral peptide from the class I H-2Kb molecule. Nature 348, 213–216.

Wali, R.K., Baum, C.L., Bolt, M.J.G., Dudeja, P.K., Sitrin, M.D., & Brasitus, T.A. (1991). Downregulation of protein kinase C activity in 1,2-dimethylhydrazine-induced rat colonic tumors. Biochim. Biophys. Acta 1092, 119–123.

Wali, R.K., Baum, C.L., Sitrin, M.D., & Brasitus, T.A. (1990). $1,25(OH)_2$ vitamin D_3 stimulates membrane phosphoinositide turnover, activates protein kinase C, and increases cytosolic calcium in rat colonic epithelium. J. Clin. Invest. 85, 1296–1303.

Ward, N.E. & O'Brian, C.A. (1988). The bile acid analog fusidic acid can replace phosphatidylserine in the activation of protein kinase C by 12-0-tetradecanoylphorbol-13-acetate *in vitro*. Carcinogenesis 9, 1451–1454.

Ways, D.K., Dodd, R.C., Bennett, T.E., Gray, T.K., & Earp, H.S. (1987). 1,25-dihydroxyvitamin D_3 enhances phorbol ester-stimulated differentiation and protein kinase C-dependent substrate phosphorylation activity in the U937 human monoblastoid cell. Endocrinol. 121, 1654–1660.

Weinberg, R.A. (1991). Tumor suppressor genes. Science 254, 1138–1146.

Weinstein, I.B. (1988). The origins of human cancer: Molecular mechanisms of carcinogenesis and their implications for cancer prevention and treatment. Canc. Res. 48, 4135–4143.

Weinstein, I.B. (1990). The role of protein kinase C in growth control and the concept of carcinogenesis as a progressive disorder in signal transduction. The Biology and Medicine of Signal Transduction, (Nishizuka, Y., ed.), pp. 307–316 Raven Press, New York.

Weinstein, I.B. (1991). Mitogenesis is only one factor in carcinogenesis. Science 251, 387–388.

Weinstein, I.B. (1991b). Cancer Prevention: Recent progress and future opportunities. Canc. Res. 51, 5080–5085.

Weyman, C.M., Taparowsky, E.J., Wolfson, M., & Ashendel, C.L. (1988). Partial downregulation of protein kinase C in C3H10T½ mouse fibroblasts transfected with the human Ha-ras oncogene. Canc. Res. 48, 6535–6541.

Wolfman, A., Wingrove, T.G., Blackshear, P.J., & Macara, I.G. (1987). Downregulation of protein kinase C and of an endogenous 80-kDa substrate in transformed fibroblasts. J. Biol. Chem. 262, 16546–16552.

Wood, A., Chang, R.L., Huang, M.T., Uskokovic, M., & Conney, A.H. (1983). 1,25-dihydroxyvitamin D_3 inhibits phorbol ester-dependent chemical carcinogenesis in mouse skin. Biochem. Biophys. Res. Comm. 116, 605–611.

Young, S., Parker, P.J., Ullrich, A., & Stabel, S. (1987). Downregulation of protein kinase C is due to an increased rate of degradation. Biochem. J. 244, 775–779.

Yu, G., Ahmad, S., Aquino, A., Fairchild, G.R., Trepel, J.B., Ohno, S., Suzuki, K., Tsuruo, T., Cowan, K.H., & Glazer, R.I. (1991). Transfection with protein kinase C-α confers increased multidrug resistance to MCF-7 cells expressing P-glycoprotein. Canc. Comm. 3, 181–189.

THE SIGNIFICANCE OF THE EXTRACELLULAR MATRIX IN MAMMARY EPITHELIAL CARCINOGENESIS

Calvin D. Roskelley, Ole W. Petersen, and Mina J. Bissell

Advances in Molecular and Cell Biology
Volume 7, pages 89–113

ISBN: 1-55938-624-X

I. INTRODUCTION

Mammary adenocarcinoma is one of the most prevalent malignancies among North American women (Willet, 1989; Henderson et al., 1991). Despite intense investigations over many decades the detailed natural history of tumor formation in the breast remains largely unknown. During reproductive life, most women develop benign hyperplastic lesions that may or may not regress with increasing age (Wellings et al., 1975; Peeters et al., 1990; Sarnelli and Squartini, 1991). It has been suggested that these lesions persist and give rise to tumors in the cancerous breast. However, there is no direct evidence that tumor formation is preceded by such a histologically recognizable pre-neoplastic stage (Ponten et al., 1990). Once formed, carcinomatous tumors are made up of disorganized, atypical cells that may or may not be hyperplastic (Egan et al., 1982; Auer et al., 1984; Taylor-Papadimitriou and Cane, 1987). From here, progression from a spatially restricted carcinoma *in situ* to the metastatic state is dependent on an initial, misguided invasion into the underlying stroma (Dupont and Page, 1985; Carter et al., 1988). Thus, during mammary tumor development changes in growth, morphogenesis and stromal invasiveness occur. Interestingly, in normal mammary gland development changes in all of these processes also take place, albeit in precisely controlled cyclical patterns. This suggests that breast cancer, like many other malignant diseases, is associated with inappropriate responses to normal tissue-specific regulatory signals (Pierce and Speers, 1988; Parchment et al., 1990). We believe that the identification of altered cellular responses to these signals, particularly those originating from the extracellular matrix, may provide clinically useful markers for breast carcinogenesis.

A strong case has been made for the importance of mutational events in breast cancer development. For example, a number of specific genetic lesions have been found in tumor biopsy specimens and tumor derived cell lines. These include amplification or activation of dominantly acting oncogenes such as *erb*-B-2, *myc* and *ras*, as well as the inactivation or loss of tumor suppressors such as the Rb and p53 genes (Horan-Hand et al., 1984; Kozbor and Croce, 1984; Slamon et al., 1987; Rochlitz et al., 1989; Thompson et al., 1990; Runnebaum et al., 1991). However, the specific genes affected vary widely between and within individual tumors. Thus, from a clinical point of view the identification of such molecular lesions remains correlative at best and the most reliable single prognostic indicator of the disease continues to be the extent of metastasis to regional lymph nodes (Gullick,

1990; Mackay et al., 1990). In fact, it is likely that no single genetic lesion, or group of lesions, is responsible for the induction of all breast tumors. Instead, varying combinations of intra- and extracellular events act together to initiate common or overlapping phenotypic changes within the gland that are crucial to the early stages of the carcinogenic process.

II. TISSUE-SPECIFIC PHENOTYPIC CHANGES IN MAMMARY CARCINOGENESIS

The phenotypic changes associated with clinical breast disease include varying degrees of aneuploidy, changes in estrogen receptor status, alterations in growth factor responsiveness, inappropriate intermediate filament expression and increased production of extracellular matrix degrading proteases (Helin et al., 1989; von Rosen et al., 1989; Basset et al., 1990; Harris and Nicholson, 1991; Thompson et al., 1991). Used in conjunction with lymph node status, these phenotypic markers provide a reasonably accurate picture of the later, progressive stages of the disease. However, it is these stages that are most often associated with poor clinical outcome, radical ablative therapies, and a high degree of morbidity. Therefore, it is highly desirable that phenotypic changes associated with the earlier inductive stage(s) of breast cancer be found. The identification of such markers would drastically improve the early diagnosis and treatment of this disease. To do so will require a close examination of tissue-specific changes that occur in the carcinogenic target cell population of the mammary gland.

It is suspected that the target cells affected in human breast cancer are derived from the epithelium that lines the terminal ducts of the mammary gland (Lippman and Dickson, 1990). This is a simple epithelium consisting of polarized cells that face a lumen apically and rest on a basement membrane basally (Hilkens et al., 1989). In histologically recognizable pre-malignant lesions the cells of this epithelium retain a modicum of normal cyto-differentiation, but they often become hyperplastic and spatially disorganized (Auer et al., 1984). The result is a multilayered epithelium that may occlude the lumen and lead to cyst formation (Wellings et al., 1975; Ponten et al., 1990). Thus, a hallmark of the carcinogenic process in the breast is the onset of inappropriate cellular responses to signals present in the surrounding microenvironment. An important component of this microenvironment is the extracellular matrix (ECM).

III. CELL-ECM INTERACTIONS DURING MAMMARY GLAND DEVELOPMENT *IN VIVO*

Mammary gland development can be crudely divided into 3 stages: an early developmental period spanning from late gestation to puberty; the functional periods of pregnancy and lactation; and the degenerative period of involution

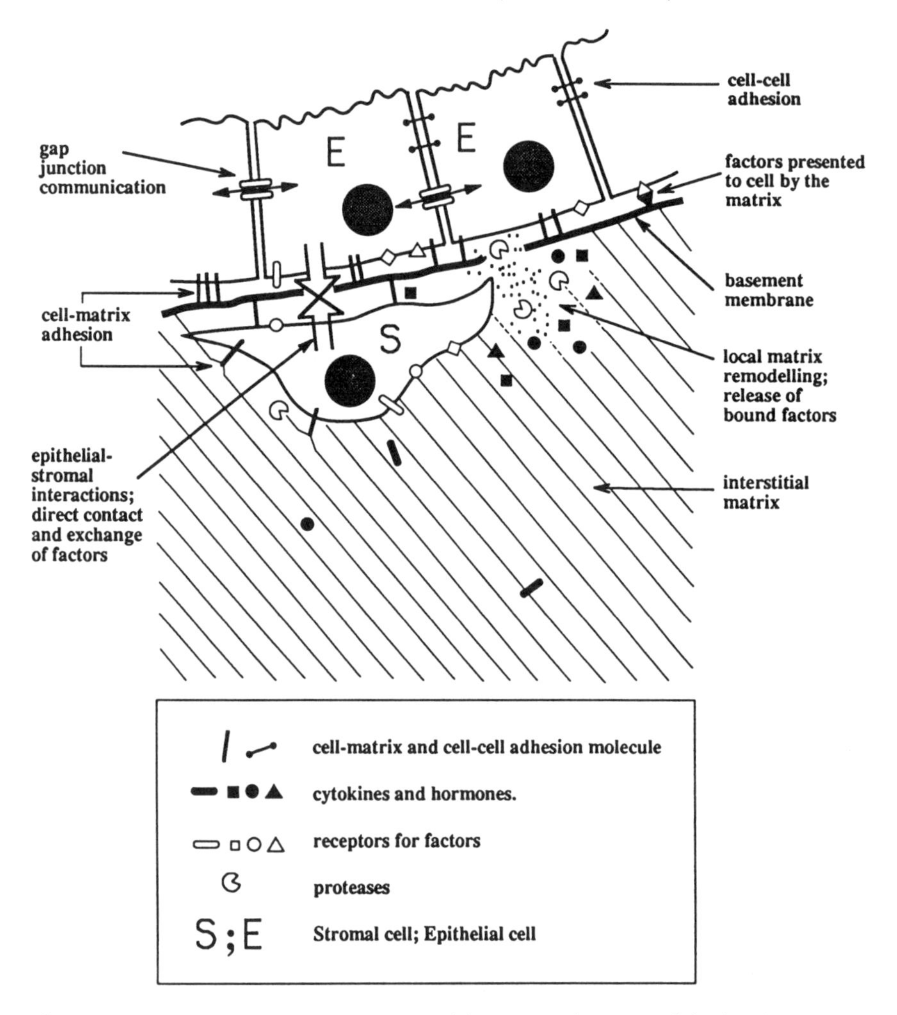

Figure 1. Diagrammatic representation of the major elements of the local microenvironment in a typical epithelial-mesenchymal organ. Interactions occur between the cells and components of the surrounding microenvironment. These include the basement membrane, stromal ECM, locally acting soluble factors and adjacent cells. In the case of the mammary gland, the actions of cycling reproductive hormones and neuroendocrine induced factors overlay this template (modified from Stoker et al., 1990).

(Knight and Peaker, 1982; Daniel and Silberstein, 1987). Throughout this development, the mammary epithelium undergoes profound changes in growth, stromal invasiveness, morphogenesis and differentiation. In addition to being triggered by ovarian, pituitary, and placental hormones, these developmental changes are dependent on dynamic and reciprocal communications between the cells and the surrounding microenvironment (Bissell et al., 1982). The microenvironment is an interactive, tissue-specific network that consists of the epithelial and stromal cells themselves, short and long range acting soluble factors, and the ECM (Figure 1, Stoker et al., 1990; Watt, 1991). Using a mouse model, where detailed observation and experimental manipulation are possible, it has been shown that the basement membrane is a significant and essential player in mammary-specific morphogenesis and gene expression (for reviews see: Bissell and Hall, 1987; Daniel and Silberstein 1987; Howlett and Bissell, 1993).

At each stage in mouse mammary gland development, interactions between the epithelium and the ECM have been identified that are important regulators of cellular behavior (Pitelka and Hamamoto, 1977; Gordon and Bernfield, 1980; Silberstein and Daniel, 1982; Talhouk et al., 1991, 1992). In addition, using specially designed microenvironments in culture, many of the normal cellular responses to specific signals emanating from the ECM have been elucidated (Emerman and Pitelka, 1977; Bissell et al., 1985; Lee et al., 1985; Blum et al., 1987; Li et al., 1987; Barcellos-Hoff et al., 1989; Streuli and Bissell, 1990). Because ECM-derived signals direct growth and differentiation, it is likely that the cellular responses to these signals represent important developmental control points that are co-opted during tumor formation. Therefore, the identification of changes in cell-ECM interactions that arise during clinical and experimental carcinogenesis should produce novel phenotypic markers of the early steps of this disease process.

In the following sections cell-ECM interactions at each stage of mammary development are briefly outlined. For each stage, cell-ECM control points that could be susceptible to carcinogenic insult are postulated.

A. Early Development

Late in gestation the female mammary gland forms as a prominent fat pad that contains simple, branching, blind-ended epithelial ducts. These ducts are surrounded by a thin sleeve of fibroblastic tissue. This basic pattern is maintained postnatally and throughout childhood. As puberty approaches, ovarian hormones and locally acting cytokines initiate a complex ductular branching. Concurrently, small ductular outpocketings known as terminal endbuds expand deep into the fatpad. This process produces the non-functional virgin gland that is maintained as a ductular scaffold until further development occurs during pregnancy (Knight and Peaker, 1982; Daniel and Silberstein, 1987).

The early branching and expansion events that occur during mammary gland development are dependent upon epithelial-mesenchymal interactions. These interactions are organospecific for morphogenesis. For example, combinations of mammary epithelium and salivary gland mesenchyme result in the development of salivary ductal patterns (Kratochwil, 1969; Sakakura et al., 1976). This indicates that the mesenchyme produces the inductive signals in this process. It appears that

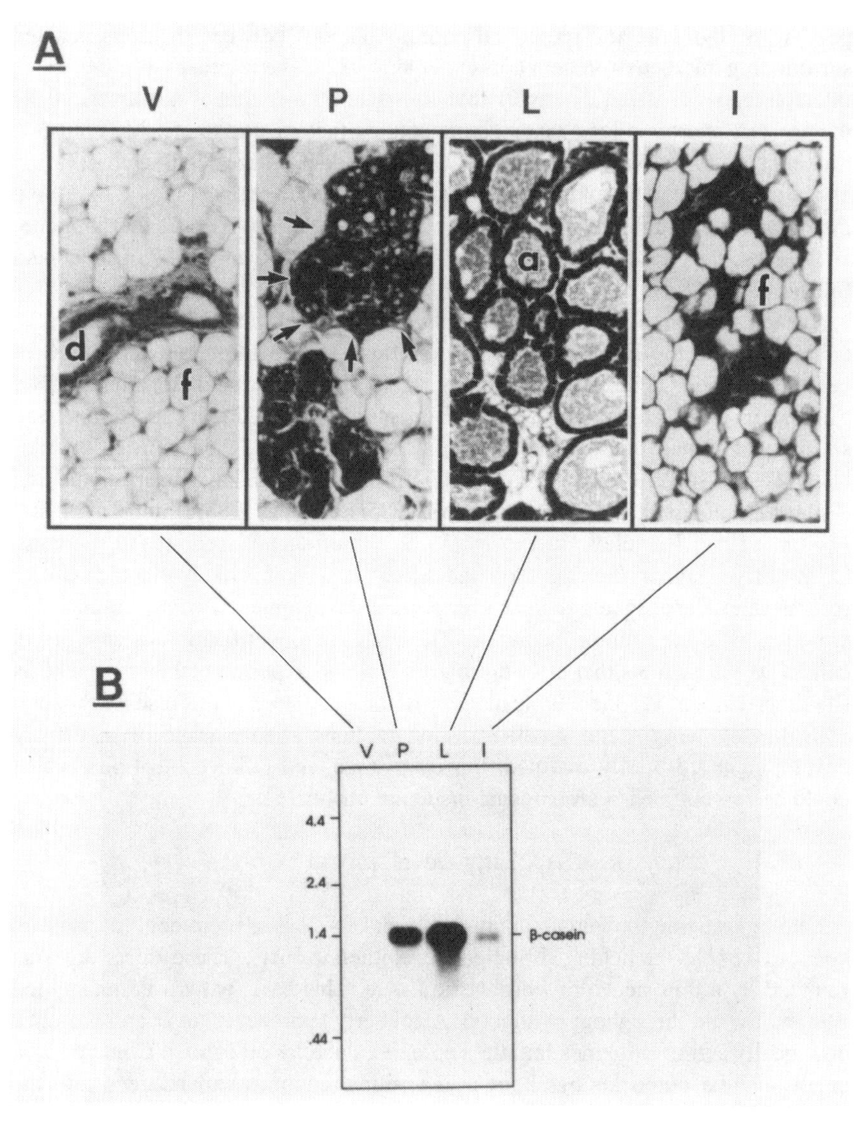

Figure 2. Morphologic (A: hematoxylin and eosin stained histologic sections) and functional (B: Northern analysis for milk protein β casein mRNA) representation of various stages of mouse mammary gland development. The virgin gland (V) contains ducts (d) that course between the adipocytes (f) of the mammary fat pad. The virgin gland is non-functional and synthesizes no β casein. During pregnancy (P) alveolar buds develop (arrows) that begin to synthesize β casein. In lactation, functional milk secreting alveoli (a) expand and obliterate most of the adipocytes. After weaning, during involution, milk production drops, alveoli degenerate and adipocytes repopulate the gland.

mammary specific mesenchymal signals are of two types. One emanates from the fibroblasts and acts to inhibit ductal expansion. The other, dependent on the presence of adipocytes, directs ductal branching and endbud development (Sakakura et al. 1982). The extracellular matrix produced by these two mesenchymal components are very different. The fibroblastic ECM contains large amounts of fibronectin and tenascin, while the adipocytes and adjacent epithelium deposit a basement membrane (BM) matrix that contains laminin, collagen IV and heparan sulphate proteoglycan (Kimata et al. 1985, Chiquet-Ehrismann et al. 1986, Sakakura et al. 1991). The importance of the type of ECM surrounding the epithelium is emphasized by experiments involving matrix modifiers. For example, TGF-β released from *in vivo* implants increases the production of fibroblastic collagen type I in the area of adipocyte-epithelial interaction. The result is an inhibition of endbud development (Daniel et al., 1989; Silberstein et al., 1990; Robinson et al., 1991). Due to its growth inhibitory effects on a number of epithelial cell types, it has been suggested that TGF-β is an inhibitor of tumorigenesis (Missero et al., 1991a,b). However, the effect of this growth factor may be dependent on the state of the target cells involved. For example, during early ductal growth and branching, which is dependent upon a stromal matrix, TGF-β could have stimulatory effects. In fact, in certain situations, where its action mimics the increased stromal matrix accumulation of wound healing, this growth factor may contribute to tumor formation (Siewicke et al., 1990). As will be discussed below, endogenous TGF-β production is dependent upon the differentiation state of mammary epithelial cells. Therefore, as alterations in differentiation arise during carcinogenesis, inappropriate expression of this matrix regulator may occur which would then have further compounding effects on growth and differentiation.

B. Pregnancy and Lactation

Triggered by increased levels of progestins, estrogen and placental lactogen, an explosive lobulo-alveolar development predominates during the early stages of pregnancy (Figure 2; Neville and Daniel, 1987). Small alveoli consisting of 50–150 cells form as outpocketings at the lateral margins of terminal ductules. In doing so they expand into the space previously occupied by adipocytes. This alveolar development is dependent upon a concurrent ductal hyperplasia that is driven by migration of the terminal endbuds to the outer margins of the fatty stroma. Throughout this process, an intimate association between the mammary epithelium and the underlying basement membrane is maintained. Interestingly, in transgenic mice which express the *myc*, *ras* or *erb*-B-2 oncogenes in a mammary-specific fashion this process is mimicked in virgin animals, and tumors form only stochastically with increasing age (Leder et al., 1986; Sinn et al., 1987; Muller et al., 1988). As these pathologies occur in animals that have never been pregnant, it is reasonable to assume that the ductal and alveolar hyperplasias arise due to changes in the local microenvironment. That a similar ductal and alveolar hyperplasia also occurs in transgenic animals overexpressing TGF-α further supports this suggestion, as this growth factor has been found to act in an autocrine fashion as a regulator of mammary epithelial cell growth (Matsui et al., 1990). Whether oncogene-associated ductal expansion occurs in response to changes in the production of autocrine signals or because of inappropriate signal processing is as yet unknown. Experimental manipulation of a culture system consisting of stable, transfectable mammary epithelial cells that remain responsive to changes in the surrounding microenvironment could be used to answer this question.

During the latter half of pregnancy alveolar expansion is accompanied by the onset of mammary-specific biochemical differentiation. This is associated with the expression of milk protein genes such as β-casein. After parturition, prolactin and growth hormone act in concert with glucocorticoids and prostaglandins to induce lactation. During lactation the epithelial cells become highly differentiated both structurally and functionally. The result is an active synthesis and vectorial secretion of milk into the alveolar lumen. Utilizing custom-designed microenvironments in culture it has been demonstrated that endogenously deposited, or exogenously added, BM components act in concert with lactogenic hormones to direct this milk production (Bissell et al., 1985; Blum et al., 1987; Li et al., 1987; Streuli and Bissell, 1990). In addition, milk production appears to be dependent on the presence of an intact basement membrane *in vivo* (Wicha et al., 1980; Talhouk et al., 1991; Talhouk, Werb and Bissell, unpublished observations). Therefore, the hypothesized dissolution of the basement membrane that occurs prior to metastatic invasion (Liotta et al., 1991) could lead to local perturbations of differentiated function in the mammary gland.

C. Involution

With the cessation of suckling at weaning, neuroendocrine signals inhibit the production of lactogenic hormones and trigger involution (Knight and Peaker, 1982). Involution is characterized by the degeneration of the secretory alveoli and regression of the gland to a stage similar to that seen in non-pregnant post-pubertal stages (Hurley, 1989). Differentiating adipocytes replace the degenerating alveoli and only a branched network of ducts and terminal ductules remain. During involution, morphologic changes in the BM occur as alveolar cells pull away from it and degenerate (Martinez-Hernandez, 1976; Warburton et al., 1982). This process appears to be regulated, at least in part, by alterations in the ratio of ECM-degrading metalloproteases and their inhibitors (Talhouk et al., 1991, 1992). These alterations occur midway through involution at a period when expression of milk proteins fall dramatically (Figure 3). Increased production of a putative metalloprotease, stromelysin-3, occurs in invasive human mammary adenocarcinomas (Bassett et al., 1990). The mouse homologue of this protease is also expressed during normal involution (Lefebvre et al., 1993). Interestingly, in both the human tumors (Wolf et al., 1993) and the involuting mouse gland (Lefebvre et al., 1993), it appears that

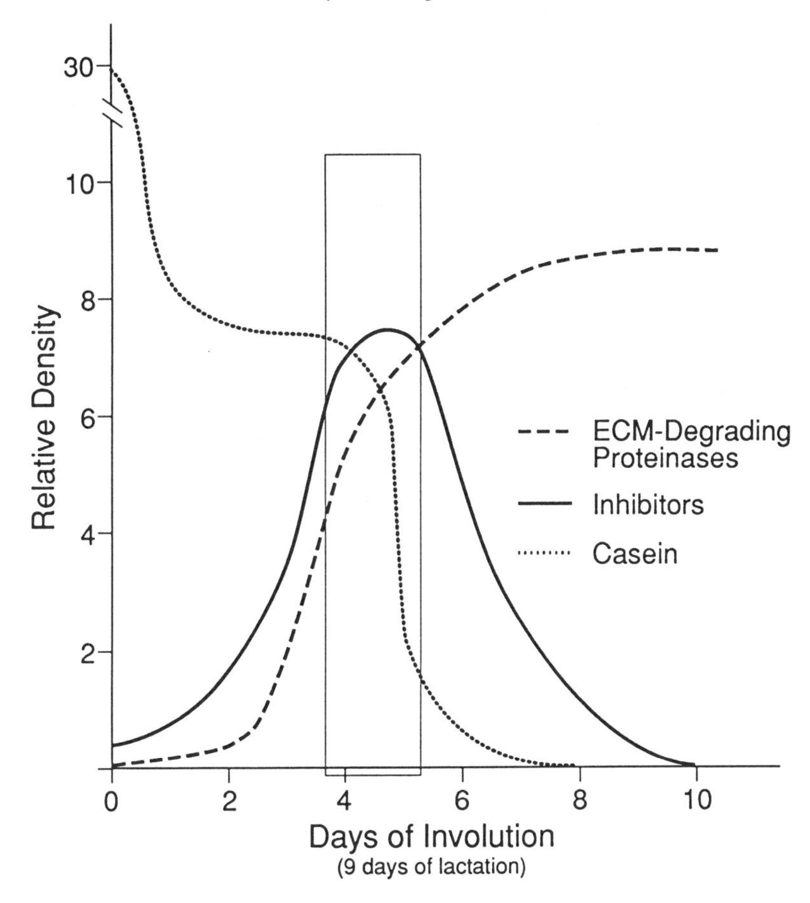

Figure 3. The kinetics of ECM-degrading proteinases, their inhibitors, and β casein production during mammary gland involution. Early in involution (days 0–4) β casein production remains high while protease levels are low. By day 5, protease levels rise concurrently with inhibitors of their action. Late in involution (days 6–10) inhibitor levels fall, protease levels remain high, and β casein production drops precipitously. Therefore, during involution the presence of active ECM-degrading proteases is associated with decreased milk protein synthesis and the subsequent alveolar degeneration (courtesy of R.S. Talhouk).

stromal cells surrounding epithelial tissues somehow become activated to produce the stromelysin-3.

In summary, ECM degradation, which is normally accelerated during the cessation of differentiated function, is also inappropriately increased during the latter stages of mammary carcinogenesis. That ECM degrading signals are associated with degeneration of the epithelium during involution and expansion of the epithelium during carcinogenesis at first glance appears to be contradictory. However, the responses to the loss of basement membrane may depend on the growth and/or differentiation state of the epithelium at the time of increased protease activity.

IV. MAMMARY EPITHELIAL CELL-ECM INTERACTIONS IN CULTURE

In a number of systems alterations in the production, deposition and degradation of the extracellular matrix are almost invariably associated with transformation in culture (for a review see: Auersperg and Roskelley, 1991). For example, in mammary cell lines acquisition of estrogen independence or introduction of the *ras* oncogene both induce changes in cell-ECM interactions that correlate with their ability to invade and metastasize (Thompson et al., 1988; Ochieng et al., 1991). However, these changes have only been observed after conversion to tumorigenicity. As has been outlined above, numerous alterations in cell-ECM interactions also occur during normal mammary gland development. Thus, an important question that remains to be answered is: what changes in the relationship between cells and the ECM are critical to the emergence of the tumor phenotype? In order to answer this question the precise nature of cell-ECM interactions must first be determined in the context of normal growth and differentiation so that transformation-induced changes can be understood in this context. These tasks can be carried out by perturbing homeostasis *in vivo* and observing the consequent changes. Alternatively, custom-designed tissue culture microenvironments can be used which induce developmental changes that mimic those that take place *in vivo*.

A. Normal Morphogenesis and Differentiation are Dependent on the Microenvironment

Morphogenesis and Function

When cultured on tissue culture plastic, mouse mammary epithelial cells from mid-pregnant mice form a non-descript epithelial monolayer. Concurrently, the cells rapidly lose tissue-specific function and they no longer respond to lactogenic hormone treatment by producing milk proteins. However, when the same cells are cultured on malleable floating collagen gels (Emerman and Pitelka, 1977) they maintain a modicum of differentiated function that is associated with the deposition of a basement membrane matrix (Streuli and Bissell, 1990). When these cells are cultured directly upon a basement membrane-derived matrix they aggregate and undergo extensive morphogenetic changes. This generates three-dimensional structures that are morphologically similar to alveoli *in vivo* (Figure 4; Bissell et al., 1985; Li et al., 1987; Streuli et al., 1991). Under these conditions, the epithelial

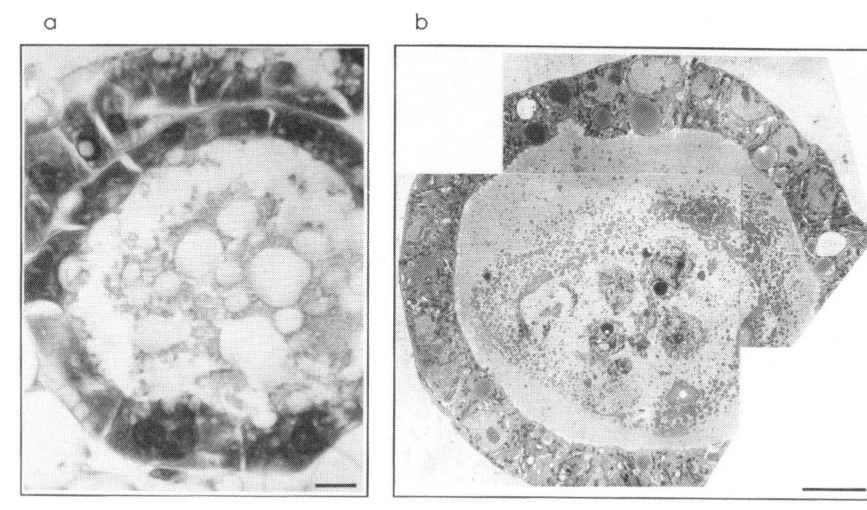

Figure 4. Transverse sections of a functional alveolus (a) and a morphologically similar 'mammosphere' (b). The latter is formed by mouse mammary epithelial cells cultured on a basement membrane-derived extracellular matrix (a: light micrograph, stained with hematoxylin and eosin. b: electron micrograph at a similar magnification. bar= 10 μm, from Streuli and Bissell, 1991).

cells form junctional complexes, polarize, and differentiate ultrastructurally. In response to lactogenic hormone treatment these mammospheres express the full range of milk proteins and secrete them in a vectorial manner into a central lumen (Figure 5; Barcellos-Hoff et al., 1989; Streuli et al., 1991).

The complete ECM-dependent differentiation of mammary epithelial cells in culture appears to be hierarchical. While cell-cell interactions are required for complete functional and structural differentiation they are not required for biochemical differentiation in the presence of a basement membrane matrix. For example, when individual cells are suspended within a basement membrane matrix they do not polarize but they do express the milk protein β-casein (Streuli et al., 1991). On the other hand, expression of whey acidic protein, a milk component that is produced in the final stages of differentiation *in vivo*, appears to require the formation of a three dimensional mammosphere (Figure 6; Chen and Bissell, 1989; Streuli and Bissell, 1991).

Molecular Mechanisms Involved in Cell-ECM Interactions

The use of a stable, transfectable cell strain has made it possible to determine regulatory elements involved in the control of milk protein gene expression. This strain, designated CID-9 (Schmidhauser et al., 1990) was isolated by differential

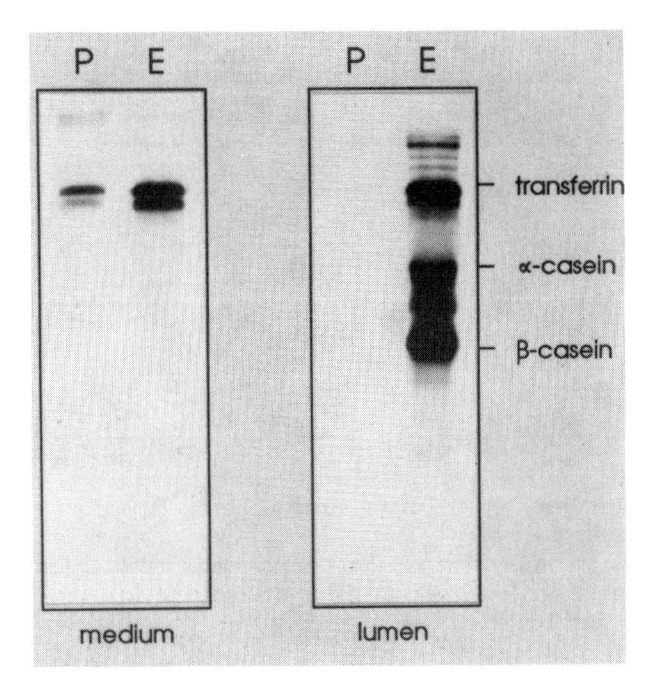

Figure 5. Milk proteins produced by mouse mammary epithelial cells cultured on tissue culture plastic (P) or on a basement membrane-derived extracellular matrix (E) in the presence of lactogenic hormones. The cells on plastic secrete small amounts of transferrin but no casein milk proteins. The alveolar like 'mammospheres' that form on basement membrane matrices secrete transferrins basally into the medium and caseins apically into their lumens. The latter is similar to the vectorial secretion that occurs *in vivo* (from Streuli and Bissell, 1991).

trypsinization of a pre-existing population, COMMA-1D (Danielson et al., 1984). CID-9 cells respond to a combination of lactogenic hormones in the presence of a basement membrane-derived ECM by undergoing both morphological and biochemical differentiation. For example, they form spheres and they produce and secrete milk proteins vectorially in response to lactogenic hormone treatment. In the case of β-casein, transcription is regulated by a 160 bp ECM-responsive enhancer (BCE-1) that is located 1.5 kilobases upstream of the gene. This *cis*-acting regulatory element was initially identified using truncated versions of the bovine β-casein promoter linked to a reporter gene (Figure 7). Subsequently, it was characterized as an enhancer by comparing the expression of stably transfected BCE-1/reporter gene constructs when the cells were maintained on either tissue culture plastic or basement membrane-derived matrix (Figure 8; Schmidhauser et al., 1992).

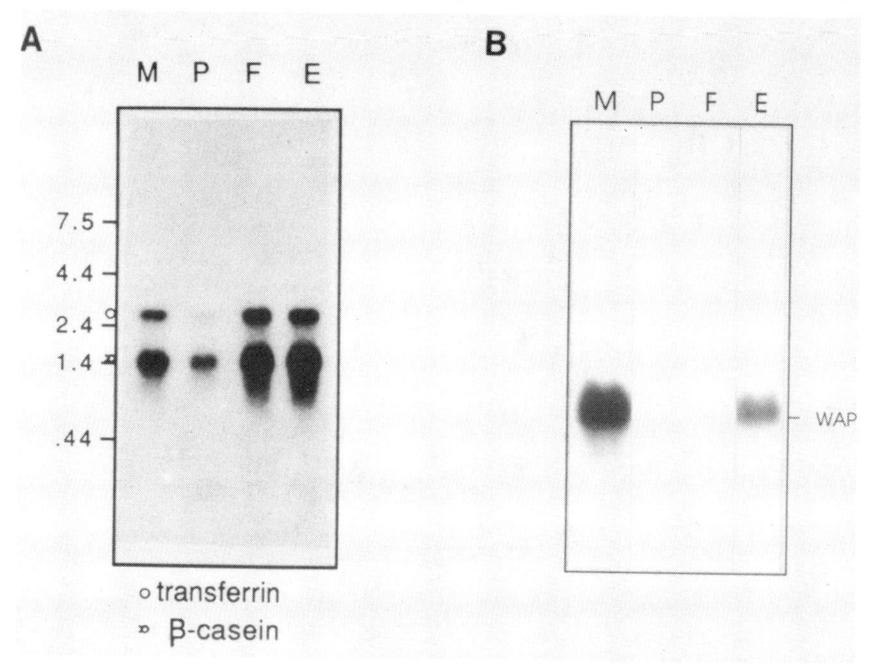

Figure 6. Northern analysis of mRNAs coding for milk proteins produced by mouse mammary epithelial cells. Transferrin and β-casein (left panel) expression occurs in mammary gland tissue (M) and is maintained when the cells are cultured on floating collagen gels (F) or on a basement membrane-derived ECM (E) but not when they are cultured on plastic (P). On the other hand, whey acidic acid protein (WAP - right panel) expression in culture occurs only on basement membrane (E) where the cells undergo a three dimensional reorganization to form alveolar like structures (from Howlett and Bissell, 1990; and modified from Chen and Bissell, 1989).

On floating gels of type I collagen, normal mammary epithelial cells synthesize and deposit their own endogenous basement membrane consisting of laminin, collagen IV, entactin and sulphated glycosaminoglycans (Parry et al., 1985; Streuli and Bissell, 1990). On tissue culture plastic the cells synthesize large quantities of BM components, but they are unable to deposit them in an organized matrix (Streuli and Bissell, 1990). Mammary epithelial cells on tissue culture plastic also produce large amounts of TGF-β, which is known to induce synthesis of ECM components. In contrast, on basement membrane matrices TGF-β expression is sharply down-regulated. Furthermore, this down-regulation occurs at the transcriptional level and it is lactogenic hormone-independent (Streuli et al., 1993). This suggests that TGF-β acts within a local feedback loop to modify matrix synthesis that is dependent on the state of the deposited basement membrane.

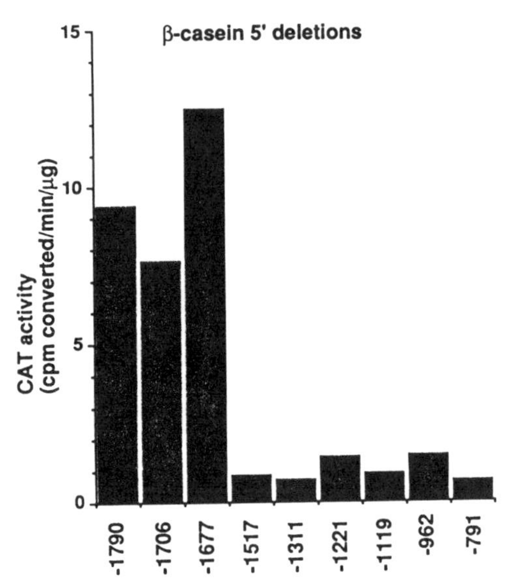

Figure 7. Deletion mutants of the 5′ regulatory sequences of the β-casein gene linked to the chloramphenicol acetyl transferase (CAT) gene were stably transfected into the ECM-responsive mouse mammary epithelial (CID-9) cells. The cells were then cultured on a basement membrane matrix in the presence of lactogenic hormones. A sharp drop in CAT activity occurs after a 160 base pair deletion located about 1.5 kilobase pairs upstream of the transcription start site. This sequence was designated BCE-1 (courtesy of C. Schmidhauser).

The interaction of mouse mammary epithelial cells with the ECM is receptor dependent. For example, antibodies directed against the β-subunit of integrins block basement membrane-dependent β-casein expression (Streuli et al., 1991). In a number of systems the association of integrins with matrix ligands appears to trigger a wide range of intracellular changes that are associated with signal transduction (for a review see: Hynes, 1992). These include: alterations in the cytoskeleton; increased phosphorylation of specific protein substrates, some of which themselves may be kinases; activation of the Na^+/H^+ antiporter; alterations in intracellular calcium metabolism; and interaction with activated growth factor receptors. Therefore, oncogene expression, which invariably impinges upon signal transduction pathways, could act to modulate integrin-mediated signaling. Such signal modulations will likely have functional consequences that can be detected in ECM-dependent differentiation assays.

Taken together, the experiments described above illustrate that the utilization of appropriate microenvironments in culture have allowed for the elucidation of mammary-specific responses to ECM at the morphogenetic, cellular and molecular

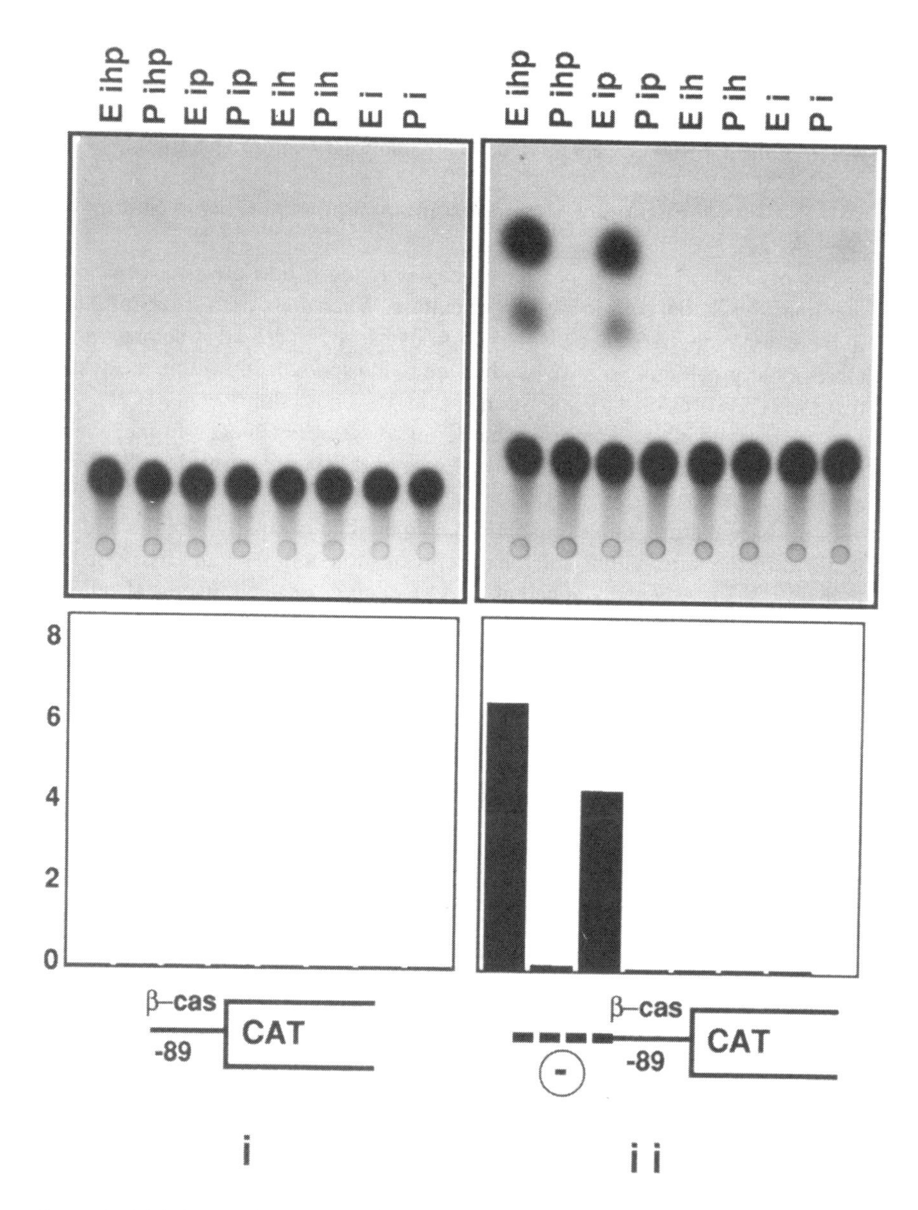

Figure 8. Thin layer chromatography and a graphic representation of CAT activity in CID-9 cells. These cells contained a fusion construct consisting of the CAT gene linked to either a minimal β-casein promoter alone (i; –89 bp to +42 bp), or with BCE-1 added in reverse orientation (ii; dotted line). The cells were then cultured on either a basement membrane ECM (E) or tissue culture plastic (P) in the presence of various combinations of lactogenic hormones (i-insulin, h-hydrocortisone, p-prolactin). The minimal β-casein promoter alone is not sufficient for CAT expression under any condition. However, BCE-1 confers responsiveness to both matrix (E) and prolactin (p) (courtesy of C. Schmidhauser).

levels. Thus, such tissue-specific culture systems will be useful in determining which of these responses are altered during mammary carcinogenesis.

B. Transformation of Mouse Mammary Epithelial Cells in Culture

The *ras* and *myc* oncogenes have been introduced into pre-senescent mouse mammary epithelial cells in monolayer culture. When these cells are re-introduced into cleared mammary fat pads they form dysplastic duct-like structures *in vivo*. Occasionally, carcinomas arise but they are found sporadically within the dysplastic tissue (Edwards et al., 1988; Strange et al., 1989; Bradbury et al., 1991). Thus, as was observed in transgenic mice harboring the same oncogenes, further changes are required for frank tumor formation (Leder et al., 1986; Sinn et al., 1987; Muller et al., 1988).

In primary culture, mouse mammary epithelial cells rapidly lose the ability to respond to the surrounding microenvironment in a complete and tissue-specific manner. This is especially apparent after passaging. As such, immortal cell lines that remain partially responsive to ECM or lactogenic hormones have been utilized in an effort to determine the effects of oncogene expression on differentiation. For example, expression of the *ras* oncogene in a cell line that forms tubular structures in collagen gels does not abrogate this morphogenetic response, although the cells do form tumors *in vivo* (Gunzburg et al., 1988). Also, introduction of the erbB-2 oncogene into a different line induces growth factor-independent biochemical differentiation in the presence of lactogenic hormones (Hynes et al., 1990; Taverna et al., 1991). However, examination of the full range of oncogene effects on mammary-specific differentiation will require the use of pre-senescent cell strains that remain fully responsive to microenvironmental signals over a number of passages in culture. This will allow the introduction of oncogenes under conditions that do not, in and of themselves, lead to the loss of differentiated function (i.e. transfection and selection). Recently, we have isolated such a cell strain that is designated CID-9 (Schmidhauser et al., 1990, 1992). Utilizing CID-9 cells the effect of oncogene expression on cell-ECM interactions is now being evaluated.

C. Normal and Malignant Human Mammary Epithelial Cells in Culture

While they are not as amenable to experimental manipulation as mouse cells, human mammary epithelial cells have also been explanted into culture in an effort to determine the phenotypic changes that emerge with carcinogenesis. In monolayer culture normal breast epithelial cells and carcinoma-derived cell lines are difficult to distinguish on a phenotypic level. For example, both cell types proliferate with similar doubling times. The only consistent changes that have been identified in tumor-derived cells include subtle alterations in intermediate filament expression, decreased expression of a novel calcium binding protein, and cellular immortalization (Guelstein et al., 1988; Stampfer and Bartley, 1988; Trask et al.,

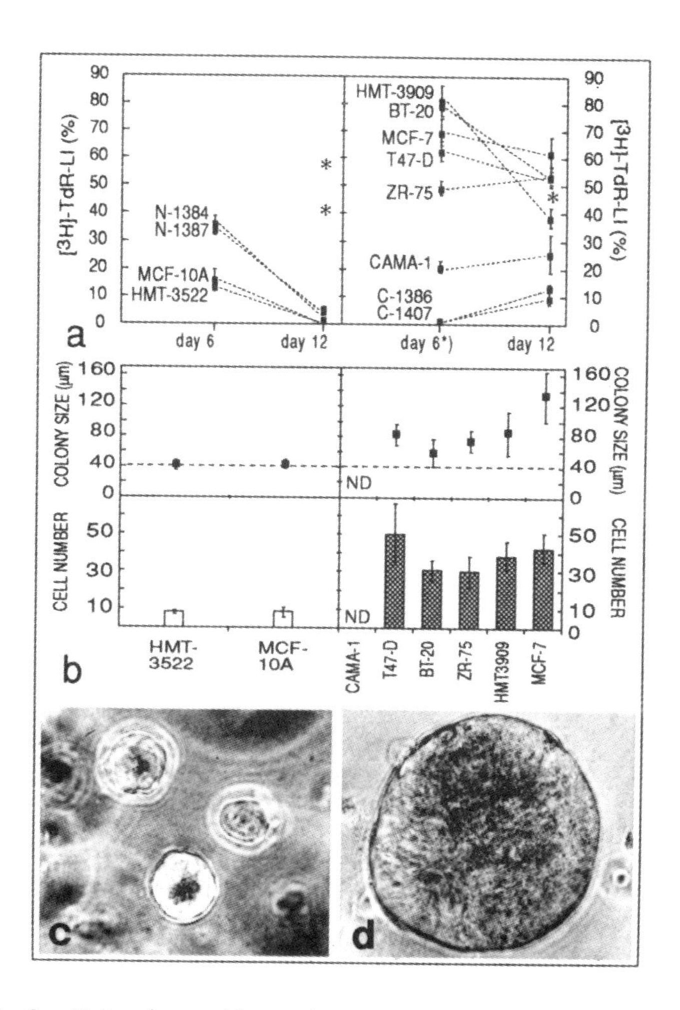

Figure 9. Sensitivity of normal human breast epithelial cells and carcinoma cells to growth regulation when they are cultured in a basement membrane-derived ECM. a) The tritiated thymidine labelling index is shown in normal cells (left panel; N-1384, N-1387-explanted cells, MCF-10A, HMT-3522- cell lines), and carcinoma cells (right panel; C-1386, C-1407- explanted cells; HMT-3909; BT-20, MCF-7, T47-D, ZR-75, CAMA-1-cell lines). Notice that between days 6 and 12 all normal cells cease dividing. In contrast, the labelling index among carcinoma cells is highly variable and they all continue to divide at day 12. b) Sphere size (μm and cell number diameters) is compared in 2 normal (left panel) and 5 carcinoma (right panel) cell lines. In all cases the carcinoma cells form larger spheres that are associated with their continued proliferation. c,d) Phase micrographs of the maximal sphere size of a normal cell line (c) and a carcinoma cell line (d) (from Petersen et al., 1992).

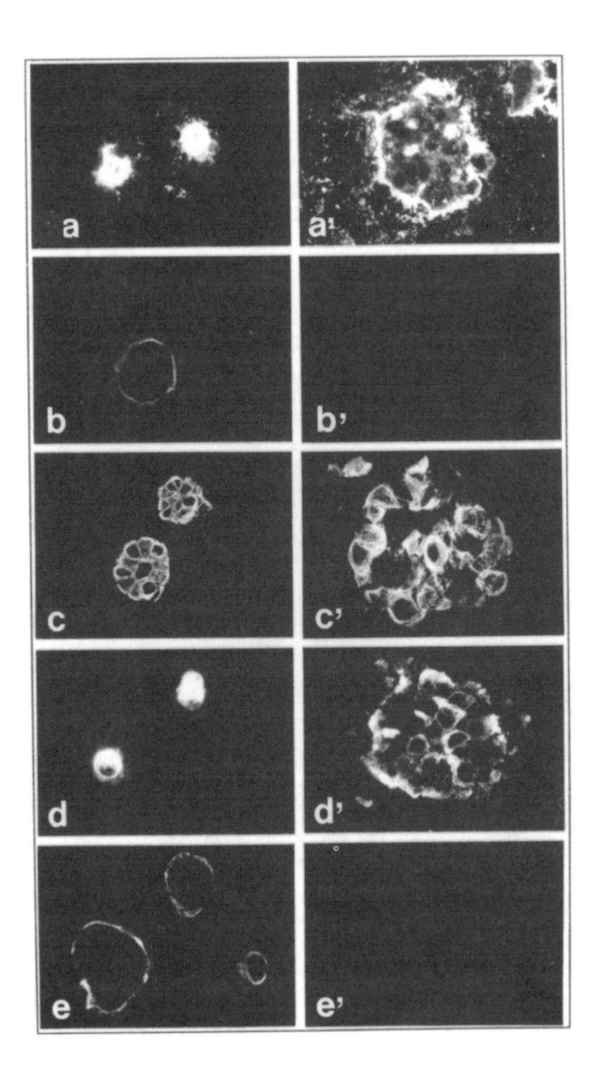

Figure 10. Immunofluorescence of frozen sections of spheres derived from normal human breast epithelial cells (left, a–e) and carcinoma cells (right a'–e') embedded in a mouse-derived basement membrane matrix. a,a') Sialomucin staining in primary cultures: Note the luminal (apical) accumulation in normal spheres and the abnormal basal accumulation in carcinoma colonies. b,b') Human collagen type IV in primary cultures: endogenous basement membrane deposition occurs in the normal spheres but not the carcinomatous colonies. c,c') Double labelling for keratin on the same sections in b,b': indicates the epithelial nature of the cells in both conditions and the basal location of the basement membrane deposition in the normal cells. d,d') Sialomucin in established lines: Similar accumulations as in the primary cultures. e,e') Human collagen type IV in established lines: Similar accumulations as in the primary cultures (from Petersen et al., 1992).

1990; Yaswen et al., 1990). When they are cultured inside extracellular matrices, luminal epithelial cells form 3 dimensional structures (Berdichevsky and Taylor-Papadimitriou, 1991; Petersen et al., 1992). In collagen gels the cells undergo a rapid morphogenesis to form ball like structures (Berdichevsky and Taylor-Papadimitriou, 1991). The latter process, which is $\alpha 2\beta 1$ integrin dependent (Berdichevsky et al., 1992), is often interrupted in cell lines derived from human breast tumor tissue (Shearer et al., 1992).

When they are embedded in a basement membrane matrix, human luminal epithelial cells form spheres that contain a lumen and the cells stop dividing when the spheres reach a size that is similar to acini *in vivo* (Petersen et al., 1992, and see Figure 9). The cells within the spheres polarize, producing mucins apically and depositing endogenous basement membrane components basally (Figure 10). This is not the case with carcinoma cell lines or with cells freshly explanted from carcinomatous tissue. These cells form large colonies within which cellular growth rates are poorly regulated (Figure 9). In addition, these colonies are structurally disorganized without a lumen, produce mucins in a non-polar fashion, and they do not deposit endogenous basement membrane components basally (Figure 10).

Thus, while it is difficult to identify differences between normal and tumor derived cells when they are growing freely in monolayer culture, fundamental, tissue-specific phenotypic differences become obvious when the cells are challenged by the growth-suppressing and differentiation-inducing potential of the basement membrane. This likely reflects an inappropriate responsiveness to tissue-specific regulatory signals that is a hallmark of the tumorigenic phenotype (Pierce and Speers, 1988; Parchment et al., 1990).

V. CONCLUSIONS

In the late stages of carcinogenesis, interactions between cells and the surrounding extracellular matrix are often altered. These changes are associated with a tumor cell's ability to dissociate from the primary mass, break through the basement membrane, invade the underlying stroma, and metastasize to distant sites (for a review see: Liotta et al., 1991). The interactions affected include adherence to, degradation of, and migration along, basement membranes and stromal matrices. Given the fact that similar alterations also occur during normal tissue development, albeit in a tightly regulated manner, it is not unreasonable to suggest that co-optation of cell-ECM interactions may occur throughout the carcinogenic process rather than being restricted only to the later metastatic stages. If such changes were to occur early in the disease process the resulting phenotypic effects would likely be tissue-specific in nature. Indeed, we have found that this is the case in the human mammary gland. In the presence of a basement membrane matrix, normal and carcinoma derived breast epithelial cells exhibit mammary-specific differences in cellular behavior that are not seen in monolayer culture. Presently, we are further defining these differences to use as markers of the pre-metastatic phenotype. In addition, a mouse cell strain that fully differentiates when cultured in custom-designed microenvironments is being subjected to carcinogenesis protocols in an effort to determine the mechanisms responsible for these altered cell-ECM interactions. By carrying out this dual experimental approach it is hoped that markers associated with, and mechanisms responsible for, the onset of the early stages of breast carcinogenesis will be elucidated. Such information will be indispensable to the further development of detailed prognostic indicators and tissue-specific treatment modalities that will help to improve the clinical outcome of this devastating disease.

ACKNOWLEDGMENTS

We thank our colleagues, Drs. Anthony Howlett, Christian Schmidhauser and Rabih Talhouk, for their helpful comments. These investigations were supported by the Health Effects Research Division, Office of Health and Environmental Research, US Departments of Energy under contract #DE-AC03-76SF00098, the Danish Cancer Society, and a gift for research from the Monsanto Corporation to MJB. CDR is a fellow of the National Cancer Institute of Canada.

REFERENCES

Auer, G., Fallenius, A., Erhardt, K., & Sundelin, B. (1984). Progression of mammary adenocarcinomas as reflected by nuclear DNA content. Cytometry 5, 420–425.

Auersperg, N. & Roskelley, C. (1991). Retroviral oncogenes: interrelationships between neoplastic transformation and cell differentiation CRC Crit. Rev. Oncogenesis 2, 125–160.

Barcellos-Hoff, M.H., Aggeler, J., Ram, T.G., & Bissell, M.J. (1989). Functional differentiation and alveolar morphogenesis of primary mammary cultures on reconstituted basement membrane. Development 105, 223–235.

Basset, P., Bellocq, J.P., Wolf, C., Stoll, I., Hutin, P., Limacher, J.M., Podhajcer, O.L., Chenard, M.P., Rio, M.C., & Chambon, P. (1990). A novel metalloproteinase gene specifically expressed in stromal cells of breast carcinomas. Nature 348, 699–704.

Berdichevsky, F. & Taylor-Papadimitriou, I. (1991). Morphological differentiation of hybrids of human mammary epithelial cell lines is dominant and correlates with the pattern of expression of intermediate filaments. Exp. Cell Res. 194, 267–274.

Berdichevsky, F., Gilbert, C., Shearer, M., & Taylor-Papadimitriou, I. (1992). Collagen induced rapid morphogenesis of human mammary epithelial cells: the role of the alpha 2 beta 1 integrin. J. Cell Sci. 102, 437–446.

Bissell, M.J., Hall, G.J., & Parry, G. (1982). How does the extracellular matrix direct gene expression? J. Theor. Biol. 99, 31–68.

Bissell, M.J., Lee, E.Y.H., Li, M.L., Chen, L.H., & Hall, H.G. (1985). Role of extracellular matrix and hormones in the modulation of tissue-specific functions in culture: mammary gland as a model for endocrine sensitive tissues. In: Benign Prostatic Hyperplasia Vol. 2. (Rogers, C.H., Coffey, D.C., Cunha, G., Grayack, J.T., Hinman, F., & Horton, R., eds.) NIH Pub '87-2881.

Bissell, M.J. & Hall, H.G. (1987). Form and function in the mammary gland: the role of extracellular matrix. In: The Mammary Gland. (Neville, M.C. & Daniel, C.W., eds.), pp. 97–146, Plenum, New York.

Blum, J.L., Zeigler, M.E., & Wicha, M.S. (1987). Regulation of rat mammary gene expression by extracellular matrix components. Exp. Cell Res. 173, 322–340.

Bradbury, J.M., Sykes, H., & Edwards, P.A. (1991). Induction of mouse mamary tumors in a transplantation system by the sequential introduction of the *myc* and *ras* oncogenes. Int. J. Cancer 48, 908–915.

Carter, C.L., Corle, D.K., Micozzi, M.S., Schatzkin, A., & Taylor, P.R. (1988). A prospective study of the development of breast cancer in 16,692 women with benign breast disease. J. Epidemiol. 128, 467–477.

Chen L-H. & Bissell, M.J. (1989). A novel regulatory mechanism for whey acidic protein gene expression. Cell Reg. 1, 45–54.

Chiquet-Ehrismann, R., Macki, E.J., Pearson, C.A., & Sakakura, T. (1986). Tenascin: an extracellular matrix protein involved in tissue interactions during fetal development and carcinogenesis. Cell 47, 131–139.

Daniel, C.W. & Silberstein, G.B. (1987). Postnatal development of the rodent mammary gland. In: The Mammary Gland: Development, Regulation and Function. (Neville, M.C. & Daniel, C.W., eds.), pp. 3–36, Plenum, New York.

Daniel, C.W., Silberstein, G.B., Van-Horn, K., Strickland, P., & Robinson, S. (1989). TGF-β_1 induced inhibition of mouse mammary ductal growth: development, specificity and characterization. Dev. Biol. 135, 20–30.

Danielson, K.G., Osborn, C.J., Durban, E.M., Butel, J.S., & Medina, D. (1984). Epithelial mouse mammary cell line exhibiting normal morphogenesis *in vivo* and functional differentiation *in vitro*. Proc. Natl. Acad. Sci. USA 81, 3756–3760.

Dupont, W.D. and Page, D.L. (1985). Risk factors for breast cancer in women with proliferative breast disease. N. Engl. J. Med. 312, 146–151.

Edwards, P.A.W., Ward, J.L. and Bradbury, J.M. (1988). Alterations of morphogenesis by the v-*myc* oncogene in transplants of mammary gland. Oncogene 2, 407–415.

Egan, R.L. (1982). Multicentric breast carcinomas: clinical-radiographic-pathologic whole organ studies and 10-year survival. Cancer 49, 1123–1130.

Emerman, J.T. & Pitelka, D.R. (1977). Maintenance and induction of morphological differentiation in dissociated mammary epithelium on floating collagen membranes. *In vitro* 13, 316–328.

Gordon, J.R. & Bernfield, M.R. (1980). The basal lamina of the postnatal mammary epithelium contains glycosaminoglycans in a precise ultrastructural organization. Dev. Biol. 74, 118–135.

Guelstein, V.I., Tchypysheva, T.A., Ermilova, V.D., Litvinova, L.V., Troyanovsky, S.M. & Bannikov, G.A. (1988). Monoclonal antibody mapping of keratins 8 and 17 and of vimentin in normal human mammary gland, benign tumors, dysplasias and breast cancer. Int. J. Canc. 42, 147–153.

Gullick, W.J. (1990). Growth factors and oncogenes in breast cancer. Prog. Growth Factor Res. 2, 1–13.

Gunzbug, W.H., Salmons, B., Schlaeffi, A., Moritz-Legrand, S., Jones, W., Sarkar, N.H., & Ullrich, R. (1988). Expression of the oncogenes mil and *ras* abolishes the *in vivo* differentiation of mammary epithelial cells. Carcinogenesis 9, 1849–1858.

Harris, A.L. & Nicholson, S. (1988). In: Breast Cancer: Cellular and Molecular Biology. (Lippman, M.E. & Dickson, R.B., eds.), pp. 93–118, Klewer Publishers, Boston.

Helin, H.J., Helle, M.J., Kallionjemi, O.P., & Isola, J.J. (1989). Immunohistochemical determination of estrogen and progesterone receptors in human breast carcinoma. Correlation with histopathology and DNA flow cytometry. Cancer 63, 1761–1767.

Henderson, R.E., Ross, R.K., & Pike, M.C. (1991). Toward the primary prevention of cancer. Science 254, 1131–1138.

Hilkens, J., Buijs, F., & Ligtenberg, M. (1989). Complexity of MAM-6, an epithelial sialomucin associated with mammary carcinomas. Canc. Res. 49, 786–793.

Horan-Hand, P., Thor, A., & Wunderlich, D. (1984). Monoclonal antibodies of pre-defined specificity detect activated ras gene expression in human mammary and colon carcinomas. Proc. Natl. Acad. Sci. USA 81, 5227–5231.

Howlett, A.R. & Bissell, M.J. (1990). Regulation of mammary epithelial cell function: a role for stromal and basement membrane matrices. Protoplasma 159, 85–95.

Howlett, A.R. & Bissell, M.J. (1992). The influence of tissue microenvironment (stroma and extracellular matrix) on the development and function of mammary epithelium. Epith. Cell Biol. 2, 79–89.

Hurley, W.L. (1989). Mammary gland function during involution. J. Dairy Sci. 72, 1637–1646.

Hynes, N.E., Taverna, D., Harwerth, I.M., Ciardiello, F., Solomon, D.S., Yamamoto, T., & Groner, B. (1990). Epidermal growth factor receptor, but not c-erbB-2, activation prevents lacogenic hormone induction of the β-casein gene in mouse mammary epithelial cells. Mol. Cell. Biol. 10, 4027–4034.

Hynes, R.O. (1992). Integrins: versatility, modulation and signaling in cell adhesion. Cell 69, 11–25.

Kimata, K., Sakakura, T., Inaquma, K., Kato, M., & Nishizuka, Y. (1985). Participation of two different mesenchymes in the developing mouse mammary gland: synthesis of basement membrane components by fat pad precursor cells. J. Embryol. Exp. Morphol. 89, 243–257.

Knight, C.H. & Peaker, M. (1982). Development of the mammary gland. J. Reprod. Fert. 65, 521–536.

Kozbor, D. & Croce, C.M. (1984). Amplification of the c-*myc* oncogene in one of five human breast carcinoma cell lines. Canc. Res. 44, 438–441.

Kratochwil, K. (1969). Organ specificity in mesenchymal induction demonstrated in the embryonic development of the mammary gland of the mouse. Dev. Biol. 20, 46–71.

Leder, A., Pattengale, P.K., Kuo, A., Stewart, T.A., & Leder, P. (1986). Consequences of widespread deregulation of the c-*myc* gene in transgenic mice: multiple neoplasms and normal development. Cell 45, 485–495.

Lee, E.Y.H., Lee, W.H., Kaetzel, C.S., Parry, G., & Bissell, M.J. (1985). Interaction of mouse mammary epithelial cells with collagen substrata: regulation of casein gene expression and secretion. Proc. Natl. Acad. Sci. USA 82, 1419–1423.

Lefebvre, O., Wolf, C., Limarhcer, J-M., Hutin, P., Wendling, C., LeMeur, M., Basset, P., & Rio, M.-C. (1993). The breast cancer-associated stromelysin 3 gene is expressed during mouse mammary gland apoptosis. J. Cell Biol. 119, 997–1002.

Li, M.L., Aggeler, J., Farson, D.A., Hatier, C., Hassell, J., & Bissell, M.J. (1987). Influence of a reconstituted basement membrane and its components on casein gene expression and secretion in mouse mammary epithelial cells. Proc. Natl. Acad. Sci. USA 84, 136–140.

Liotta, L.A., Steeg, P.S., & Stettler-Stevenson, W.G. (1991). Cancer metastasis and angiogenesis: an imbalance of positive and negative regulation. Cell 64, 327–336.

Lippman, M.E. & Dickson, R.B. (1990). Growth control of normal and malignant breast epithelium. Prog. Clinical Biol. Res. 354A, 147–178.

Mackay, J., Thompson, A.M., Coles, C., & Steel, C.M. (1990). Molecular lesions in breast cancer. Int. J. Canc. 5(S), 47–50.

Martinez-Hernandez, A., Fink, L.M., & Pierce, G.B. (1976). Removal of basement membrane in the involuting breast. Lab. Invest. 34, 455–462.

Matsui, Y., Haller, S.A., Holt, J.T., Hogan, B.L., & Coffey, R.J. (1990). Development of mammary hyperplasia and neoplasia in MMTV-TGF alpha transgenic mice. Cell 61, 1147–1155.

Missero, C., Flivaroff, E., & Dotto, G.P. (1991a). Induction of TGF-β_1 resistance by the E1A oncogene requires binding to a specific set of cellular proteins. Proc. Natl. Acad. Sci. USA 88, 3489–3493.

Missero, C., Ramon y Cajal, S., & Dotto, G.P. (1991b). Escape from transforming growth factor beta control and oncogene cooperation in skin tumor development. Proc. Natl. Acad. Sci. USA 88, 9613–9617.

Muller, W.J., Sinn, E., Pattengale, P.K., Wallace, R., & Leder, P. (1988). Single step induction of mammary adenocarcinoma in transgenic mice bearing the activated c-neu oncogene. Cell 54, 105–115.

Neville, M.C. & Daniel, C.W. (eds.) (1987). The Mammary Gland Development, Regulation and Function. Plenum Press, New York.

Ochieng, J., Basolo, F., Albini, A., Melciori, A., Watanabe, H., Elliot, J., Ray, A., Parodi, S., & Russo, J. (1991). Increased invasive, chemotactic and locomotive abilities of c-Haras-transformed human breast epithelial cells. Inv. Metas. 11, 38–47.

Parchment, R.E., Gramzinski, R.A., & Pierce, G.B. (1990). Embryonic mechanisms for abrogating the malignancy of cancer cells. Prog. Clin. Biol Res 354A, 331–344.

Parry, G., Lee, E.Y.H., Farson, D.A., Kovall, N., & Bissell, M.J. (1985). Collagenous substrata regulate the nature and distribution of glycosaminoglycans produced by differential cultures of mouse mammary epithelial cells. Exp. Cell Res. 156, 487–499.

Peeters, P.H., Verbeek, A.L., Zielhuis, G.A., Vooijs, G.P., Hendricks, J.H., & Mravunac, M. (1990). Breast cancer screening in women over age 50. Acta Radiol. 31, 225–231.

Peterson, O.W., Ronnov-Jessen, L., Howlett, A.R., & Bissell, M.J. (1992). Interaction with basement membrane serves to rapidly distinguish growth and differentiation pattern of normal and malignant human breast epithelial cells. Proc. Natl. Acad. Sci. USA 89, 9064–9068.

Pierce, G.B. & Speers, W.C. (1988). Tumors as caricatures of the process of tissue renewal: prospects for therapy by directing differentiation. Canc. Res. 48, 1996–2004.

Pitelka, D.R. & Hamamoto, S.T. (1977). Form and function in mammary epithelium: the interpretation of ultrastructure. J. Dairy Sci. 60, 643–654.

Ponten, J., Holmberg, L., Trichopoulos, D., Kalliomeni, O.P., Kuale, G., Wallgren, A., & Taylor-Papadimitriou, J. (1990). Biology and natural history of breast cancer. Int. J. Canc. 5(S), 5–21.

Robinson, S.D., Silberstein, G.B., Roberts, A.B., Flanders, D.C., & Daniel, C.W. (1991). Regulated expression and growth inhibitory effects of TGF-β_1 isoforms in mouse mammary gland development. Development 113, 867–878.

Rochlitz, C.G., Scott, G.K., & Dodson, J.M. (1989). Incidence of ras oncogene activity. Mutations associated with primary and metastatic human breast cancer. Canc. Res. 49, 357–360.

Runnebaum, I.B., Nagarajan, M., Bowman, M., Soto, D., & Sukumar, S. (1991). Mutations in p53 as potential markers for human breast cancer. Proc. Natl. Acad. Sci. USA 88, 10657–10661.

Sakakura, T., Nishizuka, T., & Dawe, C.J. (1976). Mesenchyme dependent morphogenesis and epithelium specific cytodifferentiation in mouse mammary gland. Science 194, 1439–1441.

Sakakura, T., Sakagami, Y., & Nishizuka, Y. (1982). Dual origin of mesenchymal tissue participation in mouse mammary gland embryogenesis. Dev. Biol. 91, 202–207.

Sakakura, T., Ishihara, A., & Yatini, R. (1991). Tenascin in mammary gland development: from embryogenesis to carcinogenesis. In: Regulatory Mechanisms in Breast Cancer. (Lippman, M.E. & Dickson, R.E., eds.), pp. 383–400, Kluer, Boston.

Sarnelli, R. & Squartini, F. (1991). Fibrocystic condition and at risk lesions in asymptomatic breasts: a morphologic study of post-menopausal women. Clin. Exp. Obs. Gyn. 18, 271–279.

Schmidhauser, C., Bissell, M.J., Meyers, C.A., & Casperson, G.F. (1990). Extracellular matrix and hormones transcriptionally regulate bovine β-casein 5 sequences in stably transfected mouse mammary cells. Proc. Natl. Acad. Sci. USA 87, 9118–9122.

Schmidhauser, C., Casperson, G.F., Myers, C.A., Bolten, S., & Bissell, M.J. (1992). A novel transcriptional enhancer is involved in the prolactin and ECM-dependent regulation of β-casein gene expression. Mol. Biol. Cell 3, 699–709.

Shearer, M., Bartkova, J., Bartek, J., Berdichevsky, F., Barnes, D., Millis, R., & Taylor-Papadimitriou, I. (1992). Studies of clonal cell lines developed from primary breast cancers indicate that the ability to undergo morphogenesis *in vitro* is lost early in malignancy. Int. J. Canc. 51, 602–612.

Sieweke, M.H., Thompson, N.L., Sporn, M.B., Bissell, M.J. (1990). Mediation of wound-related rous sarcoma virus tumorigenesis by TGF-β. Science 248, 1656–1660.

Sieweke, M.H., Thompson, N.L., Sporn, M.B., Bissell, M.J. (1990). Mediation of wound-related rous sarcoma virus tumorigenesis by TGF-β. Science 248, 1656–1660.

Silberstein, G.B. & Daniel, C.W. (1982). Glycosaminoglycans in the basal lamina and extracellular matrix of the developing mouse mammary duct. Dev. Biol. 90, 215–222.

Silberstein, G.B., Strickland, P., Coleman, S., & Daniel, C.W. (1990). Epithelium dependent extracellular matrix synthesis in TGF-β1 inhibited mouse mammary gland. J. Cell Biol. 110, 2209–2219.

Sinn, E., Muller, W., Pattengale, P., Tepler, I., Wallace, R., & Leder, P. (1987). Co-expression of MMTV/v-Haras and MMTV/*c-myc* genes in transgenic mice; synergistic action of oncogenes *in vivo*. Cell 49, 465–475.

Slamon, D.J., Clark, C.M., Wong, S.G., Levin, W.J., Ullrich, W.J., & McGuire, W.L. (1987). Human breast cancer: correlation of relapse and survival with amplification of the HER-2 neu oncogene. Science 235, 177–182.

Stampfer, M.R. & Bartley, J.C. (1988). Human mammary epithelial cells in culture: differentiation and transformation. Canc. Treat. Res. 40, 1–24.

Stoker, A.W., Streuli, C.H., Martins-Green, M., & Bissell, M.J. (1990). Designer microenvironments for the analysis of cell and tissue function. Curr. Opin. Cell Biol. 2, 864–874.

Strange, R., Aguila-Cordova, E., Young, L.J.T., Billy, H.T., Dandekar, S., & Cardiff, R.D. (1989). Harvey-ras mediated neoplastic development in the mouse mammary gland. Oncogene 4, 309–315.

Streuli, C.H. & Bissell, M.J. (1990). Expression of extracellular matrix components is regulated by substratum. J. Cell Biol. 110, 1405–1415.

Streuli, C.H. & Bissell, M.J. (1991). Mammary epithelial cells, extracellular matrix and gene expression. In: Regulatory mechanisms in breast cancer. (Lippman, M.E. & Dickson, R., eds.), pp. 365–381, Klewer, Boston.

Streuli, C.H., Bailey, N., & Bissell, M.J. (1991). Control of mammary epithelial differentiation; basement membrane induces tissue-specific gene expression in the absence of cell-cell interaction and morphological polarity. J. Cell Biol. 115, 1383–1395.

Streuli, C.H., Schmidhauser, C., Kobrin, M., Bissell, M.J., & Derynk, R. (1993). Extracellular matrix regulates expression of the TGF-β_1 gene. J. Cell Biol. 120, 253–260.

Talhouk, R.S., Chin, J.R., Unemori, E.N., Werb, Z., & Bissell, M.J. (1991). Proteinases of the mammary gland: developmental regulation *in vivo* and vectorial secretion in culture. Development 112, 439–449.

Talhouk, R., Bissell, M.J., & Werb, Z. (1992). Co-ordinated expression of extracellular matrix-degrading proteinases and their inhibitors regulates mammary epithelial function during involution. J. Cell Biol. 118, 1271–1282.

Taverna, D., Groner, B., & Hynes, N.E. (1991). Epidermal growth factor receptor, platelet derived growth factor receptor, and c-erbB-2 receptor activation all promote cell growth but have distinctive effects upon mouse mammary epithelial cell differentiation. Cell Growth and Diff. 2, 145–154.

Taylor-Papadimitriou, J., & Cane, E.B. (1987). Keratin expression in the mammary gland. In: The Mammary Gland (Neville, M.C. & Daniel, C., eds.), pp. 181–215. Plenum, New York.

Thompson, A., Steel, C.M., Chetty, U., Hawkins, R.A., Miller, W.R., Carter, D.C., Forrest, A.P.M., & Evans, H.J. (1990). p53 gene RNA expression and chromosome 17p allele loss in breast cancer. Brit. J. Canc. 62, 78–84.

Thompson, E.W., Reich, R., Martin, G.R., & Albini, A. (1988). Factors regulating basement membrane invasion by tumor cells. Canc. Treat. Res. 40, 239–249.

Thompson, E.W., Lippman, M.E., & Dickson, R.B. (1991). Regulation of basement membrane invasiveness in human breast cancer model systems. Mol. Cell. Endocrinol. 82, C203–C208.

Trask, D.K., Band, V., Zazchowski, D.A., Yaswen, P., Suh, T., & Sager, R. (1990). Keratins as markers that distinguish normal and tumor derived mammary epithelial cells. Proc. Natl. Acad. Sci. USA 87, 2319–2323.

Varley, J.M., Swallow, J.E., Bramman, W.J., & Whittacker, J.L. (1987). Alterations to either c-erbB-2 (neu) or *c-myc* protooncogenes in breast carcinomas correlate with short term prognosis. Oncogene 1, 423–430.

von Rosen, A., Rutqvist, L.E., Cartensen, J., Fallenius, A., Skoog, L., & Auer, G. (1989). Prognostic value of nuclear DNA content in breast cancer in relation to tumor size, nodal status and estrogen receptor content. Breast Canc. Res. Treat. 13, 23–32.

Warburton, M.J., Mitchell, D., Ormerod, E.J., & Rudland, P. (1982). Distribution of myoepithelial cells and basement membrane proteins in the resting, pregnant, lactating and involuting rat mammary gland. J. Histochem. Cytochem. 30, 667–676.

Watt, F.M. (1991). Cell culture models of differentiation. FASEB J. 5, 287–294.

Wellings, S.R., Jensen, H.M., & Marcum, R.G. (1975). An atlas of subgross pathology of the human breast with special reference to possible pre-cancerous lesions. J. Nat'l. Canc. Inst. 55, 231–239.

Wicha, M.S., Liotta, L.A., Vonderhaar, B.K., & Kidwell, W.R. (1980). Effects of inhibition of basement membrane collagen deposition on rat mammary gland development. Dev. Biol. 80, 253–266.

Willett, W. (1989). The search for the cause of breast and colon cancer. Nature 338, 389–394.

Wolf, C., Rouyer, N., Lutz, Y., Adida, C., Loriot, M., Bellocq, J-M., Chambon, P., & Basset, P. (1993). Stromelysin 3 belongs to a subgroup of proteinases expressed in breast carcinoma fibroblastic cells and possibly implicated in tumor progression. Proc. Natl. Acad. Sci. USA 90, 1843–1847.

Yaswen, P., Smoll, A., Peehl, D.M., Trask, D.K., Sager, R., & Stampfer, M.R. (1990). Down-regulation of a calmodulin-related gene during transformation of human mammary epithelial cells. Proc. Natl. Acad. Sci. USA 87, 7360–7364.

EPITHELIAL-STROMAL CELL INTERACTIONS AND BREAST CANCER

Sandra Z. Haslam, Laura J. Counterman, and Katherine A. Nummy

I. INTRODUCTION

Epithelial-stromal cell interactions are known to be required during embryonic development for normal organogenesis and the interactions between epithelial and stromal cells are believed to be reciprocal. Although many organs are functional

Advances in Molecular and Cell Biology
Volume 7, pages 115–130

ISBN: 1-55938-624-X

before birth, others undergo major morphogenetic changes and cyclic expression of functional differentiation postnatally. The mammary gland falls into this latter category. True morphogenetic changes happen cyclically during estrous or menstrual cycles, pregnancy, lactation and involution and the cyclic development and function of the mammary gland is known to be dependent upon systemic hormones (Nandi, 1958; Topper and Freeman, 1980). However, increasing evidence indicates that local epithelial-stromal cell interactions also play an important role (Cunha, 1985; Haslam, 1990; 1991; 1991a). The specific mechanisms underlying epithelial-stromal cell interactions have not been well delineated. Possible mechanisms include the physical and chemical composition of the basement membrane (Sakakura et al., 1982; Streuli et al., 1991), and/or paracrine mechanisms via soluble factors produced locally by various cell types in the gland (Haslam, 1986; Daniel and Silberstein, 1987). In this context, the abnormal cell proliferation and loss of hormonal regulation that is evident in mammary cancer is likely to be the result of a change or loss of normal regulatory interactions between mammary stromal and epithelial cells. This review will examine the roles and underlying mechanisms of epithelial-stromal cell interactions in normal mammary gland growth and development and review the properties of epithelial-stromal cell interactions in mammary cancer. Most studies have been carried out in experimental model systems using rodent tissues and provide the major source of information for this review. Although less information is available about the human breast, relevant studies of normal breast development and breast cancer will also be reviewed.

II. EPITHELIAL-STROMAL CELL INTERACTIONS IN RODENT NORMAL MAMMARY GLAND

A. Embryonic Development

By day 14 of embryonic development, two different mesenchymal components in addition to the epithelial rudiment are present in the developing mouse mammary gland. One is the fibroblast mesenchyme which directly surrounds the epithelial rudiment, and the other is an underlying adipose mesenchyme. These two mesenchymes give rise to the fibrous and adipose stromal components of the gland, respectively (Sakakura et al., 1982). Investigation of the roles of these two mesenchymes has revealed that both mesenchymal tissue types are involved in directing early gland development. The fibroblasts are involved in determining the sexual phenotype of the gland and are responsible for androgen induced regression in the male mammary gland (Durnberger et al., 1978; Heuberger et al., 1982). They may also mediate responsiveness to estrogen in the embryonic female gland. Steroid autoradiographic analysis has demonstrated that estrogen receptors are present only in the fibroblast mesenchyme before birth (Narbaitz et al., 1980).

Interestingly, injections of estrogen before birth result in epithelial malformations indicating that the fibroblast mesenchyme is capable of an estrogenic response that affects the epithelium (Raynaud and Raynaud, 1954). The fibroblasts also cause adipocyte responsiveness and the adipocytes subsequently determine the characteristic mammary pattern of ductal morphogenesis (Sakakura, 1983; Sakakura et al. 1982).

Sakakura and her colleagues have analyzed the extracellular matrix molecules produced by the different mesenchymes. The fibroblast mesenchyme produces fibronectin and tenascin (Inaguma et al., 1988), whereas the adipocyte mesenchyme produces laminin and heparan sulfate proteoglycan (Kimata et al., 1985). Sakakura has proposed that the molecular composition and structural features of the basement membrane provide the specific conditions necessary for mammary epithelium to undergo normal morphogenesis. Tenascin and fibronectin, which are derived from fibroblasts, may influence cell adhesiveness and proliferation. During embryonic development tenascin predominates and is known to reduce epithelial cell adhesiveness (Chiquet-Ehrsmann et al., 1989; Spring et al., 1989). Furthermore, the detection of epidermal growth factor-like (EGF) sequences in tenascin suggest it may also function as an autocrine and/or paracrine growth factor (Pearson et al., 1988). Both of these functions of tenascin may promote cell proliferation. Fibronectin on the other hand, predominates in the postnatal mammary gland and promotes cell adhesiveness and thus may act to stabilize the formed gland (Thiery et al., 1989). The presence of laminin and heparan sulfate proteoglycan in the basement membrane derived from the adipocytes may be responsible for determining the branching pattern of the mammary gland (Kimata et al., 1985).

In experimental studies embryonic mammary mesenchymes can contribute to mammary tumorigenesis. If fibroblast mesenchyme alone is combined with adult epithelial tissue, abnormal ductal hyperplasia results (Sakakura et al., 1982). When transplanted into virus-positive C3H mice, the hyperplasias were 12 times more likely to form tumors than normal tissue transplants (Sakakura et al., 1981). Conversely, complete mammary mesenchyme induced a known mammary tumor cell line to differentiate to a less malignant phenotype when cultured transfilter, *in vitro* (DeCosse et al., 1973).

B. Postnatal Development

The postnatal mammary stroma retains inductive activity for embryonic epithelium and postnatal epithelium can respond to mesenchymal inductive influences (Sakakura et al., 1979). These observations indicate that epithelial-stromal cell interactions continue to play an important role in postnatal mammary gland development. In the postnatal period prior to sexual maturity and pregnancy, ductal elongation and the acquisition of hormonal responsiveness are the major developmental processes (Haslam, 1988, 1989).

Ductal Elongation

Postnatal ductal elongation occurs at puberty and is the result of proliferative activity in the mammary stroma and in unique epithelial structures called end buds (Daniel and Silberstein, 1987). As the end buds proliferate and penetrate the mammary stroma they are separated from the adipocytes only by the basement membrane. As the subtending duct is formed it becomes invested by a thin sheath of fibrous connective tissue; this surrounding stroma is rich in fibroblasts (Silberstein and Daniel, 1982; Williams and Daniel, 1983). Transplantation studies have shown that an adipose stromal matrix is an absolute requirement for ductal development. Subcutaneous transplantation or transplantation to any other stromal matrix will not support normal mammary morphogenesis (Hoshino, 1978). If epithelium is transplanted into an artificial collagenous matrix within mammary adipose stroma, the epithelium will only form normal end buds and ducts when it comes in contact with the adipose tissue (Daniel et al., 1984). These observations suggest that the adipose and fibroblast stromas may have separate roles in ductal elongation. Adipose stroma promotes ductal elongation whereas the fibroblast stroma results in duct stabilization. Roles for both basement membrane composition and soluble factors are indicated in these processes. As described above, the major components associated with the basement membrane in the region of the adipose stroma are laminin, whereas fibronectin, collagen I and chondroitin sulfate are associated with fibroblast stroma and duct stabilization (Silberstein and Daniel, 1982). Thus changes in basement membrane composition and interstitial matrix may be controlled locally by stromal cells and are associated with either growth or cytodifferentiation and tissue stabilization.

Recently, it has been reported that soluble factors, epidermal growth factor (EGF), transforming growth factor α (TGF-α) and β (TGF-β) also play a role in ductal elongation (Silbertein et al., 1990; Robinson et al., 1991; Snedeker et al., 1991). EGF and TGF-α are present in the end bud epithelium and it is believed that estrogens promote ductal growth indirectly by stimulating the production of these soluble factors in epithelial cells. Ductal growth is also accompanied by DNA synthesis in unidentified stromal cells close to the growing end buds (Berger and Daniel, 1983). This proliferative activity of the stroma does not occur in the absence of epithelial proliferation. Thus soluble factors produced in the epithelium may also influence stromal proliferation indicating reciprocal interactions between epithelial and stromal cells. TGF-β appears to have an inhibitory effect on ductal elongation (Daniel et al., 1989; Silberstein et al., 1990; Robinson et al., 1991). It suppresses ductal epithelial DNA synthesis and the formation of new end buds when administered locally. TGF-β acts to increase collagen I synthesis and chondroitin sulfate synthesis in the stroma adjacent to end buds. Thus, TGF-β increases the synthesis of extracellular matrix components that are associated with tissue stabilization rather than growth. However, this occurs only if the TGF-β is applied close to the epithelium suggesting that the epithelium plays a role in directing the stromal response to TGF-β.

Acquisition of Hormonal Responsiveness

In Vivo Studies. As end buds reach the limits of the mammary stromal fat pad, they regress and ductal elongation ceases. The tip of the duct becomes ensheathed totally by fibroblast derived stroma. Further, mammary proliferation does not occur until pregnancy, at which time elevated levels of estrogen and progesterone induce ductal sidebranching and lobuloalveolar development (Topper and Freeman, 1980). Prior to the completion of ductal elongation, the mammary gland is refractory to the ability of estrogen to induce epithelial progesterone receptors and the ability of progesterone to stimulate epithelial proliferation and sidebranching (Haslam, 1988, 1989). Analysis of estrogen receptor distribution has shown that the receptors are equally distributed between the epithelial and stromal cells (Haslam and Shyamala, 1981). By contrast, only 20% of progesterone receptors are localized in stromal cells and as stated above epithelial progesterone receptors are acquired only as ductal elongation ceases.

Recently, we have investigated the role of epithelial-stromal cell interactions and the acquisition of responsiveness to estrogen and progesterone both *in vivo* and *in vitro*. We have found that if immature, non-responsive epithelium is surgically recombined with mature stroma *in vivo*, the epithelium prematurely acquires responsiveness to both estrogen and progesterone. If mature epithelium is recombined with immature stroma, responsiveness is maintained. We have also established that the premature acquisition of hormonal responsiveness is not due to a change in the overall systemic milieu after transplantation, but rather due to the local stromal environment of the adult mammary gland (Haslam and Counterman, 1991a). The nature of the mediator in stroma responsible for inducing this responsiveness is not presently known. However, recent experiments have demonstrated both that similar premature responsiveness to estrogen and progesterone could be achieved by the administration of estrogen *in vivo* or estrogen plus EGF *in vitro* in whole organ culture (Haslam, unpublished observations). This effect of EGF appears to be dissociated from its growth promoting effects, since growth was not stimulated in these experiments. Interestingly, it has also been shown that EGF has growth promoting effects only in ovariectomized mice (Coleman et al., 1988), and not in ovary intact animals (Coleman et al., 1990). In the latter case, locally administered EGF acts as a growth inhibitor. Thus in the absence of a growth promoting effect, EGF may have a differentiative function that is mediated via mammary stroma.

As reproductive function ceases in the female with the onset of menopause, the mammary gland regresses and exists as an attenuated ductal tree. Most breast cancers develop in postmenopausal women; however, the role of menopause per se in breast cancer etiology is not understood. Using the mouse mammary gland as a model system, we have investigated the status of mammary gland hormonal responsiveness during menopause. Sexually mature mice that had experienced long-term ovariectomy (5 weeks), regardless of age, exhibited a 2-fold greater epithelial cell proliferative response to estrogen; no increased response to progesterone was detected. Stromal cells exhibited a 2-fold greater proliferative response to both estrogen and progesterone. Estrogen receptor concentration was not altered by longterm ovariectomy. However, immunohistochemical analysis of the distribution of the growth inhibitor, TGF-β1 showed that after longterm ovariectomy, estrogen treatment caused a significant reduction in intracellular TGF-β in stromal cells; TGF-β levels in epithelial cells were also reduced, but to a lesser degree. (Haslam, unpublished observations). These results suggest that as a result of longterm ovariectomy and possibly menopause, the mammary gland can exhibit a heightened sensitivity to mitogenic factors such as estrogen. One possible explanation for this phenomenon could be the reduced levels of mammary growth inhibitors produced by mammary stromal cells.

In Vitro Studies. The role of epithelial-stromal cell interactions in relation to hormonally regulated growth and function has also been addressed using various cell culture systems. Epithelial cell interactions with two cell types, stromal fibroblasts and adipocytes, have been analyzed in *in vitro* models.

Using a monolayer culture system, it has been reported that mammary epithelial cells stop growing when confronted with mammary fibroblasts. However, this growth inhibition can be overcome by the addition of estrogen. Contact between the stromal cells and the epithelium was required for this response to estrogen (McGrath, 1983).

We have investigated this observation further in an attempt to get at the underlying mechanism. Specifically, we have analyzed proliferation and the progesterone receptor induction response to E *in vitro* (Haslam and Levely, 1985; Haslam, 1986). When mixed cultures containing both mammary epithelial cells and fibroblasts were treated with estrogen, there was an increase in both epithelial cell and fibroblast proliferation and epithelial progesterone receptor content. In contrast, when mammary epithelial cells or fibroblasts were grown separately neither response to estrogen was observed. Fibroblasts might influence epithelial cell behavior by modifying the substratum, analogous to the extracellular matrix and basement membrane *in vivo* or by producing soluble factors, or both. The effect on progesterone receptor inducibility appeared to be mediated predominantly through a substratum effect since coating culture dishes with collagen I could replace the fibroblast effect. In contrast, the estrogen-induced proliferative effect required the presence of live fibroblasts, in close proximity to the epithelial cells and was suggestive of soluble factor effect. Interestingly, the morphology of both epithelial cells and fibroblasts were altered when co-cultured and epithelial cells promoted E-induced fibroblast proliferation in support of the concept that there is a bidirectional interaction between the two cell populations (Haslam, 1986).

In other studies (Enami et al., 1983), conditioned media from mouse mammary fibroblasts also stimulated DNA synthesis in normal and neoplastic mouse mammary epithelial cells in culture. The growth promoting activity in conditioned medium was described as a growth factor distinct from EGF. Similar results have also been obtained using stromal cells from rat mammary glands (Rudland et al., 1979).

Interactions between mammary epithelium and adipocytes in cell culture have also been investigated. Adipocytes have been shown to stimulate DNA synthesis of mammary epithelial cells (Levine and Stockdale, 1984). The adipocyte factor is present in both conditioned medium and in substratum attached material derived from adipocytes. Adipocytes have also been reported to promote mammary epithelial cell lactational differentiation. Ultrastructural studies showed that the formation of a basement membrane was crucial for this response and cell-cell interaction was believed to be required (Levine and Stockdale, 1985; Wiens et al., 1987). However, other studies (Li et al., 1988) have shown that reconstituted basement membrane alone is sufficient for the expression of lactational function in culture. Other studies have also shown that linoleate metabolites enhance the *in vitro* proliferative and lactational responses of mouse mammary epithelial cells (Bandyopadhyay et al., 1987; Levay-Young et al., 1987). Thus, a metabolic role and the production of soluble metabolites by adipose stroma may also mediate adipocyte effects *in vitro*.

III. EPITHELIAL-STROMAL CELL INTERACTIONS IN NORMAL HUMAN BREAST

Appropriate epithelial-stromal cell interactions are likely to be required in order to express normal function in human breast. When enzymatically dispersed normal breast tissue was injected in epithelium-free mammary stroma of nude mice and appropriate growth promoting mammotropic hormones were provided, the tissue was accepted and maintained but expansive proliferation was not obtained (Sheffield and Welsch, 1988). One explanation offered for the lack of growth was an inappropriate stromal environment for human tissue. In this context, Horgan et al., (1987) have shown that human fibroblasts increase the tumor take and growth rate of MCF-7 mammary cancer cells transplanted to nude mice over tumor cells alone. Fibroblasts from benign or malignant breast tissue, or skin were equally effective in stimulating growth, whereas substitution of other cell types was not effective. Killed fibroblasts were able to increase tumor take, but not tumor growth. The authors speculate that killed fibroblasts may be effective through a substratum effect comparable to that observed in cell culture. In contrast, live fibroblasts were required for the proliferative effect and indicate that the secretion of soluble factors may be required for cell growth.

The cyclic processes of growth and differentiation of human mammary gland are similarly regulated by hormones. Morphologic changes in the breast stroma have been identified during the menstrual cycle (Vogel et al., 1981). The principal stromal change was premenstrual edema and loosening of the interlobular connective tissue. An increase in the acid mucopolysaccharide content of the intralobular stroma may be responsible for the alteration in the connective tissue structure (Ozello and Spier, 1958).

IV. EPITHELIAL-STROMAL INTERACTIONS AND MAMMARY CANCER

It is well known that the stroma of the human breast undergoes morphologic and chemical changes under pathologic conditions (Tremblay, 1979; Tamimi and Ahmed, 1986, 1987). Desmoplasia is the common host response to epithelial tumor invasion and is the result of fibroblast proliferation and excessive collagen production (Sappino et al., 1988). In the desmoplastic reaction proliferating fibroblasts express the features of smooth muscle. This is a characteristic of breast cancer and not found in normal breasts. The factor(s) responsible for smooth muscle differentiation of mammary stromal cells is not presently known. However, this process demonstrates an altered interaction between mammary epithelium and stroma.

In animal experimental model systems, aberrations in stromal morphology have also been observed. The distribution of basement membrane and connective tissue proteins in 7,12-dimethylbenzanthracene (DMBA)-induced rat mammary tumors have been examined (Omerod et al., 1985). Laminin is restricted to a thin basement membrane surrounding each duct in the normal gland and collagen I fibers are distributed throughout the stroma. In DMBA-induced tumors, each cluster of tumor cells is surrounded by interstitial collagen and an abnormally thickened basement membrane. As described earlier, abnormalities in the basement membrane and extracellular matrix may alter attachment characteristics of epithelial cells, partially freeing them from some normal restrictions of growth. In another approach, cloned mammary adenocarcinoma cells were inoculated into epithelium-free mouse mammary stroma. The host stromal response was characterized by an intense fibroblast reaction resulting in the production of large masses of collagen. The tumor cells also lacked a basement membrane again pointing to the importance of the basement membrane in regulation of epithelial-stromal cell interactions (Russo et al., 1976).

Composition of the basement membrane and extracellular matrix of normal and cancerous human breast have been recently investigated (Koukoulis et al., 1991). In the fetal, normal adult, gestational and senescent breast, ducts and acini are surrounded by a delicate layer of laminin. This pattern is retained in fibrocystic disease and benign neoplasms. However, in intraductal and invasive carcinomas, laminin was decreased or absent. The distribution of fibroblast-derived tenascin and fibronectin was also analyzed. In the adult breast, fibronectin was present in the interlobular stroma but reduced or absent in intralobular and immediate periductal matrix. However, in fibrocystic disease and carcinomas, fibronectin was associated with the epithelial structures. Tenascin in the normal breast is confined to the intralobular stroma and to a thin rim in the periductual regions. In contrast, in mammary carcinomas much greater concentrations of tenascin were observed in the matrix, preferentially localized around neoplastic aggregates (Howeedy et al., 1990). The altered pattern of tenascin deposition is considered a fibroblast response to the tumor cells. Furthermore, since tenascin appears to reduce epithelial cell adhesiveness, it may play an important role in tumor invasiveness and metastasis.

Mammary stromal cell influences on tumor cell immunogenicity and metastatic potential have also been demonstrated using a spontaneously arising mouse mammary tumor SP1 (Elliott et al., 1988). Transplants of this tumor into epithelium-free mammary stroma of syngeneic mice resulted in increased metastatic potential, whereas transplantation to the subcutis did not have this effect. Furthermore, increased expression of class I major histocompatibility antigens was observed. These observations suggest that mammary stroma provides an environment that either selects distinct tumor subpopulations, or induces a phenotypic change leading to tumor progression and the generation of metastatic subpopulations. This phenomenon has been further investigated for tissue specificity and it has been demonstrated that SP1 tumor cells grow equally well in mammary, mesenteric and ovarian fat pads as compared with the subcutis (Elliott et al., 1992). Furthermore, estrogen, progesterone and EGF receptor levels are increased in the presence of adipose tissue, and the growth of the tumor and the promotion of metastasis are estrogen dependent. Analysis of the role of growth factors and extracellular matrix components in proliferation and cell adhesion has also been carried out *in vitro* with SP1 and a stable highly metastatic variant, SPI-3M (Elliott et al., 1992a). The results suggest that proliferative response of the two cell lines to the growth factors (platelet derived growth factor BB and basic fibroblast growth factor) is dependent on the specific composition of the extracellular matrix and imply that modification of the extracellular matrix and/or surface integrin receptors may regulate responsiveness to these growth factors in this tumor model.

Recently, it has been reported that stromal fibroblasts immediately surrounding the invasive, but not *in situ* component of human breast cancer, express the stromelysin-3 gene (Basset et al., 1990). This gene encodes a new member of the family of metalloproteinases, enzymes which degrade extracellular matrix. This gene is expressed in all invasive breast cancers analyzed so far, suggesting that expression may be one of the events required for tumor metastasis and progression of breast cancer.

It has also been reported that the extracellular matrix can act as a depot for growth factors and enzymes and that the specific composition of the matrix may enhance their stability and determine their bioavailability. For example, basic fibroblast growth factors, which are important for endothelial proliferation and angiogenesis are stabilized by heparan sulfate in the extracellular matrix. Release of cellular heparanase enzymatically activates basic fibroblast growth factors. Thus, this growth factor may be involved in both tumor cell invasion and angiogenesis which often coincide with the transition from *in situ* to invasive carcinoma. Whether similar events of local growth factor concentration and stabilization by extracellular matrix play a role in normal tissue proliferation remains to be determined (Vlodavsky et al., 1990).

Familial clustering of breast cancer indicates that there is a genetic component predisposing individuals to breast cancer (Lynch et al., 1984). In general it is the epithelial cells which are designated as cancerous. However, several reports have indicated that fibroblasts derived from breast cancer may also be altered and exhibit characteristics associated with a transformed and/or fetal phenotype (Azzarone et al., 1984; Durning et al., 1984; Schor et al., 1986). Skin fibroblasts from breast cancer patients are abnormal with regard to three parameters: anchorage dependence, colony formation and increased culture density. This is in contrast to fibroblasts obtained from normal donors (Durning et al., 1984; Azzarone and Macieira-Coelho, 1987).

Skin fibroblasts from breast cancer patients also exhibit increased migratory behavior, similar to fetal fibroblasts in cell migration assays (Durning et al., 1984). More than 90% of patients with familial breast cancer and 50% with sporadic disease had fibroblasts which display fetal behavior. Furthermore, these characteristics of skin fibroblasts may precede the detection of breast cancer (Schor et al., 1986). Skin fibroblasts with fetal characteristics are present in 67% of clinically unaffected first-degree relatives of patients with hereditary breast cancer (Haggie et al., 1987). It has been proposed that retention of fetal characteristics increases the susceptibility to tumor development. Recently, it has been shown that increased fibroblast migration is due to the autocrine action of a factor which increases hyaluronic acid biosynthesis (Schor et al., 1988). These results further support the concept that aberrations in stromal function and hence abnormal epithelial-stromal cell interactions can contribute significantly to the neoplastic process.

In vitro studies have been undertaken to shed light on the mechanisms underlying abnormal epithelial-stromal cell interactions. There is evidence that the growth of human breast cancer cells is under the autocrine influence of growth factors produced by tumor cells (Dickson and Lippman, 1988). It is highly probable that paracrine influences from surrounding stromal tissue affect the growth of the epithelial cells. Adams et al. (1988, 1988a) have reported that conditioned medium from normal fibroblasts inhibited the growth of MCF-7 cells. In contrast, conditioned medium from benign and malignant-derived fibroblasts significantly enhanced the growth of the same tumor cells. One potential paracrine mechanism involves the production of interleukin-6 by breast fibroblasts, which in turn stimulates 17β-estradiol oxidoreductase activity in the tumor cells (Adams et al., 1991). This enzyme increases the availability of 17β-estradiol and can enhance the estrogen-dependent growth of mammary cancers. Recently it has also been shown that human mammary fibroblasts can secrete TGF-α and bioactive fibroblast growth factors in cell culture. Furthermore, conditioned media from fibroblast cultures were able to stimulate anchorage-independent growth of human mammary epithelial cell lines immortalized with and overexpressing *c-myc* and SV40T (Valverius et al., 1990). These results suggest that factors derived from adjacent mammary stroma might influence in a paracrine manner the phenotypic characteristics of a population of human mammary epithelial cells toward transformation.

Another paracrine mechanism, believed to be operative in normal mammary gland development and may be altered in breast cancer is the expression of TGF-β and its ability to inhibit epithelial growth. TGF-β is secreted by fibroblasts *in vivo* (Robinson et al., 1991) and *in vitro* (Colletta et al., 1990) and may be negatively regulated by estrogen (Knabbe et al., 1987). Recently it has been reported that the anti-estrogen tamoxifen significantly increases the synthesis of biologically active TGF-β by human fibroblasts *in vitro* (Colletta et al., 1990). The observation that pharmacological agents acting on mammary stromal cells to produce a growth inhibitor strongly supports a new approach to the therapy and prevention of breast cancer that focuses on the stroma.

The role of the extracellular matrix has also been investigated *in vitro* (Koa et al., 1984). When newborn skin fibroblasts are cultured on matrix derived from human mammary carcinoma cell line ZR75-1, an increased growth rate and cell density was observed. Also increased were collagen and elastin synthesis. Furthermore, direct contact between the fibroblasts and the tumor matrix were required for the mitogenic response. These effects were not observed when the fibroblasts were grown on their own preformed matrix or on matrices of other cell types, providing further evidence that cancerous epithelial cells produced extracellular matrix factor(s) which direct stromal cell behavior.

V. SUMMARY AND CONCLUSIONS

Epithelial-mesenchymal interactions are critically required for normal mammary gland morphogenesis during embryonic development. In the postnatal normal mammary gland, cyclic patterns of proliferation, morphogenesis, differentiation and regression occur cyclically and are hormonally regulated. Growing evidence supports the concept that epithelial-stromal cell interactions are required for postnatal growth and development. Most importantly there is good evidence to indicate that aberrations in epithelial stromal interactions occur during the course of the neoplastic process and involve the same mechanisms and interactive phenomena as described for normal cells. It is also noteworthy that one possible component of genetic predisposition to breast cancer in humans may be due to a stromal cell defect.

Two major mechanisms have been proposed by which stromal cells can influence epithelial cell behavior: (1) by modifying the molecular and/or physical composition of the extracellular matrix, and (2) by the production of soluble growth factor-like molecules. Furthermore, the interactions are bidirectional such that epithelial cells are clearly capable of influencing stromal cell behavior. The recent finding that the specific composition of the extracellular matrix can influence the stability and local concentration of growth factors and/or growth inhibitors provides a conceptual framework for understanding the mechanisms by which epithelial cells and stromal cells can reciprocally influence growth behavior. Epithelial and stromal cells each contribute different specific components to the extracellular matrix and probably different growth stimulators and inhibitors. Under the influence of external regulators such as systemic hormones, alterations in the composition of the extracellular matrix, the production of growth factors, and activating enzymes by specific cell types would be determined. The resulting net bioavailability of the various factors would determine the relative proliferative activity in the various cell types of the gland. Continuing progress in the identification of the precise mechanisms underlying epithelial-stromal cell interactions in the normal mammary gland should increase our understanding of the defects that occur in neoplasia. The identification of factors produced by mammary stromal cells which are growth stimulatory or growth inhibitory for mammary epithelial cells strongly supports an approach to the prevention and therapy of breast cancer that also focuses on the stroma.

REFERENCES

Adams, E.F., Newton, C.J., Tait, G.H., Braunsberg, H., Reed, M. J., & James, V. H. (1988). Paracrine influence of human breast stromal fibroblasts on breast epithelial cells: secretion of a polypeptide which stimulates reductive 17 beta-oestradiol dehydrogenase activity. Int. J. Canc. 42 (1), 119–122.

Adams, E.F., Newton, C.J., Braunsberg, H., Shaikh, N., Ghilchik, M., & James, V.H. (1988). Effects of human breast fibroblasts on growth and 17 beta-estradiol dehydrogenase activity of MCF-7 cells in culture. Breast Canc. Res. Treat. 11 (2), 165–172.

Adams, E.F., Rafferty, B., & White, M.C. (1991). Interleukin 6 is secreted by breast fibroblasts and stimulates 17 β-oestradiol oxidoreductase activity of MCF-7 cells: possible paracrine regulation of breast 17 β-oestradiol levels. Int. J. Canc. 49, 118–121.

Azzarone, B., Mareel, M., Billard, C., Scemama, P., Chaponnier, C., & Macieira-Coelho, A. (1984). Abnormal properties of skin fibroblasts from patients with breast cancer. Int. J. Canc. 33 (6), 759–764.

Azzarone, B. & Macieira-Coelho, A. (1987). Further characterization of the defects of skin fibroblasts from cancer patients. J. Cell Sci. 87 (Pt. 1), 155–162.

Bandyopadhyay, G.K., Imagawa, W., Wallace, D., & Nandi, S. (1987). Linoleate metabolites enhance the *in vitro* proliferative response of mouse mammary epithelial cells to epidermal growth factor. J. Biol. Chem. 262, 2750–2756.

Basset, P., Bellocq, J.P., Wolf, C., Stoll, I., Limacher, J.M., Podhajcer, O.L., Chenard, M.P., Rio, M.C., & Chambon, P. (1990). A novel metalloproteinase gene specifically expressed in stromal cells of breast carcinomas. Nature 348, 699–704.

Berger, J.J. & Daniel, C.W. (1983). Stromal DNA synthesis is stimulated by young, but not serially aged mouse mammary epithelium. Mech. Aging Dev. 23, 259–264.

Colletta, A.A., Wakefield, L.M., Howell, F.V., van Roozendaal, K.E.P., Danielpour, D., Ebbs, S.R., Sporn, M.B., & Baum, M. (1990). Anti-oestrogens induce the secretion of active transforming growth factor beta from human fetal fibroblasts. Br. J. Canc. 62, 405–409.

Coleman, S., Silberstein, G.B., & Daniel, C.W. (1988). Ductal morphogenesis in the mouse mammary gland: evidence supporting a role for epidermal growth factor. Dev. Biol. 127, 304–315.

Coleman, S. & Daniel, C.W. (1990). Inhibition of mouse mammary ductal morphogenesis and down regulation of the EGF receptor by epidermal growth factor. Dev. Biol. 137, 425–433.

Cunha, G.R., Bigsby, R.M., Cooke, P.S., & Sugimura, Y. (1985). Stromal-epithelial interactions in adult organs. Cell Diff. 17, 137–148.

Daniel, C.W., Berger, J.J., Strickland, P., & Garcia, R. (1984). Similar growth patterns of mouse mammary epithelium cultured in collagen matrix *in vivo* and *in vitro*. Dev. Biol. 104, 57–64.

Daniel, C.W. & Silberstein, G.B. (1987). In: The Mammary Gland: Development Regulation and Function (Neville, M., & Daniel, C.W., eds.), pp. 3–36, Plenum Press, New York.

Daniel, C.W., Silberstein, G.B., Van Horn, K., Strickland, P., & Robinson, S. (1989). TGF-β-1 induced inhibition of mouse mammary ductal growth: development, specificity and characterization. Dev. Biol. 135, 20–30.

DeCosse, J.J., Grossens, C.L., Kuzman, J.F., & Unsworth, B.R. (1973). Breast cancer: induction of differentiation of embryonic tissue. Science 181, 1057–1058.

Dickson, R.B. & Lippman, M.E. (1988). In: Breast Cancer: Cellular and Molecular Biology (Lippman, M.E., & Dickson, R., eds.), pp. 119–165, Kluwer Academic Publishing, Massachusetts.

Durnberger, H., Heuberger, B., Schwartz, P., Wasner, G., & Kratochwil, K. (1978). Mesenchyme mediated effect of testosterone on embryonic mammary epithelium. Canc. Res. 38, 4066–4070.

Durning, P., Schor, S.L., & Sellwood, R.A. (1984). Fibroblasts from patients with breast cancer show abnormal migratory behavior *in vitro*. Lancet. 2 (8408), 890–892.

Elloitt, B.E., Maxwell, M.A., Wei, W.A., & Miller, F.R. (1988). Expression of epithelial-like markers and class I major histocompatibility antigens by a murine carcinoma growing in the mammary gland and in metastases: orthotopic site effects. Canc. Res. 24, 7237–7245.

Elliott, B.E., Tam, S.-P., Dexter, D., & Chen, Z.O. (1992). Capacity of adipose tissue to promote growth and metastasis of a murine mammary carcinoma: effect of estrogen and progesterone. Int. J. Canc. 51, 416–424.

Elliott, B.E., Ostman, A., Westermark, B., & Rubin, K. (1992a). Modulations of growth factor responsiveness of murine mammary carcinoma cells by cell matrix interactions: correlation of cell proliferation and spreading. J. Cell Physiol. 152, 292–301.

Enami, J., Enami, S., & Koga, M. (1983). Growth of normal and neoplastic mouse mammary epithelial cells in primary culture: stimulation by conditioned medium from mouse mammary fibroblasts. Gann 74, 845–853.

Haggie, J.A., Sellwood, R.A., Howell, A., Birch, J.M., & Schor, S. L. (1987). Fibroblasts from relatives of patients with hereditary breast cancer show fetal-behavior *in vitro*. Lancet. 1 (8545), 1455–1457.

Haslam, S.Z., & Shyamala, G. (1981). Relative distribution of estrogen receptors and progesterone receptors among epithelial, adipose and connective tissues of normal mammary gland. Endocrinology 108, 825–830.

Haslam, S.Z. & Levely, M.L. (1985). Estrogen responsiveness of normal mouse mammary cells in primary cell culture: association of mammary fibroblasts with estrogenic regulation of progesterone receptors. Endocrinology 116 (5), 1835–1844.

Haslam, S.Z. (1986). Mammary fibroblast influence on normal mouse mammary epithelial cell responses to estrogen *in vitro*. Canc. Res. 46, 310–316.

Haslam, S.Z. (1988). Acquisition of estrogen-dependent progesterone receptors by normal mouse mammary gland. J. Ster. Biochem. 31, 9–13.

Haslam, S.Z. (1989). The ontogeny of mouse mammary gland responsiveness to ovarian steroid hormones. Endocrinology 125, 2766–2772.

Haslam, S.Z. (1990). In: Breast Cancer: Cellular and Molecular Biology (Lippman M. E. & Dickson, R., eds.), pp. 401–420, Kluwer Pub. Co.

Haslam, S.Z. (1991). In: Therapeutic Implications of the Molecular Biology of Breast Cancer (Mihich, E., ed.), John Libbey & Co., New York.

Haslam, S.Z. & Counterman, L. (1991). Mammary stroma modulates hormonal responsiveness of mammary epithelium *in vivo* in the mouse. Endocrinology 129, 2017–2023.

Heuberger, G., Fitzka, I., Wasner, G., & Kratochwil, K. (1982). Induction of androgen receptor formation by epithelial-mesenchyme interaction in embryonic mouse mammary gland. Proc. Natl. Acad. Sci. 27, 2957–2961.

Horgan, K., Jones, D.L., & Mansel, R.E. (1987). Mitogenicity of human fibroblasts *in vivo* for human breast cancer cells. Br. J. Surg. 74 (3), 227–229.

Hoshino, K. (1978). In: Physiology of Mammary Glands (Yokoyama, A., Mizuno, H., & Nagasawa, H., eds.), pp. 163–228, University Park Press, Maryland.

Howeedy, A.A., Virtanen, I., & Laitnen, L. (1990). Differential distribution of tenascin in the normal, hyperplastic and neoplastic breast. Lab. Invest. 63, 708–806.

Inaguma, Y., Kusakabe, M., Mackie, E.J., Pearson, C.A., Chiquet-Ehrismann, R., & Sakakura, T. (1988). Epithelial induction of stromal tenacin in the mouse mammary gland from embryogenesis to carcinogenesis. Dev. Biol. 128, 245–255.

Kao, R.T., Hall, J., Engel, L., & Stern, R. (1984). The matrix of human breast tumor cells is mitogenic for fibroblasts. Am. J. Pathol. 115 (1), 109–116.

Kimata, K., Sakakura, T., Inaguma, Y., Kato, M., & Nishizuka, Y. (1985). Participation of two different mesechymes in the developing mouse mammary gland: synthesis of basement membrane components by fat pad precursor cells. J. Embryol. Exp. Morphol. 89, 243–257.

Knabbe, C., Lippman, M., Wakefield, L.M., Derynk, R., & Dickson, R. B. (1987). Evidence that transforming growth factor beta is a hormonally regulated negative growth factor in human breast cancer cells. Cell 48, 417–422.

Koukoulis, G.K., Gould, V.E., Bhattacharyya, A., Gould, J.E., Howeedy, A.A., & Virtanen, I. (1991). Tenascin in normal, reactive, hyperplastic and neoplastic tissues: biologic and pathologic implications. Persp. Path. 22, 636–643.

Levay-Young, B.K., Bandyopadhyay, G.K., & Nandi, S. (1987). Linoleic acid but not cortisol, stimulates accumulation of casein by mouse mammary epithelial cells in serum-free collagen gel culture. Proc. Natl. Acad. Sci. 84, 8448–8452.

Levine, J.F. & Stockdale, F.E. (1984). 3T3-L1 adipocytes promote the growth of mammary epithelium. Exp. Cell Res. 151 (1), 112–122.

Levine, J.F. & Stockdale, F.E. (1985). Cell-cell interactions promote mammary epithelial cell differentiation. J. Cell Biol. 100 (5), 1415–1422.

Li, M.L., Aggeler, J., Farson, D.A., Hattier, C., Hassell, J., & Bissell, M.J. (1988). Influence of a reconstituted basement membrane and its components on casein gene expression and secretion in mouse mammary epithelial cells. Proc. Natl. Acad. Sci. USA 84, 136–140.

Lynch, H.T., Albano, W.A., & Heieck, A. (1984). Genetics, biomarkers and control of breast cancer: a review. Canc. Genet. Cytogenet. 13, 43–92.

McGrath, C.M. (1983). Augmentation of the response of normal mammary epithelial cells to estradiol by mammary stroma. Canc. Res. 43 (3), 1355–1360.

Nandi, S. (1958). Endocrine control of mammary gland development and function in the C3H HeCrgl mouse. J. Natl. Canc. Inst. 21, 1039–1063.

Narbaitz, K., Stumpf, W.E., & Sar, M. (1980). Estrogen receptors in mammary gland primordia of fetal mouse. Anat. Embryol. 158, 161–166.

Ormerod, E.J., Warburton, M.J., Gusterson, B., Hughes, C.M., & Rudland, P.S. (1985). Abnormal deposition of basement membrane and connective tissue components in dimethylbenzanthracene-induced rat mammary tumors: an immunocytochemical and ultrastructural study. Histochem. J. 17 (10), 1155–1166.

Ozello, L. & Spier, F.D. (1958). The mucopolysaccharides in the normal and diseased breast. Am. J. Pathol. 34, 993–1099.

Pearson, C.A., Pearson, D., Shibakara, S., Hofsteenge, J., & Chiquet-Ehrismann, R. (1988). Tenascin: cDNA cloning and induction by TGF-α EMBO J. 10, 2977–2981.

Raynaud, A. & Raynaud, J. (1954). Les diverses malformations mammaires produites chez les faetus de souris par l'action des hormones sexuelles. C. R. Soc. Biol. Paris 148, 963–968.

Robinson, S.D., Silberstein, G.B., Roberts, A.B., Flanders, K.C., & Daniel, C.W. (1991). Regulated expression and growth inhibitory effects of transforming growth factor-β isoforms in mouse mammary gland development. Development 113, 867–872.

Rudland, P.S., Bennett, D.C., & Warberton, M.J. (1979). Hormonal control of growth and differentiation of cultured rat mammary gland epithelial cells. Cold Springs Harbor Conf. Cell Prolif. 6, 677–699.

Russo, J., McGrath, C., & Russo, I.H. (1976). An experimental animal model for the study of human scirrhous carcinoma. J. Natl. Canc. Inst. 57 (6), 1253–1259.

Sakakura, T., Sakagumi, Y., & Nishizuka, Y. (1979). Persistence of responsiveness of adult mouse mammary gland to induction by embryonic mesenchyme. Dev. Biol. 72, 201–210.

Sakakura, T., Sakagumi, Y., & Nishizuka, Y. (1981). Accelerated mammary cancer development by fetal salivary mesenchyme isografted to adult mouse mammary epithelium. J. Natl. Canc. Inst. 66, 953–959.

Sakakura, T., Sakagumi, Y., & Nishizuka, Y. (1982). Dual origin of mesenchymal tissue participating in mouse mammary gland morphogenesis. Dev. Biol. 91, 202–207.

Sakakura, T. (1983). In: Understanding Breast Cancer (Rich, M.A., Hager, J.C., & Furmanski, P., eds.), pp. 261–284, Marcel Decker, New York.

Sappino, A.P., Skalli, O., Jackson, B., Schurch, W., & Gabbiani, G. (1988). Smooth-muscle differentiation in stromal cells of malignant and non-malignant breast tissues. Int. J. Canc. 41 (5), 707–712.

Schor, S.L., Haggie, J.A., Durning, P., Howell, A., Smith, L., Sellwood, R.A., & Crowther, D. (1986). Occurrence of a fetal fibroblast phenotype in familial breast cancer. Int. J. Cancer 37 (6), 831–836.

Schor, S.L., Schor, A.M., Grey, A.M., & Rushton, G. (1988). Fetal and cancer patient fibroblasts produce and autocrine migration stimulating factor not made by normal adult cells. J. Cell Sci. 90, 391–399.

Sheffield, L.G. & Welsch, C.W. (1988). Transplantation of human breast epithelia to mammary-gland-free fat-pads of athymic nude mice: influence of mammotrophic hormones on growth of breast epithelia. Int. J. Canc. 41 (5), 713–719.

Silberstein, G.B. & Daniel, C.W. (1982). Glycosaminoglycans in the basal lamina and extracellular matrix of the developing mouse mammary duct. Dev. Biol. 90, 215–222.

Silberstein, G.B., Strickland, P., Coleman, S., & Daniel, C.W. (1990). Epithelium dependent extracellular matrix synthesis in transforming growth factor-β-1 growth inhibited mouse mammary gland. J. Cell Biol. 110, 2209–2219.

Snedeker, S.M., Brown, C.F., & DiAugustine, R.P. (1991). Expression and functional properties of transforming growth factor α and epidermal growth factor during mouse mammary gland ductal morphogenesis. Proc. Natl. Acad. Sci. 88, 276–280.

Spring, J., Beck, K., & Chiquet-Ehrismann, R. (1989). Two contrary functions of tenascin: Dissection of the active sites by recombinant tenascin fragments. Cell 59, 325–334.

Streuli, C.H. & Bissell, M.J. (1991). In: Regulatory Mechanisms in Breast Cancer (Lippman, M.E. & Dickson, R., eds.), pp. 365–381, Kluwer Academic Publishers.

Tamimi, S.O. & Ahmed, A. (1986). Stromal changes in early invasive and non-invasive breast carcinoma: an ultrastructural study. J. Pathol. 150 (1), 43–49.

Tamimi, S.O. & Ahmed, A. (1987). Stromal changes in invasive breast carcinoma: an ultrastructural study. J. Pathol. 153 (2), 163–170.

Thiery, J.P., Duband, J-L., Dufour, S., Savagner, P., & Imhof, B.A. (1989). In: Fibronectins (Mosher, D.F. ed.), pp. 181–212, Academic Press, Florida.

Tremblay, G. (1979). Stromal aspects of breast carcinoma. Exp. Mol. Path. 31, 248–260.

Valverius, E.M., Ciardiello, F., Heldin, N-E., Blondel, B., Merlo, G., Smith, G., Stampfer, M., Lippman, M.E., Dickson, R.B., & Salomon, D.S. (1990). Stromal influences on transformation of human mammary epithelial cells overexpressing c-myc and SV40T. J. Cell. Physiol. 145, 207–216.

Vlodavsky, I., Korner, G., Ishai-Michaeli, R., Bashkin, P., Bar-Shavit, R., & Fuks, Z. (1990). Extracellular matrix-resident growth factors and enzymes: possible involvement in tumor metastasis and angiogenesis. Canc. & Metast. Rev. 9, 203–226.

Vogel, P.M., Georgiade, N.G., Fetter, B.F., Vogel, F.S., & McCarty, K.S. Jr. (1981). The correlation of histologic changes in the human breast with the menstrual cycle. Am. J. Pathol. 104 (1), 23–34.

Wiens, D., Park, C.S., & Stockdale, F.E. (1987). Milk protein expression and ductal morphogenesis in the mammary gland *in vitro*: hormone-dependent and -independent phases of adipocyte-mammary epithelial cell interaction. Dev. Biol. 120 (1), 245–258.

Williams, J.M. & Daniel, C.W. (1983). Mammary ductal elongation: Differentiation of myoepithelium and basal lamina during branching morphogenesis. Dev. Biol. 97, 274–290.

THE TISSUE MATRIX AND THE REGULATION OF GENE EXPRESSION IN CANCER CELLS

Kenneth J. Pienta, Brian C. Murphy,
Robert H. Getzenberg, and Donald S. Coffey

I. INTRODUCTION

Cells contain extensive and elaborate 3-dimensional interactive skeletal networks that form integral structural components of the plasma membrane, the cytoplasm, and the nucleus; the entire system has been termed the "tissue matrix" (Isaacs et al., 1981; Bissell et al., 1982; Fey et al., 1984; Pienta et al., 1991). The tissue matrix system forms a structural and functional bridge from the cell periphery to the DNA

Advances in Molecular and Cell Biology
Volume 7, pages 131–156

ISBN: 1-55938-624-X

and may be defined as the structural subcomponents of the cell that interact to organize as well as coordinate cell functions and transport, and consists of the linking of the nuclear matrix to the cytoskeleton and the extracellular matrix (see Figure 1) (Pienta et al., 1991).

The nuclear matrix is a framework scaffolding forming the superstructure of the nucleus and consists of peripheral lamins and pore complexes, an internal ribonucleic protein network, and nucleoli (Berezney and Coffey, 1974; Nelson et al., 1986). The nuclear matrix has been defined as the dynamic structural subcomponent of the nucleus that directs the 3-dimensional organization of DNA into loop domains and provides sites for the specific control of nucleic acid intranuclear and particulate transport (Getzenberg et al., 1990, 1991). The nuclear matrix is cell type

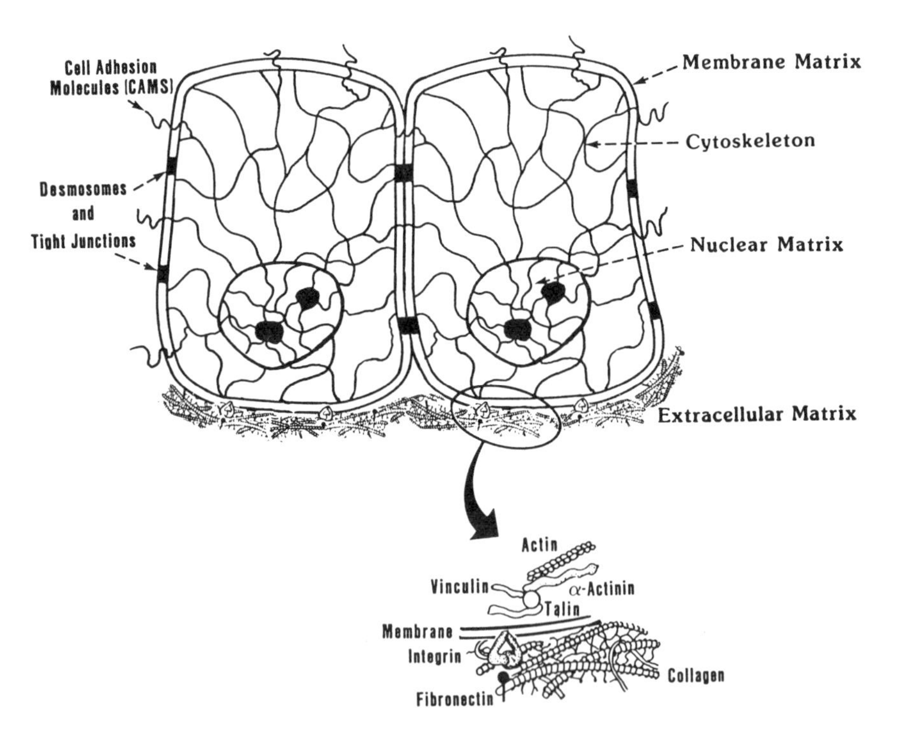

Figure 1. The tissue matrix system. This is a simplified schematic of the tissue matrix system. The tissue matrix system consists of the major insoluble networks of the cell which remain after extraction of the lipids and soluble proteins. These interconnecting networks connect the cell periphery to the DNA and are believed to play major roles in cellular communication and gene expression.

and tissue specific and has been demonstrated to be altered in cancer cells (Fey and Penman, 1988; Getzenberg et al., 1990, 1991). Several nuclear oncogenes have been shown to interact with the nuclear matrix (Klempnauer, 1988; Pienta et al., 1989).

The structural linkage of the nucleus to the cell periphery is thought to be mediated through cytoskeletal components. The cytoskeleton is composed of a structural interacting network of actin microfilaments, intermediate filaments (IF), and microtubules, and a vast array of associated proteins that modulate their polymerization and form (Hall, 1984; Pollard and Cooper, 1986; Joly et al., 1989; Cleveland, 1990; Nagle, 1990). In addition to their structural role as skeletal components, the cytoskeletal networks appear to interact with cell-signaling pathways(McNiven and Porter, 1986; Okabe and Hirokawa, 1989). This interaction with the cell-signaling pathways poises the cytoskeleton to both affect and be affected by signal transduction oncogenes.

The extracellular matrix which a cell touches has been shown to play a major role in determining how a cell functions (Gospodarowitz et al., 1978; Bissell et al., 1981; Fujita et al., 1986). Several studies have demonstrated that the extracellular matrix can alter cell morphology and affect DNA synthesis and gene expression

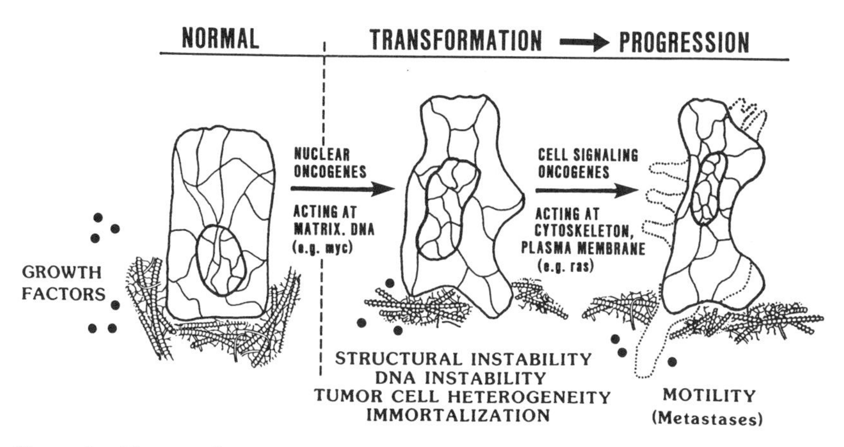

Figure 2. The transformation process viewed from a cell structure standpoint. The transformation from a normal cell to a malignant one appears to involve multiple steps, usually considered in terms of "initiation" and "progression." This process may also be viewed at the mechanism or site of action of the different oncogenes involved in tumor progression. One class of oncogenes, those acting on the nucleus (many of which interact with the nuclear matrix) e.g., myc, alter the structural stability of cells. This may alter gene expression and lead to immortalization and concomitant DNA and structural instability. A second class of oncogenes, those acting at the cell periphery of the cell, e.g., ras, induce alterations in cell signaling, extracellular matrix morphology, and induce motility into the cell. These changes also alter the ability of the cell to respond to growth factors and may impart the ability to metastasize.

(Hay, 1977; Ben-Ze'ev, 1980; Wittelsberger et al., 1981; Ingber, 1990; Pienta et al., 1991). The extracellular matrix is a complex three-dimensional array composed of proteins which include laminin, collagen IV, and entactin, as well as proteoglycans and sequestered growth factors (Prysor-Jones et al., 1985; Fujita et al., 1986; Hirano et al., 1987; Yamada, 1989). Modulation in virtually all of these elements have been found to occur in transformed cells (Pienta et al., 1989).

Cell structure and function are intimately related. Cell transformation is a multistep process involving genomic changes that can involve oncogenes acting at different sites within the cell. It appears that nuclear oncogenes such as *myc* may interact with the nuclear matrix and contribute to processes such as DNA instability and cell immortalization (Eisenman et al., 1985). A second class of oncogenes, those, such as *ras*, that act in the cytoplasm and cell periphery, alter cell signaling processes and motility and may impart the ability to metastasize and/or further alter DNA organization via structural connections to the nuclear matrix (Bar-Sagi and Feramicso, 1986; Pienta and Coffey, 1991). For example, Partin and colleagues, transfected the *v-Ha-ras* oncogene into prostate cancer cells and demonstrated that increases in cell ruffling and cell motility correlated with metastatic ability (Partin et al., 1988). Figure 2 is our schematic representation of the transformation process as viewed from a cell structure standpoint.

II. THE NUCLEAR MATRIX

The nuclear matrix has been demonstrated to play an important role in many functions of DNA, including DNA replication and the regulation of gene expression (Berezney and Coffey, 1977; Dijkwel et al., 1979; Pardoll et al., 1980; Georgiev, 1981; Fisher et al., 1982; Capco et al., 1984; Halligan et al., 1984; Bekers et al., 1985; Tubo et al., 1985) (Table 1). The nuclear matrix is the protein framework on which the DNA is organized into loop domains of approximately 60 kb (Bak et al., 1977; Laemmli, 1979; Pienta and Coffey, 1984). The bases of these loops have been identified as the regions of the DNA that are attached to the nuclear matrix and also have been identified as the location of actively transcribed genes (Cook et al., 1982; Robinson et al., 1982; Ciejek et al., 1983; Robinson et al., 1983; Intres and Donady, 1985).

In addition to the topological role of the nuclear matrix, it has also been shown to be the site of many critical cellular processes, including the site for DNA replication and has been demonstrated to possess fixed sites of DNA synthesis, named replitases, that are located at the bases of the DNA loops (Vogelstein et al., 1980; Berezney and Buchholtz, 1981; Valenzuela et al., 1983). The replitase contains enzymes involved in DNA replication, including DNA polymerase, topoisomerase, DNA methylase, dihydrofolate reductase, thymidylate synthetase, and ribonucleoside diphosphate reductase (Prem-Veer-Reddy and Pardee, 1980, 1983).

Table 1. The Nuclear Matrix is the Dynamic Structural Component of the Nucleus that Directs the Functional Organization of DNA into Loop Domains and Provides Organizational Sites for Many of the Cellular Functions Involving DNA

Reported Functions of the Nuclear Matrix	*References*
Nuclear Morphology The nuclear matrix contains structural elements of the pore complexes, lamina, internal network, and nucleoli that contribute to the overall 3-dimensional organization and shape of the nucleus.	Fisher et al., 1982; Berezney & Coffey, 1974, 1977 ; Capco et al., 1984; Fey et al., 1984; Stuurman et al., 1990; Boyd et al., 1991; Getzenberg et al., 1991
DNA Organization DNA loop domains are attached to nuclear matrix at their bases and this organization is maintained during both interphase and metaphase; nuclear matrix shares some proteins with the chromosomal scaffold including toposomerase II, an enzyme that modulates DNA topology.	Bak et al., 1977; Laemmli, 1979; Pardoll et al., 1980; Georgiev, 1981; Halligan et al., 1984; Pienta & Coffey, 1984; Bekers et al., 1985; Cockerill & Garrard, 1986; Gasser & Laemmli, 1986; Nelson et al., 1986; Ito & Sakaki, 1987; Bode & Maass, 1988; Patriotis & Djondjurov, 1988; Gasser et al., 1989; Levy-Wilson & Fortier, 1989; Stief et al., 1989; Zini et al., 1989; Cockerill, 1990; Djondjurov, 1990; Farache et al., 1990; Getzenberg et al., 1991; Pienta et al., 1991
DNA Replication The nuclear matrix has fixed sites for DNA replication, containing the replisome complex for DNA replication that includes polymerase and newly replicated DNA.	Dijkwel et al., 1979;Vogelstein et al., 1980; Berezney & Buchholtz, 1981; Prem-Veer-Reddy & Pardee, 1980, 1983; Valenzuela et al., 1983; Tubo et al., 1985;
RNA Synthesis and Transport Actively transcribed genes are associated with the nuclear matrix; the nuclear matrix contains transcriptional complexes, newly synthesized heterogeneous nuclear RNA, and small nuclear RNA; RNA-processing intermediates (splicesomes) are bound to the nuclear matrix; mRNA is transported on specific tracks within the nucleus involving nuclear matrix components.	Herman et al., 1978; Van Eekelen & Van Venrooij, 1981; Ciejek et al., 1982; Nakayasu et al., 1982; Mariman et al., 1982; Long & Och, 1983; Mariman & Van Venrooij, 1985; Harris & Smith, 1988;
Nuclear Regulation The nuclear matrix has specific sites for steroid hormone receptor binding; the composition of the nuclear matrix is cell and tissue specific. DNA viruses are synthesized in association with the nuclear matrix; the nuclear matrix is a cellular target for transformation proteins and at least one carcinogen.	Getzenberg & Coffey, 1990; Fey & Penman, 1988; Klempnauer, 1988; Eisenman et al., 1985; Robinson et al., 1983; Ciejek et al., 1983; Robinson et al., 1982; Intres & Donady, 1985; Cook et al., 1982; Staufenbiel & Deppert, 1983; Covey et al., 1984; Sarnow et al., 1982; Chou et al., 1990; Moelling et al., 1984; Verderame et al., 1983; Buckler-White et al., 1980; Younghusband, 1985; Fuerstein et al., 1988; Obi & Billet, 1991; Fey & Penman, 1984; Barrack, 1983; Barrack & Coffey, 1980; Kumari-Siri et al., 1986

Therefore, the nuclear matrix plays both a structural and functional organizing role in DNA regulation.

The nuclear matrix also plays a central role in RNA processing. Newly synthesized heteronuclear RNA and small nuclear RNA are enriched on the nuclear matrix (Herman et al., 1978; Van Eekelen and Van Venrooij, 1981; Mariman et al., 1982; Nakayasu et al., 1982; Long and Och, 1983; Harris and Smith, 1988). The nuclear matrix has also been shown to be the site of attachment for products from RNA cleavage and for RNA processing intermediates and the nuclear matrix has also been demonstrated to be the site of mRNA transcription (Ciejek et al., 1982; Mariman et al., 1985). Active genes have been found to be associated with the nuclear matrix only in cell types in which they are expressed. Genes that are not expressed in these cell types are not found to be associated with the nuclear matrix. In addition, transcription factors including the *myc* protein, the large T antigen of the SV40 virus, and E1A from adenovirus have all been found to be associated with the nuclear matrix (Sarnow et al., 1982; Staufenbiel and Deppert, 1983; Covey et al., 1984). Further investigation into the association of active genes with the nuclear matrix has revealed a DNA loop anchorage site next to the enhancer region of several genes (Cockerill and Garrard, 1986; Gasser and Laemmli, 1986; Bode and Maass, 1988; Patriotis and Djondjurov, 1988; Levy-Wilson and Fortier, 1989; Zini et al., 1989; Djondjurov, 1990; Farache et al., 1990). Termed matrix associated regions (MARs) or scaffold attached regions (SARs) (Cockerill and Garrard, 1986; Gasser and Laemmli, 1986; Bode and Maass, 1988; Patriotis and Djondjurov, 1988; Levy-Wilson and Fortier, 1989; Zini et al., 1989; Djondjurov, 1990; Farache et al., 1990), these sequences are usually approximately 200 base pairs in length, are A-T rich and contain topoisomerase cleavage sequences along with other sequences such as poly-adenylation signals (Chou et al., 1990). The MARs have also been shown to confer functional transcriptional activity in genes in which they are inserted both *in vivo* and *in vitro* and therefore, may be important carcinogen and/or mutagen targets (Ito and Sakaki, 1987; Gasser et al., 1989; Stief et al., 1989; Cockerill, 1990). The nuclear matrix has been shown to be altered in cancer cells as well as by the process of transformation. In fully transformed tumorigenic cancer cells, transformation proteins appear to be associated with the nucleus, and many of these appear to be involved with the matrix. We have demonstrated specific alterations in the dorsal rat prostate nuclear matrix protein composition that occur with transformation (Getzenberg et al., 1991). Utilizing high resolution two-dimensional gel electrophoresis, we demonstrated that specific changes in nuclear matrix protein composition occur with the acquisition of the Dunning prostate adenocarcinoma tumor phenotype. Furthermore, there are specific changes in nuclear matrix composition in the those tumor sublines with a more aggressive phenotype. The Dunning R3327 rat prostate adenocarcinoma first arose as a spontaneous tumor in a Copenhagen male rat (Smolev et al., 1977a,b). Several variants of the tumor have arisen spontaneously from the original source, but the main variants are the anaplastic lines G, AT2.1, and MAT-LyLu, and a slow-growing, androgen sensitive,

Table 2. Comparison of Nuclear Matrix Proteins Identified by Two-Dimensional Gel Electrophoresis as Being Unique to the Normal Dorsal Rat Prostate and to the Dunning Adenocarcinoma Cells

	Molecular Weight	*pI*
Normal Dorsal Prostate Specific (NDP)		
NDP-1	95,000	6.74
NDP-2	57,000	8.33
NDP-3	57,000	8.0
NDP-4	47,000	5.26
NDP-5	47,000	5.8
NDP-6	41,000	6.83
NDP-7	37,200	7.05
NDP-8	36,900	7.35
NDP-9	35,000	6.25
NDP-10	32,500	5.46
Dunning Tumor Specific		
D-1	63,000	8.55
D-2	40,000	5.91
D-3	33,000	6.97
G Specific		
G-1	55,000	6.48
G-2	52,000	6.93
AT2 and MLL Specific		
AM-1	40,000	6.73
AM-2	36,000	8.33

well differentiated line, the R3327H. The G line is a slow-growing androgen insensitive, poorly-differentiated tumor. The AT2.1 is a subline of the H tumor, which is fast-growing and anaplastic. The MAT-LyLu (MLL) subline arose spontaneously from an AT variant and is an anaplastic cell line which also metastasizes. Nuclear matrices were isolated from normal rat prostate as well as from the Dunning prostate cancer cell lines and high resolution two-dimensional electrophoresis of nuclear matrix proteins of the normal dorsal rat prostate and the Dunning R3327 G, At2.1, and MLL cell lines was performed as previously described (Getzenberg and Coffey, 1990; Boyd et al., 1991).

The results of these experiments are presented in Table 2 and Figure 3. These data indicate that the composition of the rat dorsal prostate nuclear matrix is altered between normal and cancer cells. Ten distinct proteins (NDP1-10) were present in

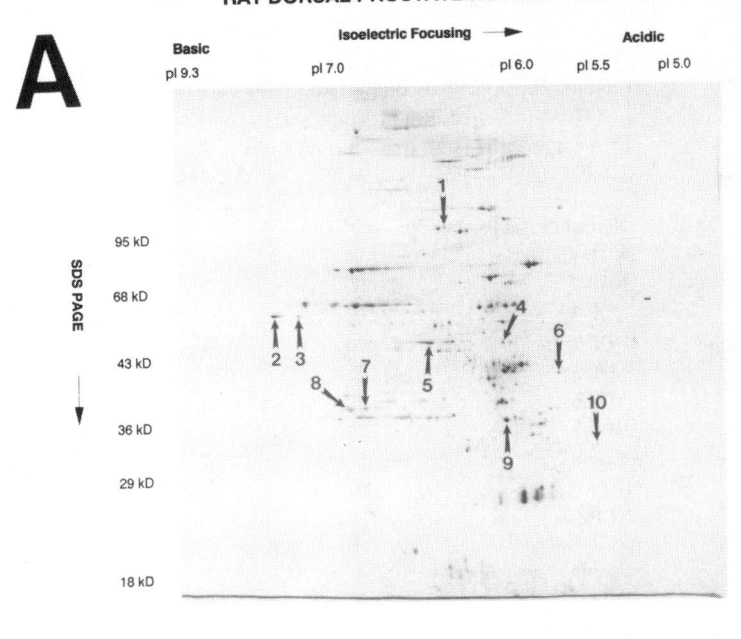
RAT DORSAL PROSTATE NUCLEAR MATRIX
A
Isoelectric Focusing
Basic
Acidic
pI 9.3
pI 7.0
pI 6.0
pI 5.5
pI 5.0
95 kD
68 kD
43 kD
36 kD
29 kD
18 kD
SDS PAGE
1
2
3
4
5
6
7
8
9
10

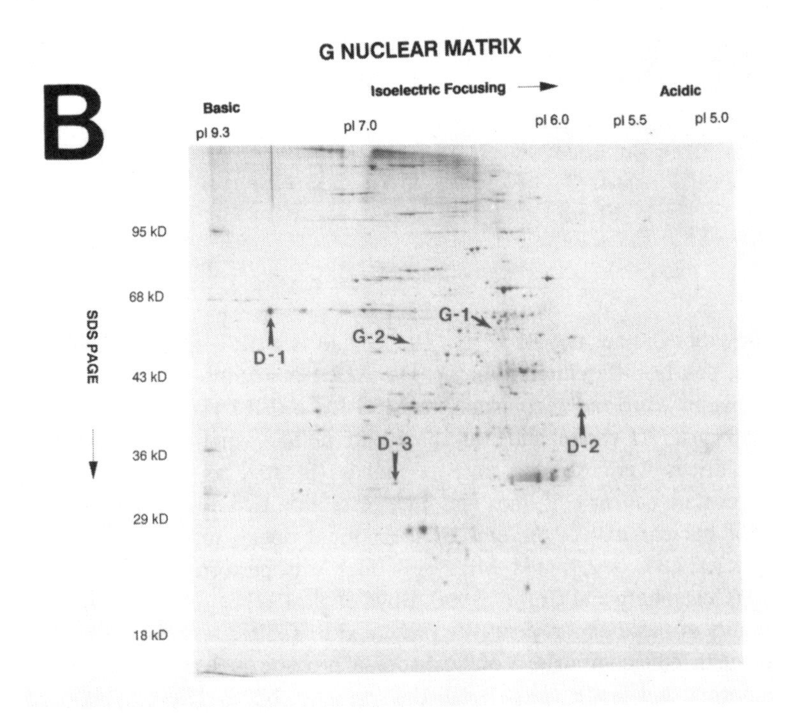
G NUCLEAR MATRIX
B
Isoelectric Focusing
Basic
Acidic
pI 9.3
pI 7.0
pI 6.0
pI 5.5
pI 5.0
95 kD
68 kD
43 kD
36 kD
29 kD
18 kD
SDS PAGE
G-1
G-2
D-1
D-2
D-3

Figure 3. High resolution two-dimensional gel electrophoresis of the nuclear matrices of normal dorsal rat prostate and the Dunning G rat prostate adenocarcinoma cell line. The nuclear matrix proteins in several Dunning cancer cell lines were examined and compared with the nuclear matrix protein composition of the dorsal prostate, the original tissue from which this tumor was derived. Utilizing the high resolution two-dimensional gel electrophoresis technique, the nuclear matrix proteins of the Dunning cell lines were found to be significantly different from the rat dorsal prostate. (**A**) A minimum of ten abundant spots were identified as unique to the rat dorsal prostate when compared with the tumor lines. (**B**) New proteins were also identified which were unique to the Dunning tumor lines, i.e., not found in the normal dorsal prostate (D1–3). Differences were also found between the different Dunning tumor cell lines (G1–2 are specific for the G subline) (Adapted from ref 9).

the normal dorsal prostate and absent in the tumor cell lines. Three proteins, D-1, D-2, and D-3, are present in all of the Dunning tumor cell lines, but are absent in the normal prostate. (A high resolution two-dimensional gel of a primary AT2.1 tumor revealed the same changes as the cell line). It is believed that these proteins may mediate alterations in function observed between normal and cancer cells. Other proteins, G-1, G-2, are expressed only in the G subline, whereas AM-1 and AM-2 are expressed only in the AT2.1 and MLL sublines. Certain nuclear matrix proteins, therefore, are specific for normal dorsal prostate, Dunning cancer cells in general, and individual Dunning cell lines. These proteins may play important roles in the process of prostate carcinogenesis.

Specific transformation events are also associated with changes in the nuclear matrix. For example, the nuclear matrix is one of the targets for retrovirus *myc* oncogene protein, the *myc* oncogene, adenovirus E1A-transforming protein, and polyoma large T antigen (Buckler-White et al., 1980; Verderame et al., 1983; Moelling et al., 1984; Younghusband, 1985). Numatrin, a nuclear matrix protein, has been associated with the induction of mitogenesis (Feuerstein et al., 1988). The carcinogen benzo[a]pyrene has been shown to bind preferentially to the nuclear matrix (Obi and Billet, 1991). Fey and Penman demonstrated that tumor promoters induce a specific morphologic signature in the nuclear matrix-intermediate filament scaffold of kidney cells (Fey and Penman, 1984). In these studies, the nuclear matrix-intermediate filament complex was profoundly reorganized in a specific manner after exposure to phorbol ester.

In an attempt to dissect the point in the transformation process where alterations in nuclear matrix protein composition occur, we have started to investigate nuclear matrix composition after specific perturbations of the cell. Transfected Kirsten-*ras* oncogene is thought to bind to GTP and disrupt cell signalling via an unknown mechanism. The nuclear matrices of normal rat kidney epithelial cells and their Kirsten-*ras* transformed counterparts were compared (Getzenberg et al., 1991). The only difference in nuclear matrix protein composition observed between these two cell lines was the loss of one protein in the K-*ras* transformed cells. The identity of this protein is currently being investigated.

To permit diversity of genomic function in various tissues, DNA must be dynamically partitioned into independent topological functional domains which must be constrained and tightly regulated. The possible reorganization of these domains would have profound effects on gene expression and may help explain the altered functions observed in cancer cells. The nuclear matrix plays a central role in DNA organization and gene expression but its role in the processes of carcinogenesis and cell transformation still remains to be explored.

III. THE CYTOSKELETON

The structural linkage of the ECM to the nucleus is thought to be mediated through cytoskeletal components. The cytoskeleton is composed of a structural interacting network of actin microfilaments (Pollard and Cooper, 1986), intermediate filaments (IF) (Klymkowsky et al., 1989), and microtubules (Purich and Kristofferson, 1984), and a vast array of associated proteins that modulate their form and function (Hall, 1984; McNiven and Porter, 1986; Perides et al., 1986; Joly et al., 1989; Okabe and Hirokawa, 1989; Scholey et al., 1989; Cleveland, 1990; Goldschmidt-Clermont et al., 1990; Matus, 1990; Nagle, 1990). This interacting network attaches to the plasma membrane and appears to be intimately involved in several cell signaling intracellular pathways (Georgetos et al., 1985). Activated growth factor receptors have been demonstrated to turn on protein kinases or lipid signals involving the inositol phosphate pathway. The initial activation of the inositol pathway in the cell periphery has been shown to produce direct interactions of the signalling components with the cytoskeleton. For example, profilin, an actin binding protein, has been demonstrated to bind the membrane phospholipid phosphatidylinositol bisphosphate, inhibiting its hydrolysis by phospholipase C (Goldschmidt-Clermont et al., 1990). Thus, components of the cytoskeleton can be involved in the regulation of cell signaling second messages.

It appears that the cytoskeleton plays a dynamic part not only in intracellular transport but as a structural modulator of cell signaling and ultimately gene expression. Its role in connecting the extracellular matrix to the nuclear matrix and

Table 3. Cytoskeletal Networks in Normal versus Transformed Rat Kidney Epithelial Cells.

Network	*Normal Rat Kidney Cell*	*Kirsten-ras Transformed Rat Kidney Cell*
Microfilaments	Fibers	Collapsed
Microtubules	Filaments	Filaments
Intermediate Filaments	Filaments	Filaments

vice versa is clearly demonstrated in the mammary epithelial system in which both cytochalasin D and colchicine inhibitors of cytoskeletal networks were shown to block the effect of laminin upregulation of gene expression by decreasing the mRNA stability of the induced genes (Langley and Cohen, 1986). The variability in the cytoskeleton components in various cell types is a potentially important mode of regulation, and it remains to be determined how common this type of regulation is within cells.

The IF system extends from the cell membrane to the nuclear matrix (Fey et al., 1984; Nagle; Klymkowsky et al., 1989). At the cell periphery, the IFs have been shown to contact tight junctions such as desmosomes, as well as to bind to the plasma membrane lipids and several of the membrane protein elements, including ankyrin, spectrin, and plectin (Wiche et al., 1983; Georgetos et al., 1985; Langley and Cohen, 1986). Traub and co-workers have suggested that the IF network interacts directly with the genome by demonstrating that vimentin, desmin, and glial fibrillary acid protein appear to bind single-stranded DNA, RNA, and histones *in vitro* (Traub, 1985; Traub et al., 1986; Vorgias and Traub, 1986). Although these results suggested that the cytoskeletal network may play an important structural role in modulating cell signal transduction and gene expression, the experiments were conducted under nonphysiological conditions. The IF cellular network has been implicated in many types of intracellular particulate transport. Cervera and colleagues demonstrated that mRNA is translated in association with the cytoskeleton, and Singer and colleagues have recently shown that cytoskeletal mRNAs are clustered in the cellular location of their protein product, suggesting a network of tracks for the directional movement of mRNA throughout the cell (Cervera et al., 1981; Singer et al., 1989).

Much of the tissue specificity of the cytoskeleton resides within the composition of the IFs. At least six classes of IFs have been identified, and each of these has been correlated with a specific tissue type (Nagle, 1990). Neurofilaments are specific in neuronal cells, and desmin is associated with striated and smooth muscle cells (Nagle, 1990). These different classes of IFs are often among the first markers to appear during early development, which allows identification of a cell type. Within the classes of IFs, each type of filament appears to be differentiation and development specific. This phenomenon has been described for the keratins (Moll et al., 1982; Van-Muijen et al., 1987) in which the different subunits of keratin combine to form mature filaments that change as a cell differentiates from the basal to apical layers of epidermis as well as for vimentin, which has been shown to be differentiation specific within the mesenchyme (Nagle, 1990).

Several investigators have demonstrated alterations in the IF network with acquisition of the cancer phenotype and with transforming events. Fey and Penman demonstrated that the IF pattern was different in normal versus cancer cells (Fey and Penman, 1988). Furthermore, they demonstrated that the IF pattern of cells can be altered by the promoter phorbol ester (Fey and Penman, 1984). Gatter and colleagues and others, have demonstrated that the expression of intermediate filaments is altered in human lung tumors (Lehto et al., 1983; Moll et al., 1983; Gatter et al., 1987). Intermediate filaments have also been demonstrated to be altered in several different cancers (Bartek et al., 1985; Ben-Ze'ev and Raz, 1985; Curschellas et al., 1987; Taylor-Papadimitriou et al., 1989; Chopra et al., 1990; Berdichevsky and Taylor-Papadimitriou, 1991). Ben Ze'ev has provided a comprehensive review of changes in the transformed cell cytoskeleton and has proposed that growth related cellular functions are regulated by signals which are transmitted through an organized cytoskeleton that has been disrupted by the transformation process (Ben-Ze'ev, 1985).

MT's play critical roles in many fundamental cellular processes, including determination and maintenance of cell shape, cell motility, and mitosis. Microtubules have been implicated as tracks for intracellular particulate movement. McNiven and Porter have demonstrated that microtubules serve as tracks to move particles of pigment within teleost melanophores (McNiven and Porter, 1988), and within neurons, organelles utilizing kinesin and dynein motors use microtubules as tracks to move bidirectionally (McNiven and Porter, 1986; McNiven and Porter, 1988; Okabe and Hirokawa, 1989; Scholey et al., 1989). In addition, MT's are conduits with core channels of approximately 10 nm and could possibly serve as a transport system for fluid or soluble signal components. Other types of intracellular particulate movement appear to be dependent on microtubules as intracellular tracks. Colchicine, which disrupts the microtubules, has been shown to inhibit the cytoplasmic transport of several different hormones within many different cell types (Lacy et al., 1972; Gautvik and Tashjian, 1973; Hall, 1984).

MT's have also been shown to vary in their composition, depending on the individual cell type where different tubulin isoforms combine to form differing microtubule types (Lewis and Cowan, 1988; Sullivan, 1988). The tubulin genes have been shown to be developmentally regulated within individual tissues (Lewis and Cowan, 1988). In addition to tissue-specific microtubule types, the cytoskeleton-associated proteins for these networks may have specific tissue functions. Different isoforms of microtubule associated proteins (maps) have recently been identified (Matus, 1990).

The microtubule network may be altered in transformed cells (Brinkley et al., 1975; Kamech and Seif, 1988), but this has been disputed by several investigators (Osborn and Webber, 1977; DeMey et al., 1978; Tucker et al., 1978; Asch et al., 1979). MT's have, however, been demonstrated to play a potential role in both cell proliferation and cell transformation. Crossin and Carney as well as others demonstrated that transient microtubule depolymerization induced DNA synthesis in resting cells (Vasiliev et al., 1971; Teng et al., 1977; Otto et al., 1979a,b; McClain and Edelman, 1980; Crossin and Carney, 1981). Griseofulvin, a tumor promotor, disrupts MT's both *in vivo* and *in vitro* (Barich et al., 1962), and vinblastine, a drug which prevents MT polymerization, increases the frequency of cell transformation (Weber et al., 1980). It appears that MT's play a role in cell proliferation and cell transformation, but much work is needed to further define the role of this cytoplasmic network in the carcinogenic process.

The actin microfilament (MF) network extends throughout the cytoplasm and is a major protein component of muscle as well as non-muscle cells (Pollard and Cooper, 1986). The MF network is involved in the regulation of cell shape, motility and intracellular transport (Wessells et al., 1971; Clarke and Spudich, 1977; Pollard and Cooper, 1986; Weeds, 1988). Consequently, MF's are poised to play a critical role in the process of cell transformation. Several investigators have demonstrated alterations of MF's in cancer cells (Friedman et al., 1984; Zachary et al., 1986; Demeure et al., 1990; Holme, 1990; Yu Rao, 1991). Yu Rao and colleagues have demonstrated low F-actin content in human bladder cancer cells (Yu Rao et al., 1991). Alterations in cellular actin architecture have also been shown in thyroid cancer, breast cancer, prostate cancer, melanoma, fibroblast, and colon cancer cell lines (Boone et al., 1979; Leavitt and Kakunaga, 1980; Raz and Geiger, 1982; Friedman et al., 1984; Zachary et al., 1986; Namba et al., 1987; Demeure et al., 1990; Holme, 1990).

The tumor promotor phorbol ester has been shown to disrupt the MF network (Demeure et al., 1990). Similarly, expression of the cell signaling oncogene *src* has been demonstrated to alter the appearance of MF's in fibroblasts (Holme, 1990). We have investigated the effect of the K-*ras* oncogene in normal rat kidney (NRK) cells and their Kirsten *ras* transformed counterparts (KNRK) (Pienta and Coffey, 1992). The transfection of Kirsten-*ras* oncogene is thought to bind to GTP and disrupt cell signaling via an unknown mechanism. Utilizing a microscopic morphometric assay to study cell structure interactions, we have demonstrated that K-*ras* causes a disruption of the actin microfilament network that is mimicked by cytochalasin D, a drug which inhibits actin polymerization. We further demonstrate that this disruption of the microfilament network by K-*ras* or cytochalasin D causes the nucleus to become dissociated from the cytoskeleton. This structural uncoupling of the nucleus may cause changes in nuclear matrix and DNA organization as we have previously proposed (Pienta et al., 1989).

At the present time, it is unclear whether or not microfilaments themselves exhibit tissue specificity. However, tropomyosin, a major actin binding protein, has several different isoforms that appear to be cell specific (Smillie, 1979; Cote, 1983; Jung-Ching et al., 1988). Other actin binding proteins such as lactinin and profilin control the rearrangement of actin and may regulate the cytoskeleton changes that occur during differentiation, cell motility and transformation (Bauer, 1981; Hendricks and Weintraub, 1981; Endo and Masaki, 1982; Cooper et al., 1985; Pollard and Cooper, 1986).

IV. THE EXTRACELLULAR MATRIX

The cytoskeletal components of the cell are placed in touch with the extracellular environment by transmembrane receptors that include the families of integrins and cell adhesion molecules (CAMs). These CAMs, such as E-cadherin (uvomorulin), may mediate specific cell-cell contact and, in doing so, link the CAMs to the intracellular matrix via the actin of the cytoskeleton (Hirano et al., 1987; Kemler et al., 1989). For example, the fibronectins of the extracellular matrix interact with specific cell surface transmembrane receptors, such as integrins, to link the extracellular matrix directly to the cytoskeleton components (Hirano et al., 1987; Ruoslahti and Giancotti, 1989; Yamada, 1989). In addition, the intermediate filament network has been demonstrated by Fey and Penman to extend from the nucleus to the cell membrane, connecting the nuclear matrix to the desmosomes, thereby forming a structural communication network extending between cells (Fey et al., 1984). Thus the cytoskeleton is directly connected from cell to cell and with the extracellular matrix through many types of these transmembrane molecules. These linkages are developmentally regulated and may determine the architecture of cell structure through contact by forming specific spatial organization sites on the membrane for ordering the cytoskeleton structure (Edelman et al., 1988). Thus, cell contact can alter cell shape through the spatial organization of the cytoskeleton.

The expression of CAM's and integrins is altered by the process of transformation and has been documented in numerous different tumor systems (Nagafuchi et al., 1987; Behrens et al., 1989; Lehmann et al., 1989; Linnemann et al., 1989; Ruoslahti and Giancotti, 1989; Moolenaar et al., 1990; Natali et al., 1990). In general, transformation results in a decreased expression of extracellular matrix (ECM) components, including fibronectin, laminin, and collagen.

What a cell touches determines what a cell does. Studies by Moscona and Folkman, as well as Gospodawawicz and colleagues have shown that the ECM controls cell differentiation and function by controlling cell shape (Folkman and Moscona, 1978; Gospodarowitz et al., 1978). Cunha and co-workers have demonstrated that the ECM can be responsible for functional differentiation in development when they showed that urogenital sinus mesenchyme induces urinary bladder epithelial cells to form prostatic epithelial cells and acini (Cunha et al., 1980). Others, including Reddi and Anderson, have observed that mature fibroblasts undergo redifferentiation to form new chondroblasts and chondrocytes when they are exposed to demineralized bone collagen matrix (Reddi and Anderson, 1976). Therefore, the contact with extracellular matrix components directs cell shape and gene function. The extracellular matrix has been shown to play both a direct and indirect role in gene expression. Bissell and co-workers have proposed that the ECM affects gene expression through interactions with transmembrane proteins and cytoskeletal elements, which in turn causes nuclear alterations that may direct gene expression (Bissell et al., 1982; Bissell and Aggeler, 1987; Bissell and Barcellos-Hoff, 1987).

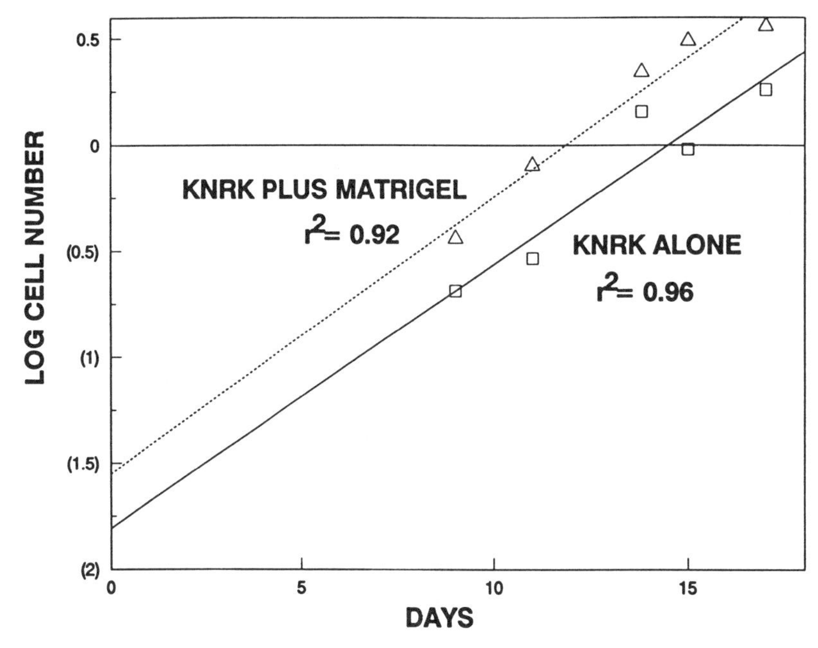

Figure 4. Growth curve of Kirsten-ras transformed normal kidney epithelial cells with and without co-inoculation with basement membrane secreted by the Engelbreth-Holm-sarcoma (EHS) tumor (matrigel). 5×10^5 KNRK cells were injected on day 0 with and without four milligrams of matrigel. The resulting tumor growth was plotted for each set of animals (eight in each group) as log cell number versus days of growth. The y-intercept of this plot represents the number of cells that originally "took" in the animal at the site of inoculation. The slope of this plot represents cell doubling time. Matrigel appears to enhance original tumor cell take but does not affect tumor cell growth rate as manifested by the similar slopes of the two growth curves.

Recently, we have demonstrated direct evidence that the ECM can directly affect nuclear matrix structure. Normal rat kidney cells (NRK) and their Kirsten-*ras* transformed counterpart (KNRK) cells were plated on tissue culture plastic or on basement membrane secreted by the Engelbreth-Holm-sarcoma (EHS) tumor (Matrigel). Plating the cells on the tumor secreted basement membrane resulted in the expression of four new nuclear matrix proteins in both the NRK and KNRK cells (Getzenberg et al., 1991). These data suggest that the ECM can directly affect nuclear structure and may suggest how the ECM regulates gene expression through the coordinate tissue matrix system (see Figure 1). A majority of studies involving the interaction of the ECM in the regulation of gene expression are done by varying the extracellular environment on which the cells are plated. The extracellular matrices used in these studies are usually either produced by other cell lines or are artificially produced using defined components. Reid and co-workers in a series of elegant experiments have defined specific biomatrices and hormonally defined media as important regulators of growth and differentiation of the liver (Enat et al., 1984; Fujita et al., 1986a,b; Muschel et al., 1986; Reid, 1990).

The ECM has been demonstrated to play a role in enhanced tumor growth. Kleinman, Martin, and colleagues as well as Pretlow and colleagues, have demonstrated that co-injection of tumor cells with Matrigel enhanced the ability of cells to grow in nude mice (Fridman et al., 1991; Grant et al., 1990; Scheutz et al., 1988; Pretlow et al., 1991). Utilizing the Kirsten-*ras* transformed normal rat kidney model, we have demonstrated that co-injection of KNRK cells with Matrigel enhanced the ability of cells to take in a nude mouse, but did not affect the subsequent growth rate of the tumor cells (see Figure 4). These experiments demonstrate that the extracellular environment of the tumor cells has a direct impact on the transformed cell's ability to interact with its host.

V. CONCLUSION

The ECM, cytoskeleton, and nuclear matrix provide an integrated structural system that is positioned to coordinate signal transduction within the cell. This tissue matrix system may regulate the tissue-specific control of gene expression within the cell by organizing the DNA into specific three-dimensional spatial conformations and by transmitting structural and functional information throughout the cell to the DNA. The exact mechanism of how cell architecture and structure may be involved in cell transformation still remains to be resolved. It does appear that the cell matrix provides specific spatial tracks for intracellular particulate transport and interacts with nuclear and cytoplasmic oncogenes. Cell transformation is a multistep process involving genomic changes that can involve oncogenes acting at different sites within the cell. Understanding this carcinogenic process will require a more complete knowledge of the interlocking matrix systems that extend from the extracellular matrix to the DNA and that govern cell shape and function.

Several investigators have proposed that the tissue matrix system can interact with the genome and affect gene expression (Bissell et al., 1981; Isaacs et al., 1981; Bissell et al., 1982; Fey et al., 1984; Fujita et al., 1986; Bissell and Aggeler, 1987; Bissell and Barcellos-Hoff, 1987; Puck et al., 1990; Reid, 1990; Pienta et al., 1991). Perhaps the most unique way of theoretically approaching the concept of the tissue matrix and how it is involved in affecting gene expression, and therefore, affecting the process of transformation, is through the idea of tensegrity. Tensegrity (*tensional integrity*) was defined in 1948 by Buckminster Fuller as a structural system composed of discontinuous compression elements connected by continuous tension cables (Fuller, 1975). Examples of tensegrity structures abound in the world and include butterfly wings and geodesic domes. Perhaps the best example of a tensegrity system is the human body, with the bones acting as compression elements and ligaments and tendons serving as tension cables. This type of tension-derived network may be a more appropriate way to view the tissue matrix system than as a rigid scaffolding system. Ingber and associates were the first to hypothesize that this principle of tensegrity could be applied to cell architecture by proposing that cell shape, and therefore, cell function, are under the control of an interacting compression-tension system (Ingber and Jamieson, 1985; Ingber and Folkman, 1989). Several investigators have now demonstrated this type of tension derived system operating in living cells. Heidemann and colleagues have shown that neurites use a tensegrity type mechanism with actin microfilaments acting as tension cables and microtubules acting as compression elements (Dennerll et al., 1988; Dennerll et al., 1989). Bereiter-Hahn hypothesized an actin-myosin fibrillar network serving as a tension producing system in eucaryotic cells and Karim and colleagues recently demonstrated this type of tensegrity system in bladder muscle cells (Bereiter-Hahn, 1987; Karim et al., 1992). The tissue matrix viewed as a tensegrity system provides an organizational structure that has the capacity to affect DNA organization and gene expression. Alterations in the tissue matrix system at any level of the organization—ECM, cytoskeleton, or nuclear matrix, have the capacity to alter gene expression. The tissue matrix of cancer cells has been altered as compared to normal cells and it is possible, even probable, that these alterations in cell skeletal structure contribute to the process of transformation.

REFERENCES

Asch, B.B., Medina, D., & Brinkley, B.R. (1979). Microtubules and actin containing filaments of normal, preneoplastic and neoplastic mouse mammary epithelial cells. Canc. Res 39, 893–907.

Bak, A.L., Zeuthen, J., & Crick, F.H.C. (1977). Higher-order structure of human mitotic chromosomes. Proc. Natl. Acad. Sci. USA 74, 1595–1599.

Bar-Sagi, D. & Feramicso, J.R. (1986). Induction of membrane ruffling and fluid-phase pinocytosis in quiescent fibroblasts by ras proteins. Science 233, 1061–1068.

Barich, L.L., Schwartz, J., & Barich, D.J. (1962). Oral griseofulvin, a cocacarcinogenic agent to methylcholanthrene induced cutaneous tumors. Canc. Res. 22, 53.

Barrack, E.R. (1983). The nuclear matrix of the prostate contains acceptor sites for androgen receptors. Endocrin. 113, 430–432.

Barrack, E.R. & Coffey, D.S. (1980). The specific binding of estrogens and androgens to the nuclear matrix of sex hormone responsive tissues. J. Biol. Chem. 255, 7265–7275.

Bartek, J., Durban, E.M., Hallowes, R.C., & Taylor-Papadimitriou, J. (1985). A subclass of luminal epithelial cells in the human mammary gland, defined by antibodies to cytokeratins. J. Cell Sci. 75, 17–33.

Bauer, H. (1981). Early changes in the distribution and organization of microfilament proteins during cell transformation. Cell 24, 175–184.

Behrens, J., Mareel, M.M., Van Roy, F.M., & Birchmeier, W. (1989). Dissecting tumor cell invasion: epithelial cells acquire invasive properties after the loss of uvomoulin-mediated cell-cell adhesion. J. Cell Biol. 108, 2435–2447.

Bekers, A.G.M., Gijzen, H.J., Taalman, R.D.F.M., & Wanka, F. (1985). Ultrastructure of the nuclear matrix from Physarum polycephalum during the mitotic cycle. J. Ultrastruct. Res. 75, 352–362.

Ben-Ze'ev, A. (1980). Protein synthesis requires cell-surface contact while nuclear events respond to cell shape in anchorage-dependent fibroblasts. Cell 21, 365–372.

Ben-Ze'ev, A. (1985). The cytoskeleton in cancer cells. Biochim. Biophys. Acta 780, 197–212.

Ben-Ze'ev, A., & Raz, A. (1985). Relationship between the organization and synthesis of vimentin and the metastatic capability of B16 melanoma cells. Canc. Res. 45, 2632–2641.

Berdichevsky, F. & Taylor-Papadimitriou, J. (1991). Morphological differentiation of hybrids of human mammary epithelial cell lines is dominant and correlates with the pattern of expression of intermediate filaments. Exp. Cell Res. 194, 267–274.

Bereiter-Hahn, J. (1987). Mechanical principles of architecture of eucaryotic cells. In: Cytomechanics. The Mechanical Basis of Cell Form and Structure. (Bereiter-Hahn, J., Anderson, O.R., & Reif, W.E., eds.), pp. 3–30. Springer-Verlag, New York.

Berezney, R. & Buchholtz, L.A. (1981). Dynamic association of replicating DNA fragments with the nuclear matrix of regenerating liver. Exp. Cell Res. 132, 1–13.

Berezney, R., & Coffey, D.S. (1974). Identification of a nuclear protein matrix. Biochem. Biophys. Res. Commun. 60, 1410–1417.

Berezney, R. & Coffey, D.S. (1977). Nuclear matrix: Isolation and characterization of a framework structure from rat liver nuclei. J. Cell Biol. 73, 616–627.

Bissell, M.J. & Aggeler, J. (1987). Dynamic reciprocity: how do the extracellular matrix and hormones direct gene expression? Prog. Clin. Biol. Res. 249, 251–262.

Bissell, M.J. & Barcellos-Hoff, M.H. (1987). The influence of extracellular matrix on gene expression: is structure the message? J. Cell Sci. Suppl. 8, 327–343.

Bissell, M.J., Hall, M.G., & Parry, G. (1982). How does the extracellular matrix direct gene expression? J. Theor. Biol. 99, 31–68.

Bissell, M.J., Matie, C., Laszlo, A., & Parry, G. (1981). The influence of the extracellular matrix on normal and transformed phenotypes in culture. J. Cell Biol. 91, 163a.

Bode, J. & Maass, K. (1988). Chromatin domain surrounding the human interferon-β-gene as defined by scaffold-attached regions. Biochem. 27, 4706–4711.

Boone, C.W., Vembu, D., White, B.J., Takeichi, N., & Paranjpe, M. (1979). Karyotypic, antigenic, and kidney-invasive properties of cell lines from fibrosarcomas arising in C3H/10T/1/2 cells implanted subcutaneously attached to plastic plates. Canc. Res 39, 2172–2178.

Boyd, J., Pienta, K.J., Getzenberg, R.H., Coffey, D.S., & Barrett, J.C. (1991). Preneoplastic alterations in nuclear morphology that accompany loss of tumor suppressor phenotype. J. Natl. Canc. Inst. 83, 862–866.

Brinkley, B., Fuller, G., & Highfield, D. (1975). Cytoplasmic microtubules in normal and transformed cells on culture: Analysis by tubulin antibody immunofluorescence. Proc. Natl. Acad. Sci. USA 72, 4981–4985.

Buckler-White, A., Humphery, G.W., & Pigiet, J., (1980). Association of polyoma T antigen and DNA with the nuclear matrix from lytically infected 376 cells. Cell 22, 37–46.

Capco, D.G., Wan, K.M., & Penman, S. (1984). Three-dimensional approaches to the residual structure of histone-depleted HeLa cell nuclei. J. Ultrastruct. Res. 87, 112–123.

Cervera, M., Dreyfuss, G., & Penman, S. (1981). Messenger RNA is translated when associated with the cytoskeletal framework in normal and VSV-infected HeLa cells. Cell 23, 113–120.

Chopra, H., Fligiel, S.E.G., Hatfield, J.S., Nelson, K.K., Diglio, C.A., Taylor, J.D., & Honn, K.V. (1990). An in vivo study of the role of the tumor cell cytoskeleton in tumor cell-platelet-endothelial cell interactions. Canc. Res. 50, 7686–7696.

Chou, R.H., Churchill, J.R., Flubacher, M.M., Mapsteon, D.E., & Jones, J. (1990). Identification of a nuclear matrix-associated region of the c-myc protooncogene and its recognition by a nuclear protein in the human leukemia HL-60 cell line. Canc. Res. 50, 3199–3206.

Ciejek, E., Tsai, M.H., & O'Malley, B.W. (1983). Actively transcribed genes are associated with the nuclear matrix. Nature (Lond.) 307, 607–609.

Ciejek, E.M., Nordstromn, J.L., Tsai, M., & O'Malley, B.W. (1982). Ribonucleic acid precursors are associated with the chick oviduct nuclear matrix. Biochem. 21, 4945–4953.

Clarke, M., & Spudich, J.A. (1977). Non-muscle contractile proteins, the role of actin and myosin in cell motility and shape determination. Ann. Rev. Biochem. 46, 797–822.

Cleveland, D.W. (1990). Microtubule mapping. Cell 60, 701–702.

Cockerill, P.N. (1990). Nuclear matrix attachment occurs in several regions of the lgH locus. Nucleic Acids Res 18, 2643–2648.

Cockerill, P.N. & Garrard, W.T. (1986). Chromosomal loop anchorage of the kappa immunoglobulin gene occurs next to the enhancer in a region containing topoisomerase II sites. Cell 44, 273–282.

Cook, P.R., Lang, J., Mayday, A., Lania, L., Fried, M., Chiswell, P.J., & Wyke, J.A. (1982). Active viral genes in transformed cells lie close to the nuclear cage. EMBO J. 1, 447–452.

Cooper, H.L., Feuerstein, N., Noda, M., & Bassin, R.H. (1985). Suppression of tropomyosin synthesis, a common biochemical feature of oncogenesis by structurally diverse retroviral oncogenes. Mol. Cell Biol. 5, 972–983.

Cote, G.P. (1983). Structural and functional properties of the nonmuscle tropomyosins. Mol. Cell Biochem. 57, 127–146.

Covey, L., Choi, Y., & Prives, C. (1984). Association of simian virus 40 T antigen with the nuclear matrix of infected and transformed monkey cells. Mol. Cell Biol. 4, 1384–1392.

Crossin, K.L. & Carney, D.H. (1981). Evidence that microtubule depolymerization early in the cell cycle is sufficient to initiate DNA synthesis. Cell 23, 61–71.

Cunha, G.R., Reese, B.A.,& Sekkingstad, M.N. (1980). Induction of nuclear androgen-binding sites in epithelium of the embryonic urinary bladder by mesenchyme of the urogenital sinus of embryonic mice. Endocrin. 107, 1767–1770.

Curschellas, E., Matter, A., & Regenass, U. (1987). Immunolocalization of cytoskeletal elements in human mammary epithelial cells. Eur. J. Canc. Clin. Oncol. 23, 1517–1527.

Demeure, M.J., Hughes-Fulford, M., Goretzki, P.E., Quan-Yang, D., & Clark, O.H. (1990). Actin architecture of cultured human thyroid cancer cells: predictor of differentiation? Surgery 108, 986–992.

DeMey, J., Janiau, M., De Brabander, M., Moens, W., & Genens, G. (1978). Evidence for unaltered structure and *in vivo* assembly of microtubules in transformed cells. Proc. Natl. Acad. Sci. USA 75, 1339–1343.

Dennerll, T.J., Joshi, M.C., Steel, J.L., Buxbaum, R.E. & Heidemann, S.R. (1988). Tension and compression in the cytoskeleton of PC-12 neurites. II. Quantitative measurements. J. Cell Biol. 107, 665–674.

Dennerll, T.J., Lamoureaux, P., Buxbaum, R.E., & Heidemann, S.R. (1989). The cytomechanics of axonal elongation and retraction. J. Cell Biol. 109, 3073–3083.

Dijkwel, P.A., Mullenders, L.M.F., & Wanka, F. (1979). Analysis of the attachment of replicating DNA to a nuclear matrix in mammalian interphase nuclei. Nucl. Acids Res. 6, 219–230.

Djondjurov, L. (1990). DNA-RNA complexes that might represent transient attachment sites of nuclear DNA to the matrix. J. Cell Sci. 96, 667–674.

Edelman, G.M. (1988). Topobiology, Basic Books, Inc., New York.

Eisenman, R.N., Tachibana, C.Y., Abrams, M.D., & Mann, S.R. (1985). V-myc and C-myc encoded proteins are associated with the nuclear matrix. Mol. Cell Biol. 5, 114–126.

Enat, R., Jefferson, D.M., Ruiz-Poazo, N., Gatmaitin, Z., Leinwald, L.A., & Reid, L.M. (1984). Hepatocyte proliferation *in vitro*: its dependence on the use of serum free hormonally defined medium and substrate of extracellular matrix. Proc. Natl. Acad. Sci. USA 81, 1411–1415.

Endo, T. & Masaki, T. (1982). Molecular properties and functions *in vitro* of chicken smooth muscle α-actinin in comparison with those of striated muscle α-actinin. J. Biochem. 92, 1457–1468.

Farache, G., Razin, S.V., Wolny, J., Moreau, J., Recillas Targa, F., & Scherrer, K. (1990). Mapping of structural and transcription-related matrix attachment sites in the α-globin gene domain of avian erythroblasts and erythrocytes. Mol. Cell Biol. 10, 5349–5358.

Feuerstein, N., Chan, P.K., & Mond, J.J. (1988). Identification of numatrin, the nuclear matrix protein associated with induction of mitogenesis, as the nucleolar protein B23. J. Biol. Chem. 263, 10608–10612.
Fey, E.G. & Penman, S. (1984)., Tumor promotors induce a specific morphological signature in the nuclear matrix-intermediate filament scaffold of Madin-Darby canine kidney (MDCK) cell colonies. Proc. Natl. Acad. Sci. USA 81, 4409–4413.
Fey, E.G. & Penman, S. (1988). Nuclear matrix proteins reflect cell type of origin in cultured human cells. Proc. Natl. Acad. Sci. USA 85, 121–125.
Fey, E.G., Wan, K.M., & Penman, S.P. (1984). Epithelial cytoskeletal framework and nuclear matrix-intermediate filament scaffold: Three-dimensional organization and protein composition. J. Cell Biol. 98, 1973–1984.
Fisher, P.A., Berrios, M., & Blobel, G. (1982). Isolation and characterization of a proteinaceous subnuclear fraction composed of nuclear matrix, peripheral lamina, and nuclear pore complexes from embryos of Drosophila melanogaster. J. Cell Biol. 92, 674–686.
Folkman, J. & Moscona, A. (1978). Role of cell shape in growth control. Nature (Lond.) 273, 345–349.
Fridman, R., Kibbey, M.C., Royce, L.S., Zain, M., Sweeney, M., Jicha, D.L., Yannelli, J.R., Martin, G.R., & Kleinman, H.K. (1991). Enhanced tumor growth of both primary and established human and murine tumor cells in athymic mice after coinjection with matrigel. J. Natl. Canc. Inst. 83, 769–774.
Friedman, E., Venderame, M., Winawer, S., & Pollack, R. (1984). Actin cytoskeletal organization loss in the benign-to-malignant tumor transition in cultured human colonic epithelial cells. Canc. Res. 44, 3040–3050.
Fujita, M., Spray, D.C., Choi, H., Saez, J., Jefferson, D.M., Hertzberg, E., Rosenberg, L.C., & Reid, L.M. (1986a). Extracellular matrix regulation of cell-cell communication and tissue specific gene expression in primary liver cultures. Prog. Clin. Biol. Res. 225, 333–360.
Fujita, M., Spray, D.C., Choi, H., Saez, J., Jefferson, D.M., Hertzberg, E., Rosenberg, L.C., & Reid, L.M. (1986b). Extracellular matrix regulation of cell-cell communication and tissue specific gene expression in primary liver cultures. Prog. Clin. Biol. Res. 226, 333–360.
Fuller, R.B. (1975). Synergetics, p. 39, Macmillan, New York.
Gasser, S.M. & Laemmli, U.K. (1986). Cohabitation of scaffold binding regions with upstream/enhancer elements of three developmentally regulated genes of D. melanogaster. Cell 46, 521–530.
Gasser, S.M., Amati, B.B., Cardenas, M.E., & Hofmann, J.F.X. (1989). Studies on scaffold attachment sites and their relation to genome function. Int. Rev. Cytol. 119, 57–96.
Gatter, K.C., Dunnill, M.S., Heryet, A., & Mason, D.Y. (1987). Human lung tumors: does intermediate filament co-expression correlate with other morphological or immunocytochemical features? Histopathology 11, 705–714.
Gautvik, K.M. & Tashjian, A.H., Jr. (1973). Effects of cations and colchicine on the release of prolactin and growth hormone by functional pituitary tumor cells in culture. Endocrin. 93, 793–799.
Georgetos, S.D., Weaver, D.C., & Marchesi, V.T. (1985). Site-specificity in vimentin-membrane interactions: Intermediate filament sub-units associate with the plasma membrane via their head domains. J. Cell Biol. 100, 1962–1967.
Georgiev, G.P. (1981). Proteins tightly bound to DNA in the regions of DNA attachment to the skeletal structures of interphase nuclei and metaphase chromosomes. Cell 27, 65–73.
Getzenberg, R.H. & Coffey, D.S. (1990). Tissue specificity of the hormonal response in sex accessory tissues is associated with nuclear matrix protein patterns. Mol. Endo. 4, 1336–1342.
Getzenberg, R.H., Pienta, K.J., Huang, E.Y.W., & Coffey, D.S. (1990). The tissue matrix: Cell dynamics and hormone action. End. Rev. 11, 399–417.
Getzenberg, R.H., Pienta, K.J., Huang, E.Y.W., & Coffey, D.S. (1991). Identification of nuclear matrix proteins in the cancer and normal rat prostate. Canc. Res. 51, 6514–6520.

Getzenberg, R.H., Pienta, K.J., Huang, E.Y.W., Murphy, B.C., & Coffey, D.S. (1991). Modifications of the intermediate filament and nuclear matrix networks by the extracellular matrix. Biochem. Biophys. Res. Comm. 179, 340–344.

Getzenberg, R.H., Pienta, K.J., Ward, W.S., & Coffey, D.S. (1991). Nuclear structure and the three-dimensional organization of DNA. J. Cell Biochem. 47, 289–299.

Goldschmidt-Clermont, P.J., Mzchesky, L.M., Baldassare, J.J., & Pollard, T.D. (1990). The actin-binding protein profilin binds to PIP2 and inhibits its hydrolysis by phosphokinase C. Science 247, 1575–1578.

Gospodarowitz, D., Greenberg, G., & Birdwell, C.R. (1978). Determination of cellular shape by the extracellular matrix and its correlation with the control of cellular growth. Canc. Res. 38, 4155–4171.

Grant, D.S., Kleinman, H.K., & Martin, G.R. (1990). The role of basement membranes in vascular development. Ann. New York Acad. Sci. 588, 61–72.

Hall, P.F. (1984). The role of the cytoskeleton in hormone action. Can. J. Biochem. Cell Biol. 62, 653–663.

Halligan, B.S., Small, D., Vogelstein, B., Hsieh, T.S., & Liu, L.F. (1984). Localization of type II DNA topoisomerase in nuclear matrix. J. Cell Biol. 99, 128a.

Harris, S.G. & Smith, M.C. (1988). SnRNP core protein enrichment in the nuclear matrix. Biochem. Biophys. Res. Commun. 152, 1383–1387.

Hay, E.D. (1977). Cell matrix interaction in embryonic induction. In International Cell Biology (Brinkley, B.R., & Porter, K.R., eds.), pp. 50–57, Rockfeller University Press, New York.

Hendricks, M. & Weintraub, H. (1981). T Gropomyosin is decreased in transformed cells. Proc. Natl. Acad. Sci. USA 78, 5633–5637.

Herman, R., Weymouth, L., & Penman, S. (1978). Heterogeneous nuclear RNA-protein fibers in chromatin depleted nuclei. J. Cell Biol. 78, 663–674.

Hirano, S., Nose, A., Hatta, K., Kawakami, A., & Takeichi, M. (1987). Calcium-dependent cell-cell adhesion molecules (cadherins): sub-class specificities and possible involvement of actin bundles. J. Cell Biol. 105, 2501–2510.

Hirano, S., Nose, A., Hatta, K., Kawakami, A., & Takeichi, M. (1987). Calcium-dependent cell-cell adhesion molecules (cadherins): sub-class specificities and possible involvement of actin bundles. J. Cell Biol. 105, 2501–2510.

Holme, T.C. (1990). Cancer cell structure: Actin changes in tumor cells-possible mechanisms for malignant tumor formation. Eur. J. Sur. Onc. 16, 161–169.

Ingber, D.E. (1990). Fibronectin controls capillary endothelial cell growth by modulating cell shape. Proc. Natl. Acad. Sci. USA 87, 3579–3583.

Ingber, D.E. & Folkman, J. (1989). Tension and compression as basic determinants of cell form and function: utilization of a cellular tensegrity mechanism. In: Cell Shape: Determinants, Regulation, and Regulatory Role (Stein, W. and Bronner F., eds.), pp. 3–31. Academic Press, San Diego.

Ingber, D.E. & Jamieson, J. (1985). Cells as tensegrity strucutes. Architectural regulation of histodifferentiation by physical forces transduced over a basement membrane. In: Gene Expression During Normal and Malignant Differentiation (Anderson, L.L., Gahmberg, C.G., & Kblom, P.E., eds.), pp. 13–32. Academic Press, New York.

Intres, R. & Donady, J.J. (1985). A constitutively transcribed actin gene is associated with the nuclear matrix in a Drosophila cell line. *In Vitro* Cell Dev. Biol. 21, 641–648.

Isaacs, J.T., Barrack, E.R., Isaacs, W.B., & Coffey, D.S. (1981). The relationship of cellular structure and function: The tissue matrix system. Prog. Clin. Biol. Res. 75A, 1–24.

Ito, T. & Sakaki, Y. (1987). Nuclear matrix association regions of rat α2 macroglobulin gene. Biochem. Biophys. Res. Commun. 149, 449–454.

Joly, J.C., Flynn, G., & Parich, D.L. (1989). The microtubule-binding fragment of MAP-2: Location of the protease accessible site and identification of an assembly-promoting peptide. J. Cell Biol. 109, 2289–2294.

Jung-Ching Lin, J., Hegmann, T.E., & Li-Chun Lin, J. (1988). Differential localization of tropomyosin isoforms in cultured nonmuscle cells. J. Cell Biol. 107, 563–572.

Kamech, N. & Seif, R. (1988). Effect of microtubule disorganizing or overstabilizing drugs on the proliferation of rat 3T3 cells and their virally induced transformed derivatives. Canc. Res. 48, 4892-4896.

Karim, O.M.A., Seki, N., Pienta, K.J., & Mostwin, J.L. (1992). The effect of age on the response of the detrusor to intracellular mechanical stimulus: DNA replication and the cell actin matrix. J. Cell Biochem. 48, 373–384.

Kemler, R., Ozawa, M., & Ringwald, M. (1989). Calcium-dependent cell adhesion molecules. Curr. Opin. Cell Biol. 1, 892–897.

Klempnauer, K.H. (1988). Interaction of *myb* proteins with nuclear matrix *in vitro*. Oncogene 2, 545–551.

Klymkowsky, M.W., Bachant, J.B., & Domingo, A. (1989). Functions of intermediate filaments. Cell Motil. Cytoskel. 13, 309–331.

Kumari-Siri, M.H., Shapiro, L.C., & Surks, M.I. (1986). Association of the 3,5,3′-triiodo-L-thyronine nuclear receptor with the nuclear matrix of cultured growth hormone-producing rat pituitary tumor cells (GC cells). J. Biol. Chem. 261, 2844–2852.

Lacy, P.E., Walker, M.M., & Fink, C.J. (1972). Perfusion of isolated rat islets *in vitro*. Participation of the microtubular system in the bi-phasic release of insulin. Diabetes 21, 987–998.

Laemmli, U.K. (1979). Levels of organization of the DNA in eukaryotic chromosomes. Pharmacol. Rev. 30, 469–476.

Langley, R.C., Jr., & Cohen, C.M. (1986). Association of spectrin with desmin intermediate filaments. J. Cell Biochem. 30, 101–109.

Leavitt, J., & Kakunaga, T. (1980). Expression of a variant form of actin and additional polypeptide changes following chemical-induced *in vitro* neoplastic transformation of human fibroblasts. J. Biol. Chem. 255, 1650–1661.

Lehmann, J.M., Riethmuller, G., & Johnson, J.P. (1989). MUC18, a marker of tumor progression in human melanoma, shows sequence similarity to the neural cell adhesion molecules of the immunoglobulin superfamily. Proc. Natl. Acad. Sci. USA 86, 9891–9895.

Lehto, V.P., Stenman, S., Miettinen, M., Dahl, D., & Virtanen, I. (1983). Expression of a neural type of intermediate filament as a distinguishing feature between oat cell carcinoma and other lung cancers. Am. J. Pathol. 110, 113–118.

Levy-Wilson, B. & Fortier, C. (1989). The limits of the DNase I-sensitive domain of the human apolipoprotein B gene coincide with the locations of chromosomal anchorage loops and define the 5′ and 3′ boundaries of the gene. J. Biol. Chem. 264, 21196–21204.

Lewis, S.A., & Cowan, N.J. (1988). Complex regulation and functional versatility of mammalian α-and β-tubulin isotypes during the differentiation of testis and muscle cells. J. Cell Biol. 106, 2023–2033.

Linnemann, D., Raz, A., & Bock, E. (1989). Differential expression of cell adhesion molecules in variants of K1735 melanoma cells differing in metastatic capacity. Int. J. Canc. 43, 709–712.

Long, B.H. & Och, R.L. (1983). Nuclear matrix, hnRNA, and snRNA in Friend erythroleukemia nuclei depleted of chromatin by low ionic strength EDTA. Bio. Cell 48, 89–98.

Mariman, E.C., Van Venrooij, W.J. (1985). The nuclear matrix and RNA-processing use of human antibodies. In Nuclear Envelope Structure and RNA Maturation (Smucker, E.G. & Clawson, G.A., eds), pp. 315–329, Alan R. Liss, Inc., New York.

Mariman, E.C.M., Van Eekelen, C.A.G., Reinders, R.J., Berns, A.J.M., & Van Venrooij, W.J. (1982). Adenoviral heterogeneous nuclear RNA is associated with the host nuclear matrix from Tetrahymena macronuclei. Biochem. 18, 1782–1788.

Matus, A. (1990). Microtubule-associated proteins. Curr. Opin. Cell Biol. 2, 10–14.
Matus, A. (1990). Microtubule-associated proteins. Curr. Opin. Cell Biol. 2, 10–14.

McClain, D.A. & Edelman, G.M. (1980). Density-dependent stimulation and inhibition of cell growth by agents that disrupt microtubles. Proc. Natl. Acad. Sci. USA 77, 2748–2753.

McNiven, M.A. & Porter, K.R. (1986). Microtubule polarity confers direction to pigment transport in chromatophores. J. Cell Biol. 103, 1547–1555.

McNiven, M.A. & Porter, K.R. (1988). Organization of microtubules in centrosome-free cytoplasm. J. Cell Biol. 106, 1593–1605.

Moelling, K., Benter, T., Bunte, T., Pfaff, E., Deppert, W., Egly, J.M., & Miyamoto, N.B. (1984). Properties of the myc-gene product: nuclear association, inhibition of transcription and activation in stimulated lymphocytes. Curr. Top. Microbiol. Immunol. 113, 198–207.

Moll, R., Franke, W.W., & Schiller, D.L. (1982). The catalog of human cytokeratins: Patterns of expression in normal epithelium, tumors and cultured cells. Cell 31, 11–24.

Moll, R., Krepler, R., & Franke, W.W. (1983). Complex cytokeratin polypeptide patterns observed in certain human carcinomas. Diff. 23, 256–269.

Moolenaar, C.E.C., Muller, E.J., Schol, D.J., Figdor, C.G., Bock, E., Bitter-Suermann, D., & Michalides, R.J.A.M. (1990). Expression of neural cell adhesion molecule-related sialoglycoprotein in small cell lung cancer and neuroblastoma cell lines H69 and CHP-212. Canc. Res. 50, 1102–1106.

Muschel, R., Khoury, G., & Reid, L.M. (1986). Regulation of insulin mRNA abundance and adenylation: dependence on hormones and matrix substrate. Mol. Cell Biol. 6, 337–341.

Nagafuchi, A., Shirayoshi, Y., Okazaki, K., Yasuda, K., & Takeichi, M. (1987). Transformation of cell adhesion properties by exogenously introduced E-cadherin cDNA. Nature 329, 341–343.

Nagle, R.B. (1990). Intermediate filaments: A review of the basic biology. Am. J. Surg. Pathol. 12 (Suppl. 1), 4–16.

Nakayasu, H., Mori, H., & Ueda, K. (1982). Association of small nuclear RNA-protein complex with the nuclear matrix from bovine lymphocytes. Cell Struct. Funct. 7, 253–262.

Namba, M., Karai, M., & Kimoto, T. (1987). Comparison of major cytoskeletons among normal human fibroblasts, immortal human fibroblasts transformed by exposure to Co-60 gamma rays, and the latter cells made tumorigenic by treatment with harvey murine sarcoma virus. Exp. Geron. 22, 179–186.

Natali, P., Nicotra, M.R., Cavaliere, R., Bigotti, A., Romano, G., Temponi, M. & Ferrone, S. (1990). Differential expression of intercellular adhesion molecule 1 in primary and metastatic melanoma lesions1. Canc. Res. 50, 1271–1278.

Nelson, W.G., Pienta, K.J., Barrack, E.R., & Coffey, D.S. (1986). The role of the nuclear matrix in the organization an dysfunction of DNA. Ann. Rev. Biophys. Biophys. Chem. 15, 457–475.

Obi, F.O. & Billet, M.A. (1991). Preferential binding of the carcinogen benzo[a]pyrene to proteins of the nuclear matrix. Carcinogenesis 12, 481–486.

Okabe, S. & Hirokawa, N. (1989). Axonal transport. Curr. Opin. Cell. Biol. 1, 91–97.

Osborn, M. & Webber, K. (1977). The display of microtubules in transformed cells. Cell 12, 561–571.

Otto, A., Zumbe, A., Gibson, L., Kubler, A.M., & De Asus, L.J. (1979a). Cytoskeleton-disrupting drugs enhance effect of growth factors and hormones on initiation of DNA synthesis. Proc. Natl. Acad. Sci. USA 76, 6435–6438.

Otto, A., Zumbe, A., Gibson, L., Kubler, A.M., & De Asua, L.J. (1979b). Microtubule-disrupting agents affect two different events regulating the initiation of DNA synthesis in Swiss 3T3 cells. Proc. Natl. Acad. Sci. USA 78, 3063–3067.

Pardoll, D.M., Vogelstein, B., & Coffey, D.S. (1980). A fixed site of DNA replication in eucaryotic cell. Cell 19, 527–536.

Partin, A.W., Isaacs, J.T., Treiger, B., & Coffey, D.S. (1988). Early cell motility changes associated with an increase in metastatic ability in rat prostatic cancer cells transfected with the v-Harvey-*ras* oncogene. Canc. Res 48, 6050–6053.

Patriotis, C. & Djondjurov, L. (1988). Tightly bound DNA-protein complexes representing stable attachment sites of large DNA loops to components of the matrix. Eur. J. Biochem. 184, 157–164.

Perides, G., Scherbarth, A., & Traub, P. (1986). Influence of phospholipids on the formation and stability of vimentin-type intermediate filaments. Eur. J. Cell Biol. 42, 268–280.
Pienta, K.J. & Coffey, D.S. (1984). A structural analysis of the role of the nuclear matrix and DNA loops in the organization of the nucleus and chromosome. J. Cell Sci. Suppl. 1, 123–135.
Pienta, K.J. & Coffey, D.S. (1991). Cellular harmonic information transfer through a tissue-tensegrity matrix system. Med. Hyp. 34, 88–95.
Pienta, K.J. & Coffey, D.S. (1992). Cytoskeletal-nuclear interactions: evidence for physical connections between the nucleus and the cell periphery and their alteration by transformation. J. Cell Biochem. 49, 357–365.
Pienta, K.J., Getzenberg, R.H., & Coffey, D.S. (1991). Cell structure and DNA organization. CRC Rev. Euk. Gene Exp. 1, 355–386.
Pienta, K.J., Murphy, B.C., Getzenberg, R.H., & Coffey, D.S. (1991). The effect of extracellular matrix interactions on morphologic transformation *in vitro*. Biochem. Biophys. Res. Comm. 179, 333–339.
Pienta, K.J., Partin, A.W., & Coffey, D.S. (1989). Cancer as a disease of DNA organization and dynamic cell structure. Canc. Res. 49, 2525–2532.
Pollard, T. & Cooper, J.A. (1986). Actin and actin binding proteins: A critical evaluation of mechanisms and functions. Ann. Rev. Biochem. 55, 987–1035.
Pollard, T. & Cooper, J.A. (1986). Actin and actin binding proteins: a critical evaluation of mechanisms and functions. Annu. Rev. Biochem. 55, 987–1035.
Prem-Veer-Reddy, G. & Pardee, A.B. (1980). Multienzyme complex for metabolic channeling in mammalian DNA replication. Proc. Natl. Acad. Sci. USA 77, 3312–3316.
Prem-Veer-Reddy, G., & Pardee, A.B. (1983). Inhibitor evidence for allosteric interaction in the replitase multi-enzyme complex. Nature (Lond.) 304, 86–88.
Pretlow, T.G., Delmoro, C.M., Dilley, G.G., Spadafora, C.G., & Pretlow, T.P. (1991). Transplantation of human prostatic carcinoma into nude mice in matrigel. Canc. Res. 51, 3814–3817.
Prysor-Jones, R.A.; Silverlight, J.J., & Jenkins, J.S. (1985). Differential effects of extracellular matrix on secretion of prolactin and growth hormone by rat pituitary tumor cells *in vitro*. Acta Endocrinol. (Copenhagen) 108, 156–160.
Puck, T.T., Krystosek, A., & Chan, D.C. (1990). Genome regulation in mammalian cells. Somatic Cell Molec. Genet. 16, 257–265. 185.
Purich, D.L. & Kristofferson, D.L. (1984). Microtubule assembly: A review of progress, principles, and perspectives. Adv. Protein Chem. 36, 133–207.
Raz, A., & Geiger, B. (1982). Altered organization of cell-substrate contacts and membrane-associated cytoskeleton in tumor cell variants exhibiting different metastatic capabilities. Canc. Res. 42, 5183–5190.
Reddi, A.M. & Anderson, W.A. (1976). Collagenous bone matrix induced endochondral ossification and hemopoiesis. J. Cell Biol. 69, 557–572.
Reid, L.M. (1990). Stem cell biology, hormone/matrix synergies and liver differentiation. Curr. Opin. Cell Biol. 2, 121–130.
Robinson, S.I., Nelkin, B.D., & Vogelstein, B. (1982). The ovalbumin gene is associated with the nuclear matrix of chicken oviduct cells. Cell 28, 99–106.
Robinson, S.I., Small, D., Izerda, R., McKnight, G.S., & Vogelstein, B.. (1983). The association of transcriptionally active genes with the nuclear matrix of the chicken oviduct. Nucl. Acids Res. 11, 5113–5130.
Ruoslahti, E., & Giancotti, G.F. (1989). Integrins and tumor cell dissemination. Canc. Cells 1, 119–126.
Sarnow, P., Mearing, P., Anderson, C.W., Reich, N., & Levine, A.J. (1982). Identification and characterization of an immunologically conserved adenovirus early region 11,000 M, protein and its association with the nuclear matrix. J. Mol. Biol. 162, 565–583.

Scheutz, E.G., Li, D., Omiecinski, C.J., Muller-Eberhard, U., Kleinman, H.K., Elswick, B., & Guzelian, P.S. (1988). Regulation of gene expression in adult rat hepatocytes cultured on a basement membrane matrix. J. Cell Physio. 134, 309–323.

Scholey, J.M., Heuser, J., Yang, J.T., & Golstein, L.S.B. (1989). Identification of globular mechanochemical heads of kinesin. Nature 338, 355–357.

Singer, R.H., Langevin, G.L., & Lawrence, J.B. (1989). Ultrastructural visualization of cytoskeletal mRNAs and their associated proteins using double-label in situ hybridization. J. Cell Biol. 108, 2343–2353.

Smillie, L.B. (1979). Structure and functions of tropomyosins from muscle and nonmuscle sources. Trends. Biochem. Sci. 4, 151–155.

Smolev, J.K., Coffey, D.S., & Scott, W.W. (1977a). Experimental models for the study of prostatic adenocarcinoma. J. Urol. 118, 216–220.

Smolev, J.K., Heston, W.D.W., Scott, W.W., & Coffey, D.S. (1977b). Characterization of the Dunning R3327H prostatic adenocarcinoma: an appropriate animal model for prostate cancer. Can. Treat. Rep. 61, 273–287.

Staufenbiel, M., & Deppert, W. (1983). Different structural systems of the nucleus are targets for SV40 large T antigen. Cell 33, 173–181.

Stief, A., Winter, D.M., Stratling, W.H., & Sippel, A.E. (1989). A nuclear DNA attachment element mediates elevated and position-independent gene activity. Nature 341, 343–345.

Stuurman, N., Meijne, A.M.L., van der Pol, A.J., de Jong, L., van Driel, R. & van Renswoude, J. (1990). The nuclear matrix from cells of different origin. J. Biol. Chem. 265, 5460–5465.

Sullivan, K.F. (1988). Structure and utilization of tubulin isotypes. Annu. Rev. Cell Biol. 4, 687–716.

Taylor-Papadimitriou, J., Stampfer, M., Bartek, J., Lewis, A., Boshell, M., Lane, E.B., & Leigh, I.M. (1989). Keratin expression in human mammary epithelial cells cultured from normal and malignant tissue: relation to *in vivo* phenotypes and influence of medium. J. Cell Sci. 94, 403–413.

Teng, M., Bartholomew, J.C., & Bissell, M.J. (1977). Synergism between anti-microtubule agents and growth stimulants in enhancement of cell cycle traverse. Nature (Lond.) 268, 739–741.

Tolle, H.G., Weber, K., & Osborne, M. (1985). Microinjection of monoclonal antibodies specific for one intermediate filament protein in cells containing multiple keratins allows insight into the composition of particular 10 nm filaments. Eur. J. Biol. 38, 234–244.

Traub, P. (1985). Are intermediate filament proteins involved in gene expression? Ann. N.Y. Acad. Sci. 455, 68–78.

Traub, P., Perides, G., Kuhn, S., & Scherbarth, A. (1986). Interaction *in vitro* of non-epithelial intermediate filament proteins with histones. Z. Naturforsch. Teil. C 42, 47–63.

Tubo, R.A., Smith, H.C., & Berezney, R. (1985). The nuclear matrix continues DNA synthesis at *in vivo* replication forks. Biochim. Biophys. Acta 825, 326–334.

Tucker, R.W., Stanford, K.K., & Frankel, F.R. (1978). Tubulin and actin in paired non-neoplastic and spontaneously transformed neoplastic cell lines *in vitro*: Fluorescent antibody studies. Cell 13, 629–642.

Valenzuela, M.S., Mueller, G.C., & Dasgupta, S. (1983). Nuclear matrix-DNA complex resulting from EcoRI digestion of HeLa nucleoids is enriched for DNA replicating forks. Nucl. Acid Res. 11, 2155–2164.

Van Eekelen, C.A.G., & Van Venrooij, W.J. (1981). hnRNA and its attachment to a nuclear protein matrix. J. Cell Biol. 88, 554–563.

Van-Muijen, G.N.P., Warnaar, S.O., & Ponec, M. (1987). Differentiation-related changes of cytokeratin expression in cultured keratinocytes and in fetal, newborn, and adult epidermis. Expl. Cell Res. 171, 331–345.

Vasiliev, J.M., Gelfand, I.M., & Guelstein, V.I., (1971). Initiation of DNA synthesis in cell cultures by Colcemid. Proc. Natl. Acad. Sci. USA 68, 977–979.

Verderame, M.F., Kohtz, D.S., & Pollack, R.E. (1983). 94,000- and 100,000-molecular-weight simian virus 40 T-antigens are associated with the nuclear matrix in transformed and revertant mouse cells. J. Virol. 46, 575–583.

Vogelstein, B., Pardoll, D.M., & Coffey, D.S. (1980). Supercoiled loops and eucaryotic DNA replication. Cell 22, 79–85.

Vorgias, C.E. & Traub, P. (1986). Nucleic acid-binding activities of the intermediate filament sub-unit proteins desmin and glial fibrillary cidic protein. Z. Naturforsch. Teil. C 41, 897–909.

Weber, K., Wheland, J., & Herzog, W. (1980). Griseofulvin interacts with microtubules both *in vivo* and *in vitro*. J. Mol. Biol. 102, 817–829.

Weeds, A. (1988). Actin binding protein-regulators of cell architecture and motility. Nature 296, 811–816.

Wessells, N.K., Spooner, B.S., Ash, J.F., Bradley, M.O., Luduena, M.A., Taylor, E.L., Wrenn, J.T., & Yamada, K.M. (1971). Microfilaments in cellular and development process. Science 171, 135–143.

Wiche, G., Krepler, R., Artlieb, U., Pytela, R., & Denk, H. (1983). Occurrence and immunologicalization of plectin in tissue. J. Cell Biol. 97, 887–901.

Wittelsberger, S.C., Kleene, K., & Penman, S. (1981). Progressive loss of shape-responsive metabolic controls in cells with increasingly transformed phenotype. Cell 24, 859–866.

Yamada, K.M. (1989). Fibronectins: structure, functions and receptors. Curr. Opin. Cell. Biol. 1, 956–963.

Younghusband, H.B. (1985). An association between replicating adenovirus DNA and the nuclear matrix of infected HeLa cells. Can. J. Biochem. Cell Biol. 63, 654–660.

Yu Rao, J., Hemstreet, G.P., Hurst, R.E., Bonner, R.B., Whan Min, K., & Jones, P.L. (1991). Cellular F-actin levels as a marker for cellular transformation, correlation with bladder cancer risk. Canc. Res. 51, 2762–2767.

Zachary, J.M., Cleveland, G., & Kwock, L., et al. (1986). Actin filament organization of the Dunning R2227 rat prostatic adenocarcinoma system: correlation with metastatic potential. Canc. Res. 46, 926–932.

Zini, N., Mazzotti, G., Santi, P., Rizzoli, R., Galanzi, A., Rana, R., & Maraldi, N.M. (1989). Cytochemical localization of DNA loop attachment sites to the nuclear lamina and to the inner matrix. Histochem. 91, 199–204.

TUMOR CELL INTERACTIONS IN CANCER GROWTH AND EXPRESSION OF THE MALIGNANT PHENOTYPE

Fred R. Miller and Bonnie E. Miller

I. INTRODUCTION

Elsewhere in this volume the genetic changes which accompany cancer development are described (Chapter 1). These changes bespeak an underlying genetic

Advances in Molecular and Cell Biology
Volume 7, pages 157–176

ISBN: 1-55938-624-X

instability in neoplastic cells. Whether or not tumors are of monoclonal origin (Fialkow, 1976; Nowell, 1976; Alexander, 1985; Iannaccone et al., 1987), a consequence of this instability is that by the time progression to a detectable stage occurs, tumors consist of multiple subpopulations of cells. Seemingly, subpopulations heterogeneous for any phenotype imaginable have been isolated from tumors of any tissue and species origin. In 1976 our laboratory derived a series of subpopulations from a single mouse mammary tumor. Our purpose was not just to demonstrate that tumors consist of heterogeneous populations, although that was our first step (Dexter et al., 1978; Heppner et al., 1978; Miller, F. and Heppner, 1979; Miller, F. et al., 1983). From the outset, our intent was to understand the role of subpopulation diversity in tumor growth and behavior. Our approach is to reconstruct heterogeneous tumors, albeit greatly simplified ones consisting of two or three defined subpopulations, and to analyze interactions capable of altering the phenotype(s) of the individual subpopulations. Initial efforts revealed immunologic interactions (Miller, F. and Heppner, 1979), altered growth rates (Miller, B. et al., 1980) sometimes resulting in clonal dominance (Miller, B. et al., 1987, 1988), altered chemotherapeutic drug responses (Miller, B. et al., 1981, 1989), and altered metastatic behavior (Miller, F., 1983), and led to the description of interactions between normal mammary stroma and epithelium with mammary tumor cells emphasizing the impact of the orthotopic site on growth and metastasis of neoplastic cells (Miller, F. et al., 1981; Miller, F., 1981). Evidence for cellular interactions in tumor biology have been reviewed in the past (Heppner et al., 1983; Heppner, 1984; Miller, F. and Heppner, 1987, 1990; Kerbel, 1990) but the relevant literature has expanded rapidly in recent months. This chapter will focus on these new reports coming from a number of laboratories now interested in tumor cell interactions.

II. ALTERATIONS IN TUMORIGENICITY: GROWTH INTERACTIONS AND CLONAL DOMINANCE

The mere fact that subpopulations separated from tumors differ widely in growth properties, some being much more aggressive than the parental heterogeneous tumor (Hauschka, 1953) and some being much less aggressive (Woodruff et al., 1982) suggests that subpopulation interactions occur within the parental tumors which modulate growth. In the former, more aggressive subpopulations must be "held in check by other elements" (Hauschka, 1953) and, in the latter, less aggressive "sublines require the cooperation of others to survive" (Woodruff et al., 1982).

Bilateral or two site protocols, in which tumor subpopulations are injected on contralateral flanks of experimental animals, have demonstrated both enhancement (Brodt et al., 1985; Caignard et al., 1985) or inhibition (Miller, B. et al., 1980; Caignard et al., 1985) of tumorigenicity. A "natural two site protocol" is represented by metastatic disease, and it is interesting to note that surgical removal of a primary tumor may result in accelerated growth of metastases (Fisher et al., 1983).

Other interactions which may not be apparent in a bilateral protocol (Newcomb et al., 1978) occur when subpopulations are physically mixed together (Nowotny and Grohsman, 1973; Newcomb et al., 1978; Butler et al., 1983). Mixtures of transformed rat prostatic fibroblast line NbF-1 and human cancer cell lines PC-3 (prostate), MDA-436 (breast), or WH (bladder) were much more tumorigenic in nude mice than were the NbF-1 or human tumor cells alone (Camps et al., 1990). In experiments in which either NbF-1 or PC-3 cells were irradiated (60Gy) prior to mixing, growth of the unirradiated line greatly exceeded that of controls in which the unirradiated cells were injected alone or with an equal number of irradiated cells of the same type (self). A study with human melanoma cell lines indicates that ability to be stimulated in growth by dermal fibroblasts increases with malignancy of the melanoma cells (Cornil et al., 1991). An extreme result of a fibroblast-tumor cell interaction is a report by U et al. (1987) in which nontumorigenic mouse fibroblasts were apparently transformed by exposure to a pituitary adenoma when the two were cocultured prior to inoculation into nude mice. Human cells were not detectable in the resulting tumors.

The net outcome of growth interactions between all subpopulations in a native, highly heterogeneous tumor is immensely complex since it is the sum of a multitude of subpopulation interactions. In studies with specific pairs of subpopulations, one may become dominant (so-called "clonal dominance") or there may be a tendency to establish a balance at a particular ratio quite unpredicted by the growth rates of the individual subpopulations grown in isolation. For example, in experiments utilizing two subpopulations derived from a single mouse mammary tumor, one subline (4TO7) strongly suppressed the growth of the other subline (168) both *in vivo* and *in vitro* (Miller, B. et al., 1988). *In vitro*, both sublines had similar doubling times when grown alone but 168 disappeared from cocultures (Miller, B. et al., 1988). *In vivo*, 168 tumors actually grew twice as fast as 4TO7 tumors, in that the volume doubling time of 168 was one half that of 4TO7 when tumors were initiated with the isolated sublines. However, tumors resulting from mixtures of the two consisted predominantly of 4TO7 cells. If initiated with a ratio of 100:1 (168:4TO7), tumors consisted of greater than 90% line 4TO7 cells by the time they reached a size of 500 mm^3. Tumors initiated with a ratio of 1000:1 (168:4TO7) averaged nearly one half 4TO7 cells at a size of 500 mm^3. In these experiments the composition of tumors was based on clonogenic cell content since 4TO7 and 168 cells were identified by their differential ability to form colonies in selective media. Interestingly, clonal dominance was only demonstrated in the 168-4TO7 combination. Other pairs of subpopulations from the same original mouse mammary tumor, including pairs in which one was either 168 or 4TO7, tended to form tumors consisting of equal cell numbers or some other pair-dependent ratio (Michelson et al., 1987; Miller, B. et al., 1989).

Clonal dominance has also been observed with line SP1 mouse mammary tumor subpopulations in a protocol utilizing a complex mixture of over 50 variants (Waghorne et al., 1988). In these experiments, tumors that had been initiated with a mixture of clones identifiable by randomly inserted pSV2*neo* genes were analyzed by restriction fragment analysis. A single clone (*neo*5) predominated in all tumors, including metastatic nodules. When tumors were initiated with a mixture of *neo*5 and the uncloned parental SP1, primary tumors were predominantly *neo*5. However, a threshold of *neo*5 in the initiating mixture was evident because *neo*5 cells were not detected in tumors resulting from mixtures of only 1% *neo*5 cells. A second metastatic variant of SP1 was obtained by transfection with activated c-Ha-*ras*. It too was dominant over parental SP1 (Waghorne et al., 1988). The parental SP1, the *neo*5 clone, and the *ras*-transfected clone (*ras*1) had nearly identical growth properties when injected into syngeneic mice.

Kerbel has proposed that clonal dominance is a feature of metastatic variants (Kerbel et al., 1988, 1989; Kerbel, 1990) and suggested that metastasis occurs after dominance has been established in the primary tumor (Theodorescu et al, 1991). (See Chapter 9 for further discussion of this concept). However, there are some caveats to this idea. Neither mammary line 168 nor 4TO7 are spontaneously metastatic, even through 4TO7 dominates 168 (see above). Furthermore, clonal dominance of one metastatic variant over another was demonstrated utilizing mixtures of the *neo*5 and the *ras*1 SP1 cells (Samiei and Waghorne, 1991). Inoculation of ratios up to 100:1 (*neo*5:*ras*1) resulted in tumors predominantly of *ras*1 cells even though both lines are metastatic and, when injected alone, have similar latency and growth rates. Similarly, a single pSV2*neo* transfected clone of the human breast carcinoma MDA-MB-435 became the dominant population in tumors formed in nude mice following inoculation of eleven pSV2*neo* clones, all of which were metastatic (Price et al., 1990). Although in this latter study the dominant clone was also the fastest growing when transfected clones were injected alone, one would have predicted better representation of other clones in these tumors if growth rates had been independently expressed. Finally, in a very recent report utilizing three tumor lines, each of which were transfected with pSV2*neo*, dominance was not demonstrable (Moffett et al., 1992). Although all injected clones were not always found in the tumors, most tumors were found to be polyclonal and clones present in metastases were not necessarily detectable by Southern analysis in the primary tumors. This may reflect metastasis by a clone making up less than 5% of the primary tumor, as that figure has been reported to be the limit detectable by Southern blot hybridization (Talmadge and Zbar, 1987).

When clonal dominance does occur, host factors may or may not play a role. In studies of growth interactions among subpopulations of a single mouse mammary tumor, we have found some in which host immunity is involved (Miller, B. et al., 1980) and some which occur *in vitro* as well as *in vivo* (Miller, B. et al., 1988). Staroselsky et al. (1990) reported that karyotypic analysis of tumors formed from mixtures of 4 nonmetastatic and one metastatic clone of the K-1735 melanoma indicated that the metastatic clone become dominant in normal syngeneic mice if the initial mixture contained 5% of the metastatic clone, but in nude mice, dominance occurred when only 1% of the initial mixture was the metastatic clone. Thus, host immunity in this system retarded establishment of dominance by the metastatic clone. In other cases, dominance may occur *in vivo*, not *in vitro*, but be independent of immune status. Thus, an androgen-independent subline (CS-2) overgrew the parental androgen-dependent mouse tumor (SC115) in male mice despite having equal growth properties when injected alone. No dominance was seen in mixtures cultured *in vitro* (Ichikawa et al., 1989).

Moffett et al. (1992) emphasize that the retention of polyclonality in tumors "has important implications for the generation of resistance to therapy" in that eradication of sensitive clones with a therapeutic agent can lead to recurrent growth as the remaining resistant clones emerge. Although this may be true, the existence of a dominant clonal population, as assessed by karyotypic marker or Southern analysis, does not mean that the tumor is homogeneous for other characteristics such as resistance to therapy or metastatic phenotype. In fact, evidence has been provided that clonal populations are less stable than mixtures for metastatic phenotype (Poste et al., 1981) and immunogenicity (Itaya et al., 1989). It is also possible that zonal heterogeneity *in vivo* creates pockets of cells which behave as isolated clones and diversify. As these areas become heterogeneous, diversification may stop until expansion of the populations again creates homogeneous pockets of sufficient size to precipitate diversification.

At any rate, tumors seem to exist for most of their "life cycle" as phenotypically heterogeneous mixtures of subpopulations regardless of whether the tumors were of monoclonal or polyclonal origin, or were clonal at some time due to overgrowth of a dominant clone.

III. ALTERATIONS IN METASTATIC BEHAVIOR

The multistage metastatic process, or "metastatic cascade," involves a variety of cell-to-cell interactions at the site of origin, transport sites (blood and lymphatics), and the metastatic site (reviewed in Miller, F. and Heppner, 1990). The impact on metastasis of interactions among normal and tumor cells in the orthotopic site have only recently been appreciated (Miller, F. et al., 1981; Tan and Chu, 1985; Naito et al., 1987, 1988; Ahlering et al., 1987; Bresalier, 1987; Elliott et al., 1988; Marikawa et al., 1988; Price et al., 1989), whereas the influence of interactions at the site of metastatic growth has been recognized as the seed and soil effect for over a century (Paget, 1889). Cells of the host immune system interact with tumor cells within the site of primary growth, during transport, and at the site of metastasis (reviewed in Miller, F., 1992).

Multicellular boluses are much more efficient at forming metastases than are single cells (Fidler, 1973; Liotta et al., 1976; Updike and Nicolson, 1986), a fact which makes it clear that tumor cell interactions influence survival during transport and arrest. Clumps of cells released from primary tumors may consist of tumor cells only, possibly aggregated due to expression of endogenous lectins (Raz and Lotan, 1987), or consist of host white cells and platelets, as well as tumor cells (Fidler et al., 1978). Increased metastatic proficiency of emboli could be due to enhancement of arrest or to increased survival of cells protected from shock and shear forces by surrounding cells. Nonmalignant cells (Fidler, 1973) or even inert particles such as glass beads may significantly enhance the number of nodules formed in the lung following i.v. injections of metastatic cells. Further, cell interactions can allow the metastasis of cells which are unable to metastasize on their own. We reported that a nonmetastatic subline (168) of a mouse mammary tumor could metastasize in the presence of a metastatic subline (410.4) of the same parental tumor. Metastatic 410.4 cells were injected i.v. into mice bearing nonmetastatic 168 tumors in the subcutis. Resulting lung nodules consisted of colony-forming 168 as well as 410.4 cells (Miller, F., 1983). Hossain et al. (1991) utilized a clone of 410.4 (4526) and a second metastatic subline (66) derived from the same parental tumor. Subline 66, injected alone, normally metastasizes to the lung but does not form nodules in the liver (Miller, F. et al, 1983). Subline 4526 forms nodules in the liver as well as in the lungs. When mixtures of 66 and 4526 were inoculated into mammary fatpads, 66 cells as well as 4526 cells were found to be present in liver metastases by Southern blot analysis (Hossain et al., 1991). Similarly, the *neo*5 clone of the SP1 mouse mammary tumor enabled nonmetastatic SP1 cells to metastasize spontaneously to the lung (Waghorne et al., 1988).

Data suggestive of similar interactions were reported by Varki et al. (1990). Eight clones were derived from a metastatic human lung carcinoma cell line, MV522, by cloning in soft agar. The ability of the eight clones to metastasize to lymph nodes and lungs from subcutaneous sites in nude mice was somewhat variable but none were as metastatic as the parental, uncloned tumor. Although these investigators concluded that cloned MV522 cells reverted to a less malignant state at a high frequency, an alternative interpretation is that interactions among some MV522 clones could increase metastatic efficiency.

The possible role of paracrine growth factors in these interactions has been investigated by a number of laboratories. The lung colonizing potential of MTLn3, a clone of the rat mammary adenocarcinoma line 13762NF, was enhanced by exposure to TGFβ (Welch et al., 1990). Production of type IV collagenase and heparinase was increased as was invasiveness of TGFβ-treated MTLn3 cells in a membrane invasion culture system. In the study of Egan et al. (1990), growth factors could enhance or inhibit lung colonization by NIH 3T3 cells in nude mice, depending on the oncogene used to transform the cells. Transfection of NIH 3T3 cells with a fusion gene of bFGF and immunoglobulin signal peptide resulted in a bFGF autocrine loop. These cells were highly efficient at colonizing the lungs. However, exogenous bFGF inhibited lung colonization by H-*ras* and v-*src* transformed 3T3 cells. CSF-1 stimulated or inhibited lung colonization by v-*fms* transformed 3T3 cells depending on whether or not the cells had been fed fresh medium before adding the exogenous growth factor (Egan et al., 1990).

G6, a clone of B16 mouse melanoma which is very inefficient at colonizing the lung after i.v. injection, became much more potent after culturing with fibroblasts from newborn mice (Tanaka et al., 1988). The effect was reversible and was not produced by exposure to conditioned media from fibroblasts. However, conditioned media from cocultures of the B16 clone and fibroblasts potentiated the lung colonizing efficiency of G6 cells not previously cocultured. Thus, a bidirectional interaction was necessary for the production of a soluble factor able to potentiate lung colonization.

These studies illustrate the complexities likely to be encountered in an analysis of paracrine modulation of metastasis.

One approach to study the mechanisms of metastasis is the use of *in vitro* assays to study isolated steps of the metastatic cascade. Tumor cell motility can be stimulated by tumor products such as bFGF (Mignatti et al., 1991), insulin polypeptides (Kohn et al., 1990), and a 55-kDa protein called autocrine motility factor (Liotta et al., 1986). TGFβ (Welch et al., 1990), EGF (Mizoguchi et al., 1991), and undefined factors in conditioned media from metastatic tumor cells (Korczak et al., 1991) can enhance *in vitro* invasive potential. EGF was found to enhance the invasiveness of one of three cell lines derived from a human squamous cell carcinoma (Mizoguchi et al., 1991). In this study there was no apparent effect on expression of type IV collagenase or urokinase plasminogen activator but EGF did enhance expression of a tissue inhibitor of metalloproteinase (TIMP). Korczak et al. (1991), derived a metastatic line, A3a, from the nonmetastatic SP1 mouse mammary tumor. The A3a cells expressed higher levels of mRNA for transin (stromelysin) and lower levels of mRNA for TIMP than did SP1 cells. Conditioned media from A3a cultures stimulated transin mRNA expression and reduced expression of mRNA for TIMP by SP1 cells. Exogenous bFGF had a similar affect. Urokinase activator mRNA expression was similar for A3a and SP1. EGF treatment induced transient expression by both followed by a return to baseline levels within 24 hours by A3a cells but SP1 expression fell to levels 3-4 times lower.

Thus, "growth" factors may affect many properties in addition to growth, including motility and invasiveness, and hence metastasis. Tumor subpopulations are heterogeneous in both the production of and the response to various growth factors. Clearly, growth factors produced by one tumor subpopulation, or by normal cells, could affect the metastatic phenotype of other subpopulations within heterogeneous tumors.

IV. ALTERATIONS IN DRUG RESPONSE

Because clones isolated from tumors express heterogenous levels of sensitivity to chemotherapeutic agents, the assumption has been that one reason for therapy failure is the elimination of the sensitive cells, leaving the drug resistant cells to grow. Our laboratory is engaged in studies testing that assumption. We have utilized our series of subpopulations derived from a single mouse mammary tumor to study interactions which alter drug responses. Initially, we expected to uncover interactions by which relatively sensitive subpopulations were made more resistant by relatively insensitive subpopulations. However, in general, we have found the opposite, in that relatively resistant subpopulations become more sensitive in the presence of relatively sensitive subpopulations. For example, we described the increased sensitivity of line 67 to methotrexate *in vitro* in the presence of line 410.4 and a host-mediated interaction in which line 410 became more sensitive to cyclophosphamide *in vivo* in the presence of line 168 (Miller, B. et al., 1981). We have developed quantitative methods to determine the subpopulation composition of clonogenic cells in mixed tumors (Miller, B. et al., 1987, 1988, 1989, 1990). Most often, we have used tumor cell lines with selectable markers that enable them to be grown to form colonies in selective media. Using this method, Miller, B. et al., (1991) found that line 66 (relatively insensitive to melphalan) became more sensitive to melphalan *in vivo* when grown in mixed tumors in the presence of the relatively sensitive line, 4TO7, but not with the equally sensitive line, 168TFAR. Transfer of melphalan sensitivity to 66 cells by 4TO7 cells was apparent only in physically mixed tumors, not in a bilateral protocol.

Other researchers have shown interactions in which there appears to be a transfer of resistance between subpopulations. Tofilon et al., (1984) demonstrated the transfer of resistance between two rat brain tumor sublines within a spheroid. The cell line more sensitive to 1,3-bis(2-chloroehtyl)-1-nitrosurea became less sensitive in the presence of the resistant line. Frankfurt et al. (1991) demonstrated a transfer of resistance to melphalan by a contact-mediated mechanism in human ovarian and lung carcinoma lines, apparently through transfer of glutathione from resistant to sensitive cells.

We also have elucidated one interaction in which resistance is transferred so that two subpopulations survive and grow in selective media which kill either subpopulation alone (Miller et al., 1986). Subpopulations 66 (wild type) and 66cl4, a thioguanine resistant, ouabain resistant variant of 66, are able to communicate via gap junctions (Miller, B. et al., 1986). Mixtures were embedded as boluses in collagen matrix and grown in HAT media with ouabain. Ouabain kills 66 cells, whereas HAT kills 66cl4 cells because these cells are deficient in the enzyme hypoxanthine guanine phosphoribosyl transferase (HGPRT-). Mixtures of communicating cells, however, grow in this medium. Excess Na^+ from the wild type cell, which accumulates because ouabain inhibits the Na^+:K^+ pump, can pass through gap junctions into the ouabain resistant cells which then get rid of the excess Na^+. The HGPRT- cells cannot salvage purines required by the block of *de novo* synthesis by aminopterin in the HAT but purine nucleotides produced by the wild type cell are available via transport through the gap junctions (Boyer and Klein, 1972; Corsaro and Migeon, 1977; Miller, B. et al., 1986). A tumor cell mass growing as a three-dimensional structure in collagen gel was extensively coupled so that a subpopulation constituting only 1% of the total could significantly rescue a mixture of cells under selective conditions (Miller, F. et al., 1990).

The transfer of resistance to melphalan described by Frankfurt et al. (1991) apparently also involved gap junctions as glutathione was transferred from gluatathione-rich resistant cells to sensitive cells.

Thus, therapy may not simply eradicate sensitive cells leaving resistant cells untouched. If resistance is transferred in mixtures, "sensitive" cells (i.e. if isolated) may remain; if sensitivity is transferred in mixtures, treatment may eliminate many "resistant" cells. We determined the response to methotrexate *in vivo* of mixtures of our mouse mammary tumor subpopulations. Response was measured both as growth delay and as a shift in tumor cell population distribution toward the more resistant line (Miller, B. et al., 1989). The cellular composition of the treated tumors did not reflect their response to therapy, implying that tumors which recur after a partial response to a drug may not contain a higher proportion of resistant cells than they did before therapy.

One must consider drug response interactions against the background of growth interactions which also occur among subpopulations. Any perturbation, such as that induced by therapeutic intervention, is likely to result in dynamic shifts in tumor composition in addition to depletion of the treatment-sensitive cells. Perhaps cases in which recurrent disease is more aggressive than the original, responsive disease, represent the unfortunate circumstance in which the treatment sensitive cells were keeping more aggressive elements in check.

V. MECHANISMS: ABERRANT HOMEOSTASIS

Tumor subpopulation interactions which alter malignancy and treatment responses are probably very similar to interactions occurring in normal tissues. Mechanisms by which interactions alter tumor behavior include autocrine/paracrine loops, contact-dependent phenomena such as intercellular communication via gap junctions, alteration of (and by) matrix or stromal cells, and modulation of host immunity.

Included in this volume are chapters pertaining to the roles of growth factors, matrix, and stroma in tumor growth (see Chapters 3, 5, and 6). Furthermore, mechanisms by which interactions may alter tumor behavior have been discussed in previous reviews (Heppner et al., 1983; F. Miller and Heppner, 1987, 1990). Thus, mechanisms will be discussed briefly only to present the concepts and a few pertinent recent findings.

VI. DIFFUSIBLE MEDIATORS

Growth factor independence appears to be important in tumor progression. Autocrine loops, in which self-stimulating factors are produced by tumor cells, are the subject of a chapter in this volume by Brattain et al. (Chapter 2). In the context of tumor subpopulation interactions, growth factors may be produced by one subpopulation and then stimulate non-producing variants. Conditioned medium from a human bladder cancer line was mitogenic for 3 of 9 clones derived from the parental line as assessed by colony formation in soft agar (Brown et al., 1990). TGFα (Ohmura et al., 1990; Chen et al., 1991; Hofer et al., 1991; Stromberg et al., 1992), bFGF (Yamanishi, 1991), IGF-I (Ohmura et al., 1990; Chen et al., 1991), and PDGF (Rozengurt et al., 1985; Nister et al., 1986; Adams et al., 1991) have all been reported to be produced by tumor cells and to be a mitogenic factor in tumor autocrine loops. Growth factor effects are not independent from each other as illustrated by the report that PDGF downregulates EGF receptor numbers (Wrann et al., 1980; Heldin et al., 1982) or affinity (Collins et al., 1983). Interactions between hormone-independent and hormone-dependent tumor cells may allow growth of hormone-dependent subpopulations in the absence of hormone or result in the hormonal stimulation of hormone-independent subpopulations. Testosterone induces an androgen-responsive clone (SC-3) of the Shionogi carcinoma 115 to release a factor which stimulates growth of SC-3 and an androgen-unresponsive subline of SC115 (Nonomura, 1988). The mitogenic factor released by androgen treated SC115 cells appears to be bFGF (Yamanishi et al., 1991). Estrogen-independent human mammary cancer line MDA-MB-231 produces a factor which stimulates growth of estrogen-responsive MCF-7 cells *in vitro* (Robinson and Jordan, 1989). Although estradiol-induced proliferation of MCF-7 cells can be inhibited by antiestrogens, such as tamoxifen, stimulation of MCF-7 replication by coculture with MDA-MB-231 was not abolished by antiestrogens (Robinson and Jordan, 1989). Thus, this type of paracrine interaction represents a potential mechanism for tamoxifen treatment failure.

Paracrine loops with inhibitory factors have also been described and may be responsible for clonal dominance phenomena. TGFβ has most frequently been implicated (Antonelli-Orlidge et al., 1989; Arteaga et al., 1990; Sing et al., 1990) but an insulin-like growth factor-binding protein was found to be produced by HT29 human colon adenocarcinoma (Culouscou and Shoyab, 1991). TGFβ may be a component of normal homeostasis. Sing et al. (1990) inhibited growth of human lymphoma lines with phorbol 12-myristate 13-acetate (PMA). The inhibitory effect of PMA was abolished by *anti*-TGFβ. Because treatment with PMA induced TGFβ receptor expression as well as the production of active TGFβ, the authors speculated that treatment restored a normal growth regulating TGFβ autocrine loop (Sing et al., 1990).

As described above, cell interactions may be necessary for the production of the regulating factor. If cell to cell contact was allowed, mixtures of endothelial cells with either pericytes or smooth muscle cells produced active TGFβ which inhibited growth of the endothelial cells (Antonelli-Orlidge et al., 1989). Conditioned media from pericytes alone, endothelial cells alone, or smooth muscle cells alone, had no effect.

Matrix and/or tissue architecture also alter growth factor production and response (see also Chapters 4, 5, and 6). Mouse mammary tumor cells have a greater growth potential in collagen gel cultures than in monolayers (Richards et al., 1983) and growth in collagen gels requires less supplementation of the media (Imagawa et al., 1982). Gospodarowicz et al. (1978) found that corneal epithelial cells respond to FGF if grown on plastic but, on collagen, respond to EGF rather than to FGF. Salomon et al. (1981) found that rat mammary cells were stimulated by EGF in monolayer on plastic or type I collagen but not when grown on type IV collagen. The authors speculated that EGF induced production of type IV collagen and, therefore, was unnecessary for optimal growth on type IV collagen. Ervin et al. (1989) reported that conditioned media from a normal human mammary cell line inhibited growth of human mammary tumor cells *in vitro* in monolayer culture. Our laboratory found that normal mouse mammary epithelium and stroma also inhibited growth of mouse mammary tumor cells in monolayer culture (Miller, F. et al., 1989). Because *in vivo* mouse mammary tumor growth is facilitated by implantation into mammary glands containing normal epithelium and stroma (Miller, F. et al., 1981), the biological significance of this inhibitory interaction *in vitro* was suspect. By utilizing a system in which both normal mammary cells and mammary tumor cells grow as 3-dimensional structures in collagen gels, we were able to demonstrate marked stimulation of tumor cell growth in the presence of normal mammary cells in a configuration which excluded cell to cell contact (Miller, F. et al., 1989).

VII. CONTACT-DEPENDENT INTERACTIONS

Cell to cell contact has long been recognized as an important element in homeostasis. Indeed, loss of contact inhibition has been considered a hallmark of transformation for years. However, growth of transformed cells can often be inhibited by normal cells *in vitro* (Mehta et al., 1986; Yoshikura, 1989; Martin et al., 1991). The role of gap-junction mediated intercellular communication in transformation, promotion, and progression to a metastatic phenotype has been the subject of numerous investigations. Loewenstein (1979) suggested that regulatory molecules may be transmitted between cells via gap junctions. Because communication occurs very efficiently between some tumor subpopulations and poorly between others (Miller, B. et al., 1983), growth signals transmitted via gap junctions may be nonrandomly distributed throughout a tumor. Control of tissue in which only a few cells respond directly to a diffusible factor might be accomplished by gap junction coupling of receptor-positive and receptor-negative cells. An *in vitro* model which demonstrated this possibility was devised by Lawrence et al. (1978), using a mixture of rat ovarian granulosa cells and mouse myocardial cells. In isolated cultures, noradrenaline stimulated only myocardial cells and follicle stimulating hormone stimulated only ovarian granulosa cells, but in mixed cultures, both cell types responded to either hormone.

Loss of intercellular communication competence has been implicated during essentially all steps of initiation, promotion, and progression of neoplasia. Cells transformed by oncogenes may lose communication competence (Mesnil et al., 1986; Azarnia and Loewenstein, 1987; Vanhamme et al., 1989; Nicolson et al., 1990). Transformation with H-*ras* reduced communication between transformed epithelial cells but not between transformed 3T3 fibroblasts (Vanhamme et al., 1989). Tumor promoters, such as 12-0-tetradecanoylphorbol-13-acetate (TPA), inhibit gap junction mediated communication (Yotti et al., 1979; Schindler et al., 1987). Suppression of intercellular communication by TPA was found to be more effective for transformed clones than for nontransformed clones of mouse epidermal JB6 cells (Miki et al., 1990). However, in other models progression may be accompanied by a resistance to uncoupling agents. In a system which is unaffected by TPA, ouabain more effectively uncoupled normal mouse mammary epithelial cells than it did mouse mammary tumor cells (Miller, F. et al., 1986). Hormones may function as promoters for initiated, hormone responsive cells. Because testosterone, but not estradiol, uncoupled two human transitional cell carcinoma lines, Kihara et al. (1990) suggested that the androgen might be an important factor in the higher incidence of urothelial cancer in males.

Sequential loss of coupling was reported during progression from 1) normal rat ovarian granulosa cells to 2) immortalized cells to 3) contact independent cells to (4) cells tumorigenic in nude mice (Stein et al., 1991). Eghbali et al. (1991) found that transfection of cDNA for the rat liver gap junction protein connexin 32 into a communication deficient human hepatoma line coordinately increased communication competence and decreased growth rate of transfected tumor cells in nude mice.

The loss of communication competence by metastatic cells has been reported (Nicolson et al., 1988; Ren et al., 1990). After transfecting a benign clone of rat 13762NF with *neo*-resistance or H-*ras* genes, intercellular communication was initially high but decreased with passage (Nicolson et al., 1990). Transfectants, clones, and subclones expressed a range of communication competence, tumorigenicity, and metastatic capacity but correlation between communication competence and metastatic capacity was poor (Nicolson et al., 1990).

VIII. THE HOST IN SUBPOPULATION INTERACTIONS

It is clear that stroma/mesenchyme and matrix interactions are important factors in cancer growth (Chapters 4 and 5). Alteration of stroma and/or extracellular matrix by one tumor subpopulation may subsequently cause an alteration in a second subpopulation's behavior. A human colon carcinoma grown on matrix produced by a well differentiated colon line was found to be more sensitive to inhibition by TGFβ (Levine et al., 1989). Tumor cells may induce expression of growth factor receptors (Funa et al., 1990) or the production of proteinases (Basset et al., 1990) by stromal cells. If active in matrix degradation, the production of a metalloproteinase by cancer stromal cells (Basset et al., 1990) may enhance the metastatic capacity of all tumor cell subpopulations present. The stromal component tenascin may be induced by epithelial cells (Inagama et al., 1988) and may participate in tumor-mesenchymal interactions (Chiquet-Ehrismann et al., 1989). Tenascin may also be immunosuppressive (Ruegg et al., 1989). Immune interactions include innocent bystander effects (Nagai et al., 1983), one-way cross-reactivity in which one subpopulation is sensitive to immunity induced by a second subpopulation (Miller, F. and Heppner, G. 1979), and preferential stimulation of suppressor cells by some subpopulations (Naor, 1983). Urban et al. (1984) described a series of variants of an ultraviolet light-induced mouse fibrosarcoma which demonstrated a hierarchy of response. An antigenic variant did not induce a response in the presence of an "immunodominant" antigen on a second subpopulation (Urban et al., 1984). Murine tumors may suppress immunity by releasing GM-CSF which expands the population of immunoregulatory macrophages (Tsuchiya et al., 1988; Fu et al., 1990). A recent study has found that lymphocytes from tumor-bearing mice showed an altered response to prolactin (Biswas and Chattopadhyay, 1992). Lymphocytes from both normal and tumor-bearers responded to the mitogen ConA and glucocorticoid suppressed the response. However, prolactin restored responsiveness to ConA only for the normal cells, not for those from tumor-bearers (Biswas and Chattopadhyay, 1992).

IX. SUMMARY

In recent years, a number of laboratories have begun to investigate how interactions between genetically distinct subpopulations existing within tumors can alter aspects of tumor behavior such as growth, metastasis, and response to chemotherapy. In some instances, these interactions have been shown to take place through diffusible mediators such as known autocrine and paracrine growth factors. Contact-mediated mechanisms appear to facilitate other interactions. Factors provided by the tumor host such as extracellular matrix and/or stromal cells and factors provided by the immune system have also been shown to play a role in some instances of tumor cell interactions. In many cases, these interactions appear to take place by mechanisms similar to those known to occur in normal tissue, which appear to control tissue homeostasis. It should not be surprising if normal mechanisms of cellular interaction continue to operate in tumors, albeit in aberrant form.

The existence of such interactions, however, reinforces the concept that tumors are *tissues*, not *collections* of "autonomous" cells.

ACKNOWLEDGMENTS

The authors' studies have been supported by grants CA28366 and CA27419 from the National Cancer Institute. We gratefully acknowledge 15 years of "interactions" with our colleague, Gloria Heppner. We thank Ms. Peterson for manuscript preparation.

REFERENCES

Adams, E.F., Todo, T., Schrell, V.M.H., Thierauf, P., White, M.C., & Fahlbusch, R. (1991). Autocrine control of human meningioma proliferation: secretion of platelet-derived growth-factor-like molecules. Int. J. Cancer 49, 398–402.

Arteaga, C.L., Coffey, R.J., Jr., Dugger, T.C., McCutchen, C.M., Moses, H.L., & Lyons, R.M. (1990). Growth stimulation of human breast cancer cells with anti-transforming growth factor β antibodies: evidence for negative autocrine regulation by transforming growth factor β. Cell Growth Diff. 1, 367–374.

Alexander, P. (1985). Do cancers arise from a single transformed cell or is monoclonality of tumors a late event in carcinogenesis. Br. J. Cancer 51, 453–457.

Antonelli-Orlidge, A., Saunders, K.B., Smith, S.R., & D'Amore, P.A. (1989). An activated form of transforming growth factor β is produced by cocultures of endothelial cells and pericytes. Proc. Natl. Acad. Sci. USA 86, 4544–4548.

Azarnia, R. & Loewenstein, W.R. (1987). Polyomavirus middle T antigen down regulates junctional cell-to-cell communication. Mol. Cell. Biol. 7, 946–950.

Basset, P., Bellocq, J.P., Wolf, C., Stoll, I., Hutin, P., Limacher, J.M., Podhajcer, O.L., Chenard, M.P., Rio, M.C., & Chambon, P. (1990). A novel metalloproteinase gene specifically expressed in stromal cells of breast carcinomas. Nature 348, 699–704.

Biswas, R., & Chattopadhyay, V. (1992). Altered prolactin response of the lymphocytes of tumor-bearing mice. Int. J. Cancer 50, 93–98.

Boyer, P.D., & Klein, W.L. (1972). Energy-coupling mechanisms in transport. In: Membrane Molecular Biology (Fox, C.F., & Klein, W.L., eds.) pp. 323–344, Sinaur Associates Inc., Stanford, CT.

Brodt, P., Parhar, R., Sankar, P., & Lala, P.K. (1985). Studies on clonal heterogeneity in two spontaneously metastasizing mammary carcinomas of recent origin. Int. J. Cancer 35, 265–273.

Brown, J.L., Russell, P.J., Philips, J., Wotherspoon, J., & Raghavan, D. (1990). Clonal analysis of a bladder cancer cell line: an experimental model of tumor heterogeneity. Br. J. Cancer 61, 370–376.

Butler, W.B., Toenniges, M.M., & Hillman, R.M. (1983). *In vivo* complementation between clones of the human breast cancer cell line MCF-7. Proc. Amer. Assoc. Cancer Res. 24, 35.

Caignard, A., Martin, M.S., Michel, M.F., & Martin, F. (1985). Interaction between two cellular subpopulations of a rat colonic carcinoma when inoculated to the syngeneic host. Int. J. Cancer 36, 273–279.

Camps, J.L., Chang, S.-M., Hsu, T.C., Freeman, M.R., Hong, S.-J., Zhau, H.E., von Eschenbach, A.C., & Chung, L.W.K. (1990). Fibroblast-mediated acceleration of human epithelial tumor growth *in vivo*. Proc. Natl. Acad. Sci. 87, 75–79.

Chen, S.-C., Chou, C.-K., Wong, F.-H., Chang, C., & Hu, C.-P. (1991). Overexpression of epidermal growth factor and insulin-like growth factor-I receptors and autocrine stimulation in human esophageal carcinoma cells. Canc. Res. 51, 1898–1903.

Chiquet-Ehrismann, R., Kalla, P., & Pearson, C.A. (1989). Participation of tenascin and transforming growth factor-β in reciprocal epithelial-mesenchymal interactions of MCF-7 cells and fibroblasts. Canc. Res. 49, 4322–4325.

Collins, M.K.L., Sinnett-Smith, J.W., Rozengurt, E. (1983). Platelet-derived growth factor treatment decreases the affinity of the epidermal growth factor receptors of Swiss 3T3 cells. J. Biol. Chem. 258, 11689–11693.

Cornil, I., Theodorescu, D., Man, S., Herlyn, M., Jambrosic, J., & Kerbel, R.S. (1991). Fibroblast cell interactions with human melanoma cells affect tumor cell growth as a function of tumor progression. Proc. Natl. Acad. Sci. 88, 6028–6032.

Corsaro, C.M., & Migeon, B.R. (1977). Comparison of contact-mediated communication in normal and transformed human cells in culture. Proc. Natl. Acad. Sci. 74, 4476–4480.

Culouscou, J.-M., & Shoyab, M. (1991). Purification of a colon cancer cell growth inhibitor and its identification as an insulin-like growth factor binding protein. Canc. Res. 51, 2813–2819.

Dexter, D.L., Kowalski, H.M., Blazar, B.A., Fligiel, Z., Vogel, R., & Heppner, G.H. (1978). Heterogeneity of tumor cells from a single mouse mammary tumor. Cancer Res. 38, 3174–3181.

Egan, S.E., Jarolim, L., Rogelj, S., Spearman, M., Wright, J.A., & Greenberg, A.H. (1990). Growth factor modulation of metastatic lung colonization. Anticanc. Res. 10, 1341–1346.

Eghbali, B., Kessler, J.A., Reid, L.M., Roy, C., & Spray, D.C. (1991). Involvement of gap junctions in tumorigenesis: transfection of tumor cells with connexin 32 cDNA retards growth *in vivo*. Proc. Natl. Acad. Sci. 88, 10701–10705.

Ervin, P.R., Jr., Kaminski, M.S., Cody, R.L., & Wicha, M.S. (1989). Production of mammastatin, a tissue-specific growth inhibitor, by normal human mammary cells. Science 244, 1585–1587.

Fialkow, P.J. (1976). Clonal origin of human tumors. Biochem. Biophys. Acta 458, 283–321.

Fidler, I.J. (1973). The relationship of embolic homogeneity, number, size and viability of experimental metastasis. Eur. J. Cancer 9, 223–227.

Fidler, I.J., Gersten, D.M., & Hart, I.R. (1978). The biology of cancer invasion and metastasis. Adv. Cancer Res. 38, 149–250.

Fisher, B., Gunduz, N., & Saffer, E.A. (1983). Influence of the interval between primary tumor removal and chemotherapy on kinetics and growth of metastases. Canc. Res. 43, 1488–1492.

Frankfurt, O.S., Seckinger, D., & Sugarbaker, E.V. (1991). Intercellular transfer of drug resistance. Canc. Res. 51, 1190–1195.

Fu, Y., Watson, G., Jimenez, J.J., Wang, Y., & Lopez, D.M. (1990). Expansion of immunoregulatory macrophages by granulocyte-macrophage colony-stimulating factor derived from a murine mammary tumor. Canc. Res. 50, 227–234.

Funa, K., Papanicolaou, V., Juhlin, C., Rastad, J., Akerstrom, G., Heldin, C.-H., & Oberg, K. (1990). Expression of platelet derived growth factor β-receptors on stromal tissue cells in human carcinoid tumors. Canc. Res. 50, 748–753.

Gospodarowicz, D., Greenburg, G., & Birdwell, C.R. (1978). Determination of cellular shape by the extracellular matrix and its correlation with the control of cellular growth. Canc. Res. 38, 4155–4171.

Hauschka, T.S. (1953). Methods of conditioning the graft in tumor transplantation. J. Natl. Canc. Inst. 14, 723–726.

Heldin, C.H., Wasteson, A., & Westermark, B. (1982). Interactions of platelet-derived growth factor with its fibroblast receptor. J. Biol. Chem. 257, 4216–4221.

Heppner, G.H., Dexter, D.L., DeNucci, T., Miller, F.R., & Calabresi, P. (1978). Heterogeneity in drug sensitivity among tumor cell subpopulations of a single mouse mammary tumor. Canc. Res. 38, 3758–3763.

Heppner, G.H., Miller, B.E., & Miller, F.R. (1983). Tumor subpopulation interactions in neoplasms. Biochim. Biophys. Acta 695, 215–226.

Heppner, G.H. (1984). Tumor heterogeneity. Canc. Res. 44, 2259–2265.

Hofer, D.R., Sherwood, E.R., Bromberg, W.D., Mendelsohn, J., Lee, C., & Kozlowski, J.M. (1991). Autonomous growth of androgen-independent human prostatic carcinoma cells: Role of transforming growth factor α. Canc. Res. 51, 2780–2785.

Hossain, A., Sarkar, A., & Sarkar, N.H. (1991). Mixed inocula of mouse mammary tumor cell subpopulations result in changes of organ-specific metastasis. Clin. Expl. Metast. 9, 501–515.

Iannaccone, P.M., Weinberg, W.C., & Deamant, F.D. (1987). On the clonal origin of tumors: a review of experimental models. Int. J. Canc. 39, 778–784.

Ichikawa, T., Akimoto, S., Hayata, I., & Shimazaki, J. (1989). Progression and selection in heterogeneous tumor composed of androgen-responsive Shionogi carcinoma 115 and its autonomous subline (Chiba subline 2). Canc. Res. 49, 367–371.

Imagawa, W., Tamooka, Y., & Nandi, S. (1982). Serum-free growth of normal and tumor mouse mammary epithelial cells in primary culture. Proc. Natl. Acad. Sci. 79, 4074–4077.

Inagama, Y., Kusakabe, M., Mackie, E.J., Pearson, C.A., Chiquet-Ehrismann, R., & Sakakura, T. (1988). Epithelial induction of stromal tenascin in the mouse mammary gland: from embryogenesis to carcinogenesis. Dev. Biol. 128, 245–255.

Itaya, T., Judde, J.-G., Hunt, B., & Frost, P. (1989). Genotypic and phenotypic evidence of clonal interactions in murine tumor cells. J. Natl. Canc. Inst. 81, 664–668.

Kerbel, R.S., Waghorne, C., Korczak, B., Lagarde, A., & Breitman, M.L. (1988). Clonal dominance of primary tumours by metastatic cells: genetic analysis and biological implications. Canc. Surv. 7, 597–629.

Kerbel, R.S., Cornil, I., & Korczak, B. (1989). New insights into the evolutionary growth of tumors revealed by Southern gel analysis of tumors genetically tagged with plasmid or proviral DNA insertions. J. Cell Sci. 94, 381–387.

Kerbel, R.S. (1990). Growth dominance of the metastatic cancer cell: cellular and molecular aspects. Adv. Canc. Res. 55, 87–132.

Kihara, K., Fukui, I., Higashi, Y., & Oshima, H. (1990). Inhibitory effect of testosterone on gap junctional intercellular communication of human transitional cell carcinoma cell lines. Canc. Res. 50, 2848–2852.

Kohn, E.C., Francis, E.A., Liotta, L.A., & Schiffman, E. (1990). Heterogeneity of the motility responses in malignant tumor cells: a biological basis for the diversity and homing of metastatic cells. Int. J. Canc. 46, 287–292.

Korczak, B., Kerbel, R.S., & Dennis, J.W. (1991). Autocrine and paracrine regulation of tissue inhibition of metalloproteinases, transin, and urokinase gene expression in metastatic and nonmetastatic mammary carcinoma cells. Cell Growth Diff. 2, 335–341.

Lawrence, T.S., Beers, W.H., Gilula, N.B. (1978). Transmission of hormonal stimulation by cell-to-cell communication. Nature 272, 501–503.

Levine, A.E., Black, B., & Brattain, M.G. (1989). Effects of N,N-dimethylformamide and extracellular matrix on transforming growth factor-β binding to a human colon carcinoma cell line. J. Cell Physiol. 138, 459–466.

Liotta, L.A., Kleinerman, J., & Saidal, G.M. (1976). The significance of hematogenous tumor cell clumps in the metastatic process. Canc. Res. 36, 889–894.

Liotta, L.A., Mandler, R., Murano, G., Katz, D.A., Gordon, R.K., Chiang, P.K., & Schiffman, E. (1986). Tumor autocrine motility factor. Proc. Natl. Acad. Sci. 83, 3302–3306.

Loewenstein, W.R. (1979). Junctional intercellular communication and the control of growth. Biochim. Biophys. Acta 560, 1–65.

Martin, W., Zempel, G., Hulser, D., & Willecke, K. (1991). Growth inhibition of oncogene-transformed rat fibroblasts by cocultured normal cells: relevance of metabolic cooperation mediated by gap junctions. Canc. Res. 51, 5348–5354.

Mehta, P.R., Bertram, J.S., & Loewenstein, W.R. (1986). Growth inhibition of transformed cells correlates with their junctional communication with normal cells. Cell 44, 187–196.

Mesnil, M., Montesano, R., & Yamasaki, H. (1986). Intercellular communication of transformed and non-transformed rat liver epithelial cells. Modulation by TPA. Exp. Cell Res. 165, 391–402.

Michelson, S., Miller, B.E., Glicksman, A.E., & Leith, J.T. (1987). Tumor micro-ecology and competitive interactions, J. Theor. Biol. 128, 233–246.

Mignatti, P., Morimoto, T., & Rifkin, D.B. (1991). Basic fibroblast growth factor released by single, isolated cells stimulates their migration in an autocrine manner. Proc. Natl. Acad. Sci. 88, 11007–11011.

Miki, H., Yamadori, I., Heine, S., Riggs, C.W., & Rice, J.M. (1990). Effect of 12-O-tetradecanoylphorbol-13-acetate on intercellular communication in various clones of mouse epidermal JB6 cells. Canc. Res. 50, 1324–1329.

Miller, B.E., Miller, F.R. Leith, J., & Heppner, G.H. (1980). Growth interaction *in vivo* between tumor subpopulations derived from a single mouse mammary tumor. Canc. Res. 40, 3977–3981.

Miller, B.E., Miller, F.R., & Heppner, G.H. (1981). Tumor heterogeneity and drug sensitivity: Interactions between tumor subpopulations affecting their sensitivity to the antineoplastic agents cyclophosphamide and methotrexate. Canc. Res. 41, 4378–4381.

Miller, B.E., Roi, L.D., Howard, L.M., & Miller, F.R. (1983). Quantitative selectivity of contact-mediated intercellular communication in a metastatic mouse mammary tumor line. Canc. Res. 43, 4102–4107.

Miller, B.E., McInerney, D., Jackson, D., & Miller, F.R. (1986). Metabolic cooperation between mouse mammary tumor subpopulations in three-dimensional collagen gel cultures. Canc. Res. 46, 89–93.

Miller, B.E., Miller, F.R., Wilburn, D., & Heppner, G.H. (1987). Analysis of tumor cell composition of tumors composed of paired mixtures of mammary tumor cell lines. Br. J. Canc. 56, 561–569.

Miller, B.E., Miller, F.R., Wilburn, D., & Heppner, G. (1988). Dominance of a mammary tumor subpopulation line in mixed, heterogeneous tumors. Canc. Res. 48, 5747–5753.

Miller, B.E., Aslakson, C.J., & Miller, F.R. (1990). Efficient recovery of clonogenic stem cells from solid tumors and occult metastatic deposits. Inv. Metast. 10, 101–112.

Miller, B.E., Miller, F.R., & Heppner, G.H. (1989). Therapeutic perturbation of the tumor ecosystem in reconstructed heterogeneous mammary tumors. Canc. Res. 49, 3747–3753.

Miller, B.E., Machemer, T., Lehotan, M., & Heppner, G.H. (1991). Tumor subpopulation interactions affecting melphalan sensitivity in palpable mouse mammary tumors. Canc. Res. 51, 4378–4387.

Miller, F.R. (1981). Comparison of metastasis of mammary tumors growing in the mammary fatpad versus the subcutis. Inv. Metast. 1, 220–226.

Miller, F.R. (1983). Tumor subpopulation interactions in metastasis. Invasion Metastasis 3, 234–242.

Miller, F.R. (1993). Immune mechanisms in the sequential steps of metastasis. CRC Crit. Rev. Onc. 4, 293–311.

Miller, F.R., & Heppner, G.H. (1979). Immunologic heterogeneity of tumor cell subpopulations from a single mouse mammary tumor. J. Natl. Canc. Inst. 63, 1457–1463.

Miller, F.R., Medina, D., & Heppner, G.H. (1981). Preferential growth of mammary tumors in intact mammary fatpads. Canc. Res. 41, 3863–3867.

Miller, F.R., Miller, B.E., & Heppner, G.H. (1983). Characterization of metastatic heterogeneity among subpopulations of a single mouse mammary tumor: Heterogeneity in phenotypic stability. Inv. Metast. 3, 22–31.

Miller, F.R., & Heppner, G.H. (1987). Interaction of mammary tumor subpopulations. In: Cellular and Molecular Biology of Experimental Mammary Cancer (Medina, D., Kidwell, W., Heppner, G., & Anderson, E., eds.) pp. 141–162, Plenum Publishing Corp., New York.

Miller, F.R., McEachern, D., & Miller, B.E. (1989). Growth regulation of mouse mammary tumor cells in collagen gel cultures by diffusible factors produced by normal mammary gland epithelium and stromal fibroblasts. Canc. Res. 49, 6091–6097.

Miller, F.R., & Heppner, G.H. (1990). Cellular interactions in metastasis. Canc. Metast. Rev. 9, 21–34.

Miller, F.R., McEachern, D., & Miller, B.E. (1990). Efficiency of communication between tumor cells in collagen gel cultures. Br. J. Canc. 62, 360–363.

Mizoguchi, H., Komiyama, S., Matsui, K., Hamanaka, R., Ono, M., Kiue, A., Kobayashi, M., Shimizu, N., Welgus, H.G., & Kuwano, M. (1991). The response to epidermal growth factor of human maxillary tumor cells in terms of tumor growth, invasion and expression of proteinase inhibitors. Int. J. Canc. 49, 738–743.

Moffett, B.F., Baban, D., Bao, L., & Tarin, D. (1992). Fate of clonal lineages during neoplasia and metastasis studied with an incorporated genetic marker. Canc. Res. 52, 1737–1743.

Nagai, A., Zbar, B., Terata, N., & Horis, J. (1983). Rejection of retrovirus-infected tumor cells in guinea pigs: effect on bystander tumor cells. Canc. Res. 43, 5783–5788.

Naor, D. (1983). Coexistence of immunogenic and suppressogenic epitopes in tumor cells and various types of macromolecules. Cancer Immunol. Immunother. 16, 1–10.

Newcomb, E.W., Silverstein, S.C., & Skilagi, S. (1978). Malignant mouse melanoma cells do not form tumors when mixed with cells of a malignant subclone: relationships between plasminogen activator expression by the tumor cells and the host's immune response. J. Cell Physiol. 95, 169–171.

Nicolson, G.L., Dulski, K.M., & Trosko, J.E. (1988). Loss of intercellular junctional communication correlates with metastatic potential in mammary adenocarcinoma cells. Proc. Natl. Acad. Sci. 85, 473–476.

Nicolson, G.L., Gallick, G.E., Dulski, K.M., Spohn, W., Lembo, T.M., & Tainsky, M.A. (1990). Lack of correlation between intercellular junctional communication, $p21^{rasEJ}$ expression, and spontaneous metastatic properties of rat mammary cells after transfection with c-H-ras^{EJ} or neo genes. Oncogene 5, 747–753.

Nister, M., Heldin, C.-H., & Westermark, B. (1986). Production of a platelet-derived growth factor-like protein and expression of corresponding receptors in a human malignant glioma. Canc. Res. 46, 332–340.

Nonomura, N., Nakamura, N., Uchida, N., Noguchi, S., Sato, B., Sonoda, T., & Matsumoto, K. (1988). Growth-stimulatory effect of androgen-induced autocrine growth factor(s) secreted from Shionogi carcinoma 115 cells on androgen-unresponsive cancer cells in a paracrine mechanism. Canc. Res. 49, 4904–4908.

Nowell, P.C. (1976). The clonal evolution of tumor cell populations. Science 194, 23–28.

Nowotny, A., & Grohsman, J. (1973). Mixed tumor challenge of strain specific and nonspecific TA3 mouse ascites mammary adenoarcinoma. Int. Arch. Allergy 44, 434–440.

Ohmura, E., Okada, M., Onoda, N., Kamiya, H., Tsushima, T., & Shizume, K. (1990). Insulin-like growth factor I and transforming growth factor α as autocrine growth factors in human pancreatic cancer cell growth. Canc. Res. 50, 103–107.

Paget, S. (1889). The distribution of secondary growths in cancer of the breast. Lancet 1, 571–573.

Poste, G., Doll, J., & Fidler, I.J. (1981). Interactions among clonal subpopulations affect stability of the metastatic phenotype in polyclonal populations of B16 melanoma cells. Proc. Natl. Acad. Sci. 78, 6226–6230.

Price, J.E., Bell, C., & Frost, P. (1990). The use of a genotypic marker to demonstrate clonal dominance during the growth and metastasis of a human breast carcinoma in nude mice. Int. J. Canc. 45, 968–971.

Raz, A., & Lotan, R. (1987). Endogenous galactoside binding lectins: a new class of functional tumor cell surface molecules related to metastasis. Canc. Metast. Rev. 6, 433–452.

Ren, J., Hamada, J.-I., Takeichi, N., Fujikawa, S., & Kobayashi, H. (1990). Ultrastructural differences in junctional intercellular communication between highly and weakly metastatic clones derived from rat mammary carcinoma. Canc. Res. 50, 358–362.

Richards, J., Pasco, D., Yang, J., & Nandi, S. (1983). Comparison of the growth of normal and neoplastic mouse mammary cells on plastic, on collagen gels and in collagen gels. Exp. Cell Res. 146, 1–14.

Robinson, S.P., & Jordon. V.C. (1989). The paracrine stimulation of MCF-7 cells by MDA-MB-231 cells: possible role in antiestrogen failure. Eur. J. Canc. Clin. Oncol. 25, 493–497.

Rozengurt, E., Sinnett-Smith, J., & Taylor-Papadimitriou, J. (1985). Production of PDGF-like growth factor by breast cancer cell lines. Int. J. Canc. 36, 247–252.

Ruegg, C.R., Chiquet-Ehrismann, R., & Alkan, S.S. (1989). Tenascin, an extracellular matrix protein, exerts immunomodulatory activities. Proc. Natl. Acad. Sci. 86, 7437–7441.

Salomon, D.S., Liotta, L.A., & Kidwell, W.R. (1981). Differential response to growth factor by rat mammary epithelium plated on different collagen substrata in serum-free medium. Proc. Natl. Acad. Sci. 78, 382–386.

Samiei, M., & Waghorne, C.G. (1991). Clonal selection within metastatic SP1 mouse mammary tumors is independent of metastatic potential. Int. J. Canc. 47, 771–775.

Schindler, M., Trosko, J.E., & Wade, M.H. (1987). Fluorescence photobleaching assay of tumor promoter 12-O-tetradecanoylphorbol 13-acetate inhibition of cell-cell communication. Meth. Enzymol. 141, 439–447.

Sing, G.W., Ruscetti, F.W., Beckwith, M., Keller, J.R., Ellingsworth, L., Urba, W.J., & Longo, D.L. (1990). Growth inhibition of a human lymphoma cell line: induction of a transforming growth factor-β-mediated autocrine negative loop by phorbol myristate acetate. Cell Growth & Diff. 1, 549–577.

Staroselsky, A., Pathak, S., & Fidler, I.J. (1990). Changes in clonal composition during *in vivo* growth of mixed subpopulations derived from the murine K-1735 melanoma. Anticanc. Res. 10, 291–296.

Stein, L.S., Stoica, G., Tilley, R., & Burghardt, R.C. (1991). Rat ovarian granulosa cell culture: A model system for the study of cell-cell communication during multistep transformation. Canc. Res. 51, 696–706.

Stromberg, K., Collins, T.J., IV, Gordon, A.W., Jackson, C.L., & Johnson, G.R. (1992). Transforming growth factor-α acts as an autocrine growth factor in ovarian carcinoma cell lines. Canc. Res. 52, 341–347.

Talmadge, J.E., & Zbar, B. (1987). Clonality of pulmonary metastases from the bladder 6 subline of the B16 melanoma studied by Southern hybridization. J. Natl. Canc. Inst. 78, 315–320.

Tanaka, H., Mori, Y., Ishii, H., & Akedo, H. (1988). Enhancement of metastatic capacity of fibroblast-tumor cell interaction in mice. Canc. Res. 48, 1456–1459.

Theodorescu, D., Cornil, I., Sheehan, C., Man, S., & Kerbel, R.S. (1991). Dominance of metastatically competent cells in primary murine breast neoplasms is necessary for distant metastatic spread. Int. J. Canc. 47, 118–123.

Tofilon, P.J., Buckley, N., & Deen, D.F. (1984). Effect of cell-cell interactions on drug sensitivity and growth of drug-sensitive and -resistant tumor cells in spheroids. Science 226, 862–864.

U, H.S., Kelley, P., Ashbaugh, S., Tatsukawa, K., & Werner, R. (1987). Tumorigenicity in the nude mouse of cocultures derived from two nontumorigenic cell types, human pituitary adenomas and mouse C3H 10 T1/2 fibroblasts. Canc. Res. 47, 5678–5683.

Updike, T.V., & Nicolson, G.L. (1986). Malignant melanoma cell lines selected *in vitro* for increased homotypic adhesion properties have increased experimental metastatic potential. Clin. Exp. Metast. 4, 273–284.

Urban, J.L., Van Waes, C., & Schreiber, H. (1984). Pecking order among tumor-specific antigens. Eur. J. Immunol. 14, 181–187.

Vanhamme, L., Rolin, S., & Szpirer, C. (1989). Inhibition of gap-junctional intercellular communication between epithelial cells transformed by the activated H-ras-1 oncogene. Exp. Cell Res. 180, 297–301.

Varki, N.M., Tseng, A., Vu, T.P., & Estes, L.A. (1990). Cloned low metastatic variants from human lung carcinoma metastases. Anticanc. Res. 10, 637–644.

Waghorne, C., Thomas, M., Lagarde, A., Kerbel, R.S., & Breitman, M.L. (1988). Genetic evidence for progressive selection and overgrowth of primary tumors by metastatic cell subpopulations. Canc. Res. 48, 6109–6114.

Welch, D.R., Fabra, A., & Nakajima, M. (1990). Transforming growth factor β stimulates mammary adenocarcinoma cell invasion and metastatic potential. Proc. Natl. Acad. Sci. 87, 7678–7682.

Woodruff, M.F.A., Ansell, J.D., Forbes, G.M., Gordon, J.C., Burton, D.I., & Micklem, H.S. (1982). Clonal interactions in tumours. Nature, 299, 822–824.

Wrann, M., Fox, C.F., & Ross, R. (1980). Modulation of epidermal growth factor receptors on 3T3 cells by platelet-derived growth factor. Science 210, 1363–1365.

Yamanishi, H., Nonomura, N., Tanaka, A., Nishizawa, Y., Terada, N., Matsumoto, K., & Sato, B. (1991). Proliferation of Shionogi carcinoma 115 cells by glucocorticoid-induced autocrine heparin-binding growth factor(s) in serum-free medium. Canc. Res. 51, 3006–3010.

Yoshikura, H. (1989). Suppression of focus formation by bovine papillomavirus-transformed cells by contact with non-transformed cells: involvement of sugars and phosphorylation. Int. J. Canc. 44, 885–891.

Yotti, L.P., Chang, C.C., & Trosko, J.E. (1979). Elimination of metabolic cooperation in Chinese hamster cells by a tumor promoter. Science 206, 1089–1091.

EFFECTS OF CLASS I MHC GENE PRODUCTS ON THE IMMUNOBIOLOGICAL PROPERTIES OF BL6 MELANOMA CELLS

Misoon Kim and Elieser Gorelik

Advances in Molecular and Cell Biology
Volume 7, pages 177–204

ISBN: 1-55938-624-X

I. CLASS I H-2 ANTIGEN AND IMMUNOGENICITY OF BL6 MELANOMA CELLS

Bl6 melanoma is the most popular and the most investigated experimental tumor model. Various low and high metastatic lines were selected in order to investigate tumor cell properties responsible for metastatic spread and growth (Fidler et al., 1981; Nicolson, 1982). B16 melanoma cell lines are weakly immunogenic and relatively low levels of immune response can be generated by various immunization procedures (Fidler and Kripke, 1980).

Several years ago we attempted to isolate a highly immunogenic variant of B16 melanoma that can be utilized for immunization of tumor-excised mice resulting in eradication of postoperative tumor metastasis. These studies were performed using B16F10BL6 cells that were selected by Hart from the Bl6F10 line by their ability to penetrate the bladder wall *in vitro*. Six sequential selective procedures were performed and the resulting highly invasive variant, was termed bladder six (BL6) (Hart, 1979). To obtain a highly immunogenic variant of BL6 line we used an approach developed by Boon (1983), and found to be highly efficient in augmentation of the immunogenicity of various murine tumor cell lines.

BL6 melanoma cells were treated *in vitro* twice with N-methyl-N-nitro-N′-nitrosoguanidine (MNNG). A new cell variant was isolated and termed BL6T2 (Gorelik et al., 1985). The immunogenic property of BL6T2 cells was tested by transplanting into the immunocompetent syngeneic C57BL/6, or T-cell deficient nude, or immunosuppressed (x-irradiated) C57BL/6 mice. BL6T2 melanoma cells inoculated s.c. (1×10^5 cells/mouse) failed to develop progressively growing tumors in about 70% of C57BL/6 mice, whereas no tumor rejection was found after inoculation of these cells into nude or immunosuppressed (x-irradiated) C57BL/6 mice. Mice which rejected the first tumor inoculation of BL6T2 cells were resistant to high challenge doses ($1 \times 10^6 - 1 \times 10^7$) of BL6T2 cells. However, mice immune to BL6T2 were only partially resistant to the original BL6 melanoma cells. BL6 tumor cells were rejected only in 10% of mice and in the rest growth of BL6 tumor was substantially inhibited but not prevented (Gorelik et al., 1985).

In all previous studies performed in different laboratories the isolated immunogenic tumors treated with MNNG shared antigen specificity with the parental tumors and were able to induce strong crossprotective immunity against challenge injections of the parental tumor (Boon, 1983). The observed relatively low crossprotection of the parental BL6 melanoma in our experiments could be a result of induction by MNNG of new antigens missing in the parental BL6 melanoma cells or due to the lack of class I MHC antigen expression required for recognition by the anti-tumor CTL that developed in mice which rejected BL6T2 cells.

To test this, we analyzed expression of class I MHC antigens by the parental B16 and various isolated sublines (Table 1). We found that the parental B16 and low metastatic B16F1 melanoma cells expressed relatively low levels of H-2K^b and H-2D^b antigens. High metastatic B16F10 and BL6 melanoma lines expressed very

Table 1. H-2 Antigen Expression on the Cell Surface of B16 Melanoma Cell Variants

Tumor	*Characteristics*	*% Positive Cells*	
		$H\text{-}2K^b$	$H\text{-}2D^b$
B16	Initial population	31.6[a]	26.9
B16F1[b]	Low metastatic subline	20.6	24.5
B16F10[b]	Selected on high lung colonization	3.5	9.1
BL6[c]	Selected from B16F10 on high invasiveness	1.2	10.1
BL6TI	One treatment with MNNG	50.2	79.2
BL6T2	Two treatments with MNNG	47.4	76.5
BL6T3	Three treatments with MNNG	46.7	76.9

Notes: [a]Using monoclonal antibodies and flow cytometry analysis, the expression of $H\text{-}2K^b$ and $H\text{-}2D^b$ antigens on the cell surface of B16 melanoma cell variants was investigated.

[b]Melanoma lines selected by Dr. J. Fidler.

[c]Melanoma line selected by Dr. I. Hart.

low, if any, serologically detectable class I H-2 antigens. After single or multiple treatment of BL6 cells with MNNG, a substantial and stable increase in expression of both $H\text{-}2K^b$ and $H\text{-}2D^b$ antigens was observed (Table 1). Thus, increase in tumor cell immunogenicity after MNNG treatment of BL6 melanoma cells is a result of up-regulation of class I MHC antigen expression. Low crossprotection of BL6 melanoma in mice-rejected BL6T2 cells is due to their low expression of class I MHC molecules required for T-cell recognition. Analysis of H-2 antigen expression by the individual clones isolated from the BL6T2 melanoma line revealed that all immunogenic (*tum*$^-$) clones had high levels of class I H-2 antigen expression. Clones with low expression of H-2 molecules were nonimmunogenic. High expression of class I H-2 antigens cannot be a marker for melanoma cell immunogenicity. Indeed, some BL6T2 clones with high H-2 antigen expression were weakly immunogenic, suggesting that expression of class I MHC antigen is essential but not sufficient for tumor cell rejection.

Although H-2 antigen expression in some clones was not sufficient for their rejection in normal mice, it makes them more recognizable and rejectable in the preimmunized mice. BL6T2 and its clone 39 with high levels of H-2 antigen expression were highly immunogenic and rejectable in normal and immune mice (Table 2). Clone 3 expressed high levels of H-2 antigens but was rejected only in 12% of normal mice, but its growth was prevented in 100% of immune mice that rejected clone 39 of BL6T2 melanoma. In comparison, clone 47 with low levels of $H\text{-}2K^b$ antigen expression was rejected in only 40% of immune mice. B16Fl melanoma cells usually showed higher levels of H-2 antigen expression than BL6

Table 2. Effect of H-2 Antigen Expression on Immunogenicity and Crossreactivity of B16 Melanoma Cells

	% Positive Cells		*% of Tumor Rejection in Mice*	
Tumor Cells	*H-2K^b^*	*H-2D^b^*	*Normal*	*Immune*
BL6T2	41.2	84.6	71	100
BL6T2 clone 39	50.6	82.6	94	100
BL6T2 clone 3	56.5	86.6	12	100
BL6T2 clone 47	9.2	48.0	0	40
B16F1	17.3	23.5	0	50
BL6	2.1	6.7	0	10

Notes: [a]C57BL/6 mice were inoculated i.f.p. with 3×10^5 cells of clone 39. Mice that were resistant to challenging injections of 5×10^5 and 1×10^6 cells of clone 39 were considered as immune mice. The ability of normal and immune C57BL/6 mice to reject 5×10^4 cells of B16 melanoma lines was determined.

[b]Percent of cells expressing H-$2K^b$ and H-$2D^b$ antigens was analyzed using flow cytometric analysis.

cells and in parallel they had higher percentage of rejectability in immune mice (Table 2).

To analyze further the role of class I H-2 molecules in regulation of the immunogenicity of the BL6 melanoma, the experiments with gene transfection were performed. Clone BL6-8 was isolated from BL6 line and was found to be H-$2K^{b-}$, H-$2D^{b+}$. Northern blot analysis show no message for H-2K gene in these cells (Tanaka et al., 1988). This clone was transfected with plasmid containing the genomic fragment of H-$2K^b$ gene and cotransfected with *neo^r^* gene. In parallel the BL6-8 clone was transfected with class II H-$2IA^k$ gene and *neo^r^* gene or *neo^r^* gene alone (Tanaka et al., 1988). Tumor cells expressing high levels of the transfected class I H-$2K^b$, but not class II H-$2KA^k$ genes were rejected in the syngeneic mice (Tanaka et al., 1988).

To test what types of effector cells are involved in rejection of H-$2K^{b+}$ melanoma cells, tumor cells were inoculated into C57BL/6 mice in which subpopulations of $CD4^+$, $CD8^+$ or NK cells were depleted. C57BL/6 mice received i.p. 0.2 ml of ascites containing anti-L3T4 or anti-Lyt2.2 mAb. To deplete NK cells, mice were treated with anti-asialo GMl antibodies. Depletion of $CD4^+$ or NK cells did not affect the ability of mice to reject H-$2K^b$ transfected CL8-1 melanoma cells or H-$2K^{b+}$ MNNG-treated T2-39 clone. Only depletion of CD8 T-lymphocytes resulted in progressively growing CL8-1 or T2-39 tumors. These data indicate that H-$2K^b$ molecules play an important role in recognition and elimination of BL6 melanoma cells by T cell-mediated immunity. $CD8^+$ T lymphocytes are paramount for rejection of H-$2K^{b+}$ BL6 melanoma cells.

II. CLASS I H-2 MOLECULES AND NK SENSITIVITY OF BL6 MELANOMA CELLS

Expression of class I MHC is important for recognition of "self" and "nonself" in the body and elimination of "nonself" by T cell-mediated immunity. Recently, it was proposed that tumor cells that lack class I MHC ("no self") can be more efficiently killed by NK cells than tumor cells with high expression of class I MHC molecules (Karre et al., 1986; Ljunggren and Karre, 1990). It was postulated that MHC gene products play a role in regulation of tumor cell sensitivity to NK cell-mediated lysis and actually down- regulate the NK sensitivity of tumor cells (Karre et al., 1986; Ljunggren and Karre, 1990). Indeed, numerous tumor lines, especially of lymphoid origin show negative association between MHC antigen expression and NK sensitivity (Harel-Bellan et al., 1986; Karre et al., 1986; Storkus et al., 1987; Ljunggren and Karre, 1990). However, in some studies with nonlymphoid tumor cell lines, a positive correlation between MHC antigen expression and NK sensitivity was found (Sewada et al., 1985; Kenyon and Raska, 1986). E1A gene products of adenovirus type 12 inhibited the expression of class I H-2 antigen and increased the resistance to NK cell-mediated lysis of the transformed fibroblasts, whereas cells transformed with the E1A gene of adenovirus type 5 had high levels of H-2 antigen expression and increased sensitivity to NK cell cytotoxicity (Sewada et al., 1985; Kenyon and Raska, 1986). Similarly, we found that treatment of BL6 melanoma cells with MNNG led to increased expression of class I H-2 antigens and sensitivity to NK cells (Gorelik et al., 1988).

Previous attempts to assess the effect of transfected MHC genes on NK sensitivity of tumor cells have also produced divergent results. Numerous experiments with class I MHC gene transfection into human and murine tumor cells showed reduction in tumor cell sensitivity to NK cell-mediated lysis (see for review Ljunggren and Karre, 1990). However, in many instances, transfection of various human or murine MHC tumor cells with a class I MHC gene did not significantly affect their sensitivity to lysis by NK cells (Nishmura et al., 1988; Leiden et al., 1989). This discrepancy in the effect of MHC transfected genes on the susceptibility of the various tumor cell lines to NK lysis could be due to multiple reasons. Numerous factors and mechanisms are probably involved in regulation of tumor cell sensitivity to NK cell cytotoxicity (Herberman et al., 1986). NK resistance of different tumor cell lines is probably based on varying mechanisms and only some of them could be affected by MHC gene products. Furthermore, different class I MHC genes might have different effects on the NK sensitivity of tumor cells (Storkus et al., 1989). The level of MHC antigen expression might also be an important factor in determining the level of protection of tumor cells from NK cell destruction (Storkus et al., 1989).

In addition, divergence in the results could also be based on differences in the assays used for the evaluation of NK sensitivity of tumor cells. Short (4 h) and long(10 to 24 h) cytotoxicity assays have been used for testing the lysability of the lymphoid and solid tumor cells, respectively. It might be that different effector cells are involved in cytotoxicity in the different assays against different target cells.

Two types of lymphoid effector cells, NK and NC cells, are known to be able to lyse tumor cells in an MHC-nonrestricted manner without any activation or presensitization (Lattime et al., 1981, 1982; Herberman et al., 1986). NK and NC cells differ phenotypically and by their mechanisms of tumor cell lysis (Lattime et al., 1981, 1982; Herberman et al. 1986). NK cells are a morphologically distinguishable population of large granular lymphocytes (LGLs) that lyse tumor cells by the exocytosis of cytolytic granules (Herberman et al., 1986). NK cells are phenotypically characterized a, $CD3^-$, $CD16^+$ and $CD56^+$. In mice they express determinants recognizable by anti-asialo GMl or NK1.1, NK1.2 antibodies (Herberman et al., 1986). NC activity was found to be associated with various cell types such as T, B, and NK cells (Lattime et al., 1981, 1982). Whereas NK cells may lyse highly susceptible target cells within 4 h, NC cells only kill susceptible targets after a relatively long time of exposure (12 to 24 h) and their activity can be blocked by anti-TNF antibodies (Ortaldo et al., 1986).

In our studies with MNNG-treated BL6T2 we found that these cells were lysed by normal spleen cells only after prolonged (18 h) incubation (Gorelik et al., 1988). H-$2K^{b+}$ BL6T2 cells were sensitive and H-$2K^{b-}$ BL6 or BL6T2 cells were resistant to natural cell-mediated cytotoxicity (NCMC) (Gorelik et al., 1988). To obtain more direct evidence about the involvement of class I H-2 antigen in regulation of sensitivity of BL6 melanoma cells to NCMC, we used the H-$2K^{b-}$ clone BL6-8 that was transfected with class I H-$2K^b$ or class II H-$2IA^k$ gene (Tanaka et al., 1988).

The original clone BL6-8 as well as clones transfected with *neo*^{r} gene were resistant to lysis by spleen cells. Class II IA^k gene-transfected BL6-12 and BL6-22 clones did not show increased sensitivity to lysis. In contrast, transfection with class I H-$2K^b$ gene substantially increased tumor cell sensitivity to NCMC. The higher sensitivity of class I transfected BL6 melanoma cells also was observed when spleen cells of allogeneic BALB/c or C3H/HeN mice were used as effector cells (Gorelik et al., 1990).

Thus, these data indicate that natural effector cells do not recognize the polymorphic part of the H-$2K^b$ molecules expressed by the transfected BL6 cells, but rather kill the target cells in an MHC-nonrestricted manner. However, the expression of H-$2K^b$ gene products increased tumor cell sensitivity to NCMC.

With stimulation of NK cell activity with poly I:C, lysis of H-$2K^{b+}$, but not H-$2K^{b-}$, BL6 melanoma clones were increased. Elimination of NK cell activity by pretreatment of mice with anti-asialo GM1 serum or NK1.1, mAbs resulted in significant but partial reduction in spleen cell cytotoxicity against H-$2K^{b+}$ BL6 melanoma cells, suggesting that NK cells are involved in lysis of these cells.

Inasmuch as lysis of BL6 melanoma cell variants has been manifested mainly after 18 h of incubation with effector cells, it was unclear whether this lysis was mediated by NK and/or NC cells. To assess the relative importance of NK and NC cells in lysis of H-$2K^{b+}$ and H-$2K^{b-}$ BL6 melanoma cells, the following experi-

Table 3. Sensitivity of H-2K^b Transfected CL8-1 Melanoma Cells to NK and NC Cell-Mediated Cytotoxicity

	% Cytotoxicity		
Effector Cells	*CL8-1*	*YAC-1*	*WEHI-164*
Normal spleen ($NK^+ NC^+$)	37.9	60.7	49.8
Normal spleen+ anti-TNF Abs ($NK^+ NC^-$)	20.4	57.1	6.1
NK-depleted spleen ($NK^- NC^+$)	14.3	12.9	47.7
NK-depleted spleen+anti-TNF Abs ($NK^- NC^-$)	5.1	10.3	5.2

Note: [a]C57BL/6 mice were treated with anti-asialo GM1 serum (0.2 ml, dilution 1:20) 3 and 1 days before assay. Cytotoxic activity of the spleen cells of normal or NK cell-depleted mice was tested against ^{51}Cr-labeled target cells in an 18h cytotoxicity assay in the presence or absence of anti-TNF antibodies (2000nu/well). Effector:target ratio is 100:1.

mental approach was used (Gorelik et al., 1990). The cytotoxic activity of normal spleen cells (NK^+, NC^+) was compared with other groups: normal spleen cells in the presence of anti-TNF antibodies (NK^+, NC^-), NK-depleted spleen cells (NK^-, NC^+) and NK-depleted spleen cells with NC activity blocked by anti-TNF antibodies (NK^-, NC^-).

The results presented in Table 3 show that in the presence of anti-TNF antibodies the cytotoxic activity of normal spleen cells against CL8-1 H-2K^b transfected melanoma cells was significantly reduced suggesting the involvement of NC cells in lysis of these cells. NK depletion also reduced spleen cell cytotoxicity against CL8-1 melanoma cells. However, these reductions in spleen cell cytotoxicity were always partial. More complete inhibition of spleen cell cytotoxicity was observed when NK cell activity of spleen cells was depleted and NC activity was blocked in the presence of anti-TNF antibodies. Similar data were obtained when spleen cell cytotoxicity of C57BL/6 and NK deficient beige mice was tested in the presence and absence of anti-TNF antibodies (Gorelik et al., 1990). These data demonstrate that H-2K^b expression resulted in increase of BL6 melanoma cells sensitivity to both NK and NC cell-mediated cytotoxicity.

To test whether other allelic H-2K genes are capable of affecting melanoma cell sensitivity to NK/NC cell cytotoxicity we transfected BL6 melanoma cells with the allelic H-2K^d gene. The selected clones in addition to expression of H-2K^d gene showed high expression of the endogenous H-2K^b gene (Gorelik et al., 1991). All these clones were found highly sensitive to normal spleen cell cytotoxicity which was mediated by both NK and NC cells.

We analyzed H-2K^b expression among individual clones isolated from the parental BL6 melanoma. Fifteen examined clones were H-2K^{b-} and one clone

(BL6-29) was found to be expressing high level of the endogenous H-2K^b gene. All tested H-2K^{b-} clones were completely resistant and the H-2K^{b+} clone was sensitive to normal spleen cytotoxicity. Similar data were obtained for clones isolated from MNNG-treated BL6T2 melanoma expressing or nonexpressing the endogenous H-2K^b gene (Gorelik et al., 1988). These data indicate that increase in sensitivity of BL6 melanoma cells to NCMC is due to H-2K^b gene expression rather than to gene insertion or some nonspecific influences associated with gene transfection techniques.

III. EFFECT OF H-2K^b GENE EXPRESSION ON BL6 MELANOMA CELLS SENSITIVITY TO TNF LYSIS

NC cells lyse sensitive target cells via the production of TNF since anti-TNF antibodies can block NC cell-mediated lysis (Ortaldo et al., 1986). Inasmuch as NC cell cytotoxicity is mediated by TNF produced by the effector cells, it might be expected that an increase in the sensitivity of H-2K^{b+} BL6 melanoma cells to NC cell-mediated lysis would be due to their increase in sensitivity to lysis by TNF. To test this possibility, we compared the sensitivity of H-2K^{b+} and H-2K^{b-} BL6 melanoma clones to the cytotoxic action of human rTNF. The data obtained demonstrate that H-2K^{b-} clones, including those that were transfected with H-2IA^k gene, were resistant to lysis by rTNF. In contrast, clones transfected with H-2K^{b+} gene manifested apparent sensitivity to TNF-mediated lysis (Gorelik et al., 1990).

It was demonstrated that tumor cells lysis by TNF could be augmented by inhibitors of protein or RNA synthesis such as cycloheximide and actinomycin D respectively (Vilcek and Lee, 1991). We tested whether cycloheximide (CHI) could further increase TNF sensitivity of H-2K^{b+} BL6 melanoma cells and whether TNF resistance of H-2K^{b-} melanoma cells might be overcome by CHI treatment (Table 4). The data obtained indicate that CHI could substantially increase sensitivity of H-2K^{b+} CL8-1 melanoma cells to TNF lysis. In the presence of TNF (1–1000 u/well) and CHI (1 ug/ml) the cytotoxicity increased from 18–27% to 58–82%. Thus, TNF sensitivity of these cells was similar to WEHI-164 cells. TNF lysis of BL6-8 and Class II H-2IA^k transfected clone BL6-12 was significantly less and did not increase in the presence of CHI. Similar data were obtained when cytotoxic activity of TNF in the presence of CHI or actinomycin D was tested against other H-2K^{b+} and H-2K^{b-} B16 melanoma clones.

TNF sensitivity of BL6-8 cells transfected with the allelic H-2K^d gene was also tested in the presence or absence of CHI (1 ug/ml). Four clones isolated after transfection of BL6-8 with *neo*^{r} gene showed low sensitivity to TNF similar to the parental BL6-8 clone. In the presence of CHI their lysability by TNF did not increase (cytotoxicity ranged between 5–10%). In contrast, clones transfected with H-2K^d that expressed transfected H-2K^d and endogenous H-2K^b genes became sensitive to TNF and their sensitivity substantially increased in the presence of low

Table 4. Effect of Cycloheximide on TNF Sensitivity of BL6-8 Melanoma Cells Transfected with H-2K^b or H-2IA^k Genes

	% Cytotoxicity							
	TNF, u/ml				TNF+Cycloheximide			
Tumor Cells	*1000*	*100*	*10*	*1*	*1000*	*100*	*10*	*1*
BL6-8	8	7	10	5	7	9	10	10
CL8-1	27	25	20	18	82	80	75	58
BL6-12	13	12	6	9	18	12	13	5
WEHI-164	40	26	20	15	80	83	70	42

Note: [a]Clones CL8-1 and BL6-12 were isolated after transfection of BL6-8 clone with Class I H-2K^b or Class II H-2IA^k gene, respectively. ^{51}Cr-labeled melanoma cells were incubated with human rTNF-α (1–1000u/ml) and cycloheximide (1 ug/ml). Percent specific tumor cell lysis was determined after 18 h of incubation. For comparison, lysis of the TNF-sensitive WEHI-164 was also determined.

concentrations of CHI (1 ug/ml). Lysis of eight H-2K^d transfected clones by TNF (1000 u/well) in the presence of CHI ranged from 60 to 90%. We also compared the ability of TNF to lyse H-2K^{b-} and H-2K^{b+} clones isolated from the parental BL6 melanoma. All 4 tested H-2K^{b-} clones were resistant to TNF and CHI failed to increase their sensitivity to TNF lysis (cytotoxicity was below 10%). The H-2K^{b+} BL6-29 clone was highly sensitive to TNF and CHI and cytotoxicity was higher than 70%. Similarly H-2K^{b+} clones isolated from MNNG-treated BL6T2 cells were highly sensitive whereas H-2K^{b-} clone from the same line was resistant to TNF lysis.

The mechanisms responsible for increased TNF sensitivity in clones expressing H-2K^b gene remain unknown. Our preliminary data indicate that BL6 melanoma cells that express the transfected or endogenous H-2K^b gene had twice higher numbers of TNF receptor 1 and twice more of ^{125}I-TNF was found to be internalized and metabolized by H-2K^{b+}, than H-2K^{b-} melanoma clones. TNF sensitivity of H-2K^{b+} clones was also associated with the ability of TNF or TNF plus CHI to activate phospholipase A_2, whereas in H-2K^{b-} clones TNF failed to activate phospholipase A_2. It is possible that TNF resistance of H-2K^{b-} BL6 melanoma cells is due to the existence of the block in the transduction of the TNF-induced second signal.

It is unclear how the expression of H-2K^b molecules can affect the sensitivity of BL6 melanoma to TNF cytotoxicity. The TNF-alpha encoding gene is in the MHC loci 70 kb proximal to the H-2D gene (Muller et al., 1987). It was demonstrated that TNF is capable of stimulating MHC gene expression (Collins et al., 1986). However, it seems unlikely that TNF can directly recognize Class I MHC gene products. It is more plausible that the effect of the H-2K^b gene on TNF sensitivity of BL6 melanoma cells is indirect. We have hypothesized that H-2K^b gene expression could alter some tumor cell properties that in turn make tumor cells more sensitive to NCMC and TNF lysis (Gorelik et al., 1991).

IV. NONIMMUNOLOGICAL EFFECTS OF H-2K^b GENE IN BL6 MELANOMA CELLS

There are numerous experimental evidences indicating that MHC molecules might have wider biological significance beyond their interaction with the elements of the immune system. For example, MHC molecules have been demonstrated to play an important role in control of expression of receptors for insulin (Chvatchko et al., 1983; Due et al., 1986; Kittur et al., 1987), glucagon (Lafuse and Edidin, 1980), luteinizing hormone, beta-adrenergic receptor (Solano et al., 1988) and epidermal growth factor (EGF) receptor (Schreiber et al., 1984).

Using various congenic strains of mice that differ only in H-2 genes, it was found that the MHC gene products could influence some hormone-related phenomena such as resistance to glucocorticoid-induced cleft palate in embryos (Bonner and Tyan, 1983), levels of glucocorticoid receptor in the lungs (Goldman and Katzumata, 1986) and estrogen receptor in uterus (Palumbo and Vladutiu, 1979). Overexpression of class I MHC molecules in the pancreatic beta cells of transgenic mice could be responsible for the disfunction of the pancreatic cells and development of insulin-dependent diabetes mellitus (Allison et al., 1988). MHC molecules are involved in the regulation of liver adenylate cyclase activity and cell-to-cell contacts (Lafuse and Edidin, 1980). Furthermore, MHC molecules could interact with endogenous and exogenous substances (Class and Van Rood, 1985) and viruses (Helenius et al., 1978) and are also implicated in chemosensory recognition (Yamazaki et al., 1982).

Transgenic mice with disrupted beta$_2$-microglobulin gene did not express class I H-2K^b and had very low levels of H-2D^b antigens and were found to lack mature CD8 T lymphocytes. No obvious malformations, morphological or biochemical defects were found (Koller et al., 1990). However, recently it was described that mice deficient in class I MHC antigens developed insulin-dependent diabetes mellitus (Faustman et al., 1991). Further characterization of these mice is required. Lack of obvious alterations in these transgenic mice cannot exclude the possible nonimmunological effects of class I MHC especially taking into consideration that cell surface expression of class I MHC was achieved by disrupting of ß$_2$ microglobulin gene that did not effect expression of class I MHC genes and intracellular generation of their products. Furthermore, it is quite possible that nonimmunological effects of MHC genes might be different in normal and malignant cells. Indeed, in tumor cells with numerous chromosomal alterations the expression of MHC gene could affect expression and function of some non-MHC genes and thus affect biological properties of tumor cells.

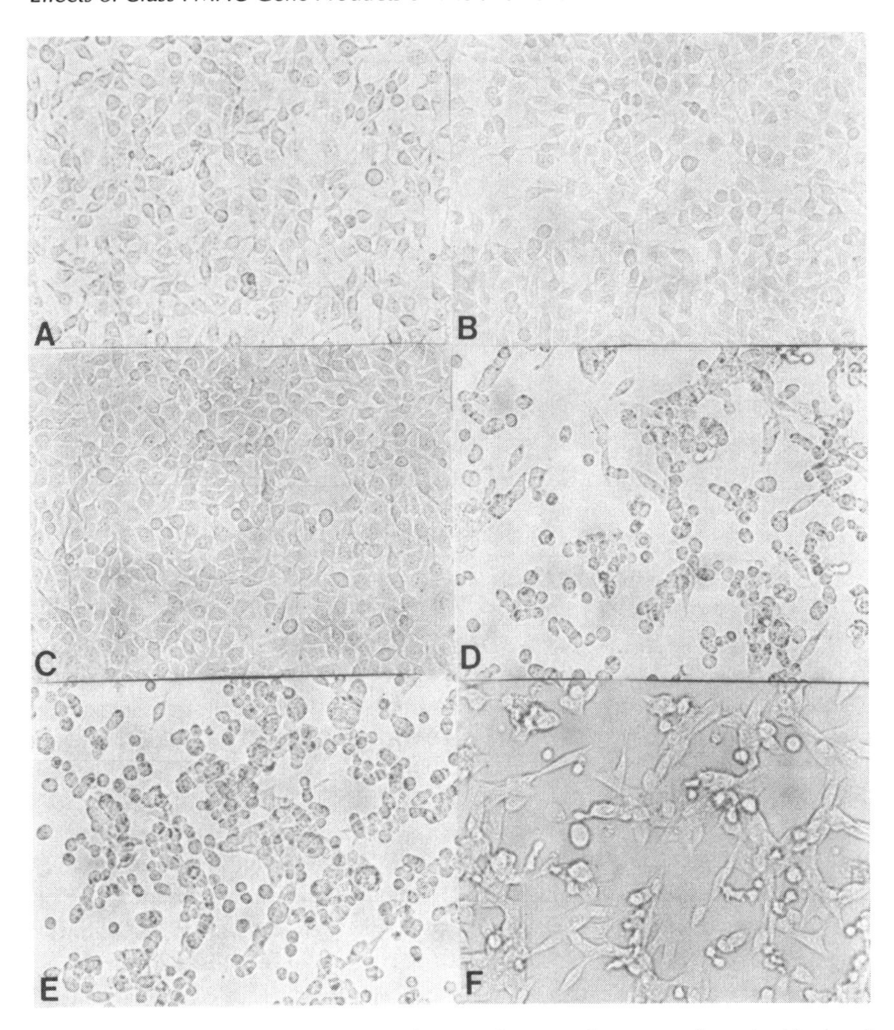

Figure 1. Morphologic appearance of BL6 melanoma clones transfected with class I and class II H-2 genes. A: BL6-8 parental clone, B: BL6-9 clone transfected with *neo*^{r} gene, C: BL6-12 transfected with H-2IA^k gene, D and E: CL8-1 and CL8-2 clones transfected with H-2K^b gene, F: T2-39 clone selected from MNNG-treated BL6T2 line.

Our data show that expression of the transfected or endogenous class I H-2K^b gene was associated with numerous alterations in cell characteristics of BL6 melanoma (Gorelik et al., 1991). All clones of BL6 melanoma expressing H-2K^b gene lost the ability to produce melanin both *in vitro* and *in vivo*, whereas transfected class II H- 21A^k or *neo*^{r} gene did not affect melanogenesis in these cells. Transfection of BL6 melanoma cells with *neo*^{r} or H-21A^k genes did not change

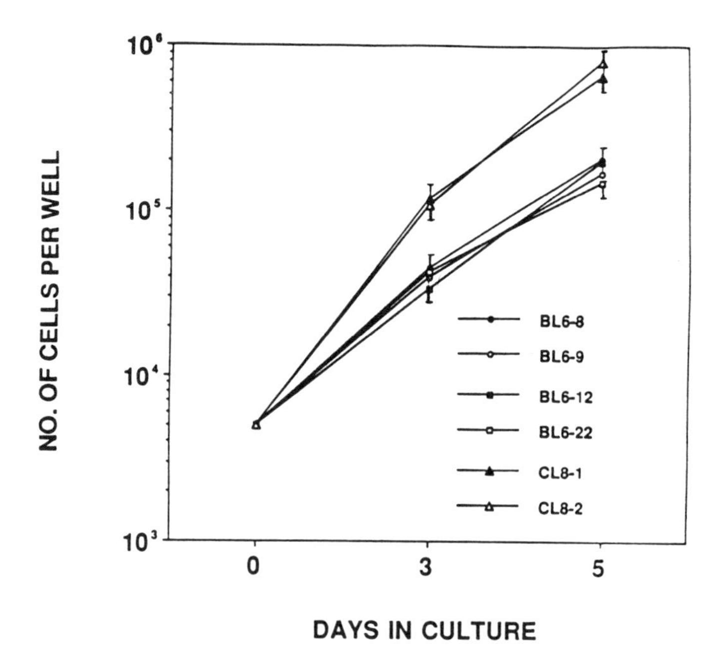

Figure 2. Kinetics of cell proliferation of the BL6 melanoma cells transfected with H-2 genes. Tumor cells were seeded into 24 well plate (0.5×10^4 per well). At different time intervals, cells were harvested and numbers of cells per well were determined.

their morphological appearance. All these *in vitro* cultured cells are strongly adherent fibroblast-like cells. In contrast, obvious changes were observed in BL6 cells transfected with H-2K^b gene and expressing high levels of H-2K^b antigens (Figure 1). The majority of cultured CL8-1 and CL8-2 cells are round and nonadherent with some aggregated in clumps, whereas other cells are adherent but with fewer protruding processes than are seen with the other clones. (Figure 1).

In addition, H-2K^b transfected CL8-1 and CL8-2 cells showed 3–4 times higher rates of proliferation in comparison to the parental BL6-8 and other H-2K^b low clones transfected with *neo*^{r} or class II H-2IA^k genes (Figure 2).

Monoclonal antibodies MM2-9B6 and MM2-3C6 were produced by immunization of C57BL/6 mice with B16F10 melanoma cells and were found to react with various low and high metastatic sublines of B16 melanoma (Leong et al., 1988). These antibodies also reacted with 2 other melanomas (JB/RH and JB/MS) that originated in C57BL/6 mice, but not with melanomas of other strains of mice or with a variety of different histological types of normal or malignant cells of C57BL/6 mice (Leong et al.,1988). Thus, the antigen recognizable by these antibodies appear to be specific for melanomas of C57BL/6 mice.

It might be that this melanoma associated antigen (MAA) is the target of the host immune response. It was of interest to test whether the expression of MAA increased in the highly immunogenic clones. High levels of expression of the MAA recognized by MM2-9B6 monoclonal antibodies were found on the cell surface of BL6-8, *neo*^{r} transfected BL6-9 and H-2 IA^k transfected BL6-12 and BL6-22 clones. In contrast, H-2K^b transfected clones CL8-1 and CL8-2 had virtually no detectable reactivity with these monoclonal antibodies (Figure 3).

Thus, high levels of expression of class I but not class II H-2 genes was associated with the disappearance of MAA on murine BL6 melanoma cells. Therefore, this antigen can not be responsible for increased immunogenicity of H-2K^b transfected BL6 melanoma cells.

Previously we found that MNNG-treated H-2K^{b+} BL6T2 melanoma cells differ from the original B16, B16F1, B16F10 and BL6 melanoma cell lines by their ability to bind SBA lectin. Therefore, it was of interest to test whether appearance of SBA lectin binding sites is a result of MNNG treatment or whether it is associated with appearance of H-2K^b molecules. Thus, we investigated whether transfection of class I or class II H-2 genes might result in alteration of the cell surface carbohydrates.

Using a panel of 15 biotinylated lectins and avidin-phycoerythrin complex, the percent of positive cells and the intensity of the expression of the lectin-binding sites on the cell surface of the tumor cell lines were determined by flow cytometric analysis. Differences among the lines were demonstrated for binding of several lectins. Clones CL8-1 and CL8-2 transfected with H-2K^b gene as well as clone T2-39 from the BL6T2 melanoma line reacted strongly with SBA, GSIB$_4$ and PNA lectins, whereas class II IA^k and *neo*^{r} gene transfected clones showed no reactivity with these lectins. (Figure 4). Although the H-2K^{b-} BL6-8, BL6-9, BL6-12 and BL6-22 clones did not bind SBA and PNA, *in vitro* treatment of these clones with neuraminidase uncovered binding sites specific for PNA and SBA. These data indicate that the failure of class I H-2K^{b-} clones to bind SBA and PNA appears to be due to sialylation of the carbohydrates specific for the binding of SBA and PNA. Neuraminidase treatment did not increase binding of GS1B$_4$ lectin by H-2K^{b-} clones, suggesting that GS1B$_4$ binding sites on the surface of H-2K^{b+} clones was not due to their unmasking but rather induction *de novo* of GS1B$_4$ binding cell surface carbohydrates.

GS1B$_4$ lectin is specifically bound to α-galactosyl epitope (Goldstein et al., 1981). These epitopes were found on a wide variety of murine normal and malignant cells and also in extracellular matrix as a major carbohydrate component of laminin (Shibata et al., 1982). α-Galactosyl epitope synthesis is mediated by α1,3-galactosyltransferase in the Golgi apparatus (Roth, 1987). Analysis of the expression of galactosyltransferase gene revealed that the H-2K^{b-} BL6-8 cells do not express this gene whereas in H-2K^b transfected CL8-1 melanoma cells high expression of this gene was found (Uri Galili, personal communication). It was suggested that expression of α1,3-galactosyltransferase might compete with α1.2-

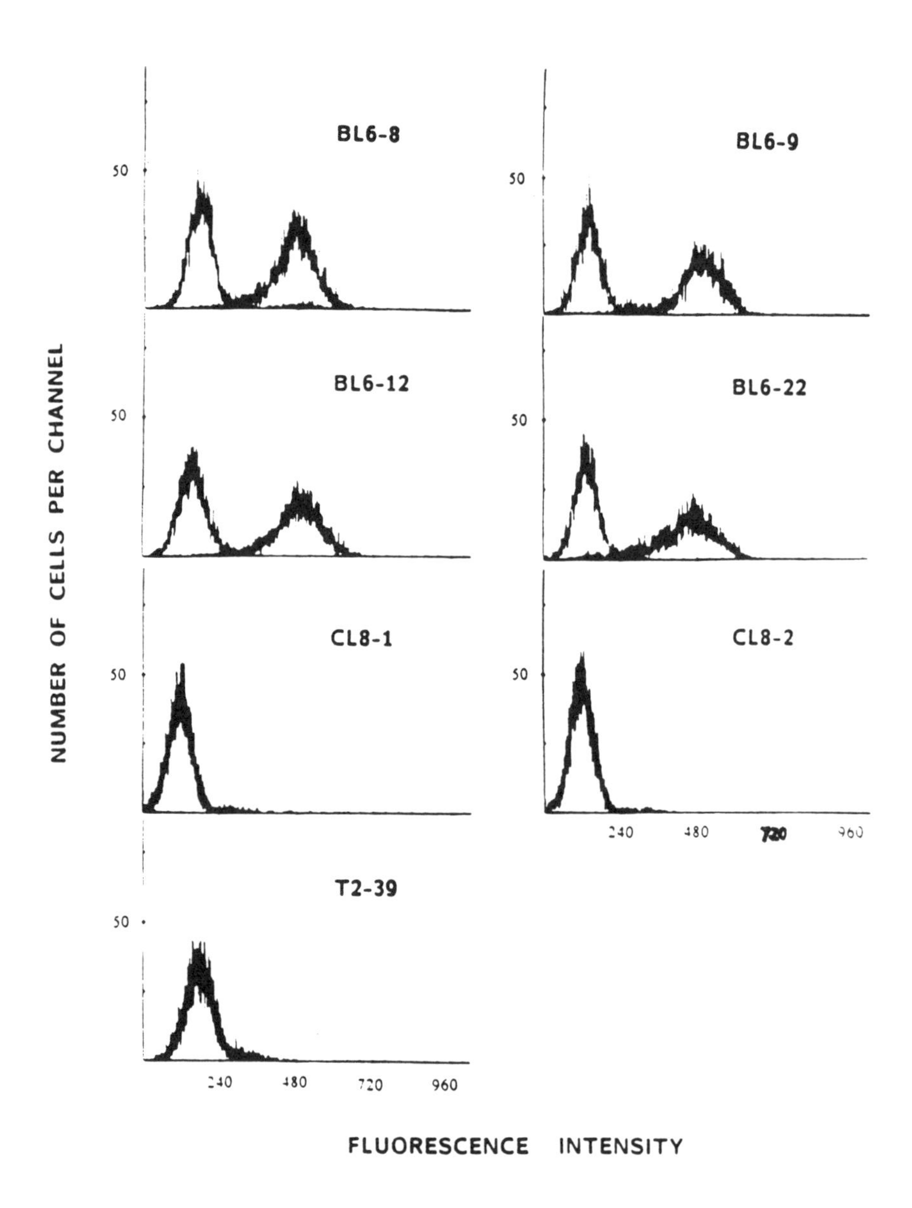

Figure 3. Expression of melanoma associated antigen (MAA) by BL6 melanoma clones transfected with class I H-2K^b or class II H-2IA^k genes. Expression of melanoma associate antigen was analyzed using MM2-9B6 (IgG) monoclonal antibodies and flow cytometric analysis.

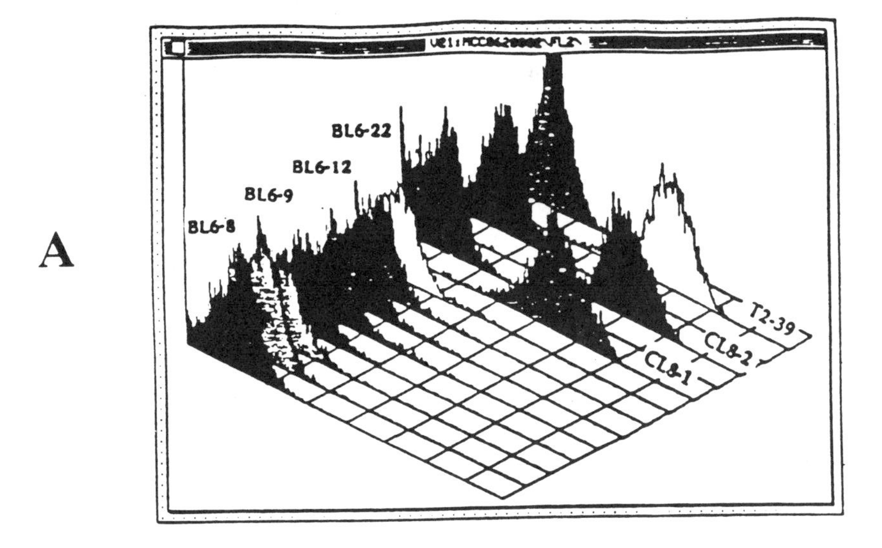

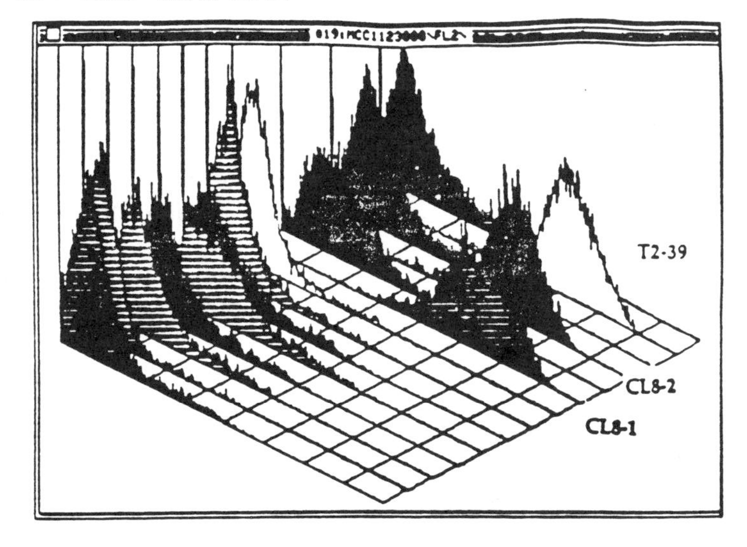

Figure 4. Expression of GS1B_4 (A) and SBA (B) lectin binding carbohydrates by BL6 melanoma cells transfected with class I or class II H-2 genes. BL6-8 clone transfected with *neo*^{r} gene (clone BL6-9), H-2IAk gene (BL6-12 and BL6-22) or H-2K^b gene (CL8-1 and CL8-2) or MNNG-treated H-2K^{b+} clone T2-39 were incubated with 10 ug/ml of biotinylated GSTB_4 (A) or SBA (B) lectin alone or in the presence of 0.2M the specific inhibitory sugar (a-D-Gal for GS1B_4 and GalNAc for SBA lectin). The log red signal of the cells was analyzed using a FACstar flow cytometer.

sialyltransferase for capping of the common substrate N-acetyllactosamine receptor. As a result it will be galactosylated rather than sialylated (Smith et al., 1990). It might explain our observation that appearance of α-galactosyl epitopes binding $GS1B_4$ lectin was parallel with reduction in sialylation and unmasking of SBA and PNA binding carbohydrates.

Thus, expression of the transfected $H\text{-}2K^b$ gene resulted in various phenotypic changes of BL6 melanoma cells. In order to test whether the changes in the BL6 melanoma phenotype were uniquely associated with the $H\text{-}2K^b$ gene or with allelic H-2K genes as well, we transfected the BL6-8 clone with the $H\text{-}2K^d$ gene. In parallel BL6- 8 cells were transfected with *neo*^{r} gene alone using the Lipofectin reagent (Gorelik et al., 1991). Ten *neo*^{r} gene transfected clones were investigated and all were $H\text{-}2K^{b-}$ and expressed MAA but they did not bind $GS1B_4$, SBA or PNA lectins. As was mentioned above these characteristics are common for all $H\text{-}2K^{b-}$ BL6 melanoma clones. Twenty five clones transfected with the $H\text{-}2K^d$ gene were isolated. These clones expressed relatively low levels of the transfected $H\text{-}2K^d$ gene with 5–53% positive cells and high levels of the endogenous $H\text{-}2K^b$ genes (75–95% of these cells were $H\text{-}2K^{b+}$). Thus, transfection with $H\text{-}2K^d$ gene resulted in the induction of expression of the endogenous $H\text{-}2K^b$ gene. $H\text{-}2K^d$ transfected clones lost MAA expression but became highly reactive with $GS1B_4$, SBA and PNA lectins. These phenotypic changes are probably associated with the expression of the endogenous $H\text{-}2K^b$ gene rather than the transfected, but lowly expressed, $H\text{-}2K^d$ gene. Indeed, clone 100 expressed a relatively high level of the transfected $H\text{-}2K^d$ gene and manifested the same phenotypic changes as $H\text{-}2K^d$ low clone 118, that is, low MAA expression and high expression of the lectin binding sites paralleling the high expression of endogenous $H\text{-}2K^b$ antigen (Figure 5). All these phenotypic changes were identical to those described for the clones expressing high levels of the transfected $H\text{-}2K^b$ gene.

For further confirmation of the association between $H\text{-}2K^b$ gene expression and alteration of the biological properties of BL6 melanoma cells, we studied an additional 13 clones isolated from the MNNG-treated BL6T2 line. Previously we demonstrated that the BL6T2 line contains preferentially $H\text{-}2K^{b+}$ clones with low proportion of $H\text{-}2K^{b-}$ clones (Gorelik et al., 1985). By testing 13 individual clones from the BL6T2 line we found that only one clone (T2-18) did not express $H\text{-}2K^b$ antigen. This clone displayed high levels of MAA expression identified by MM2-9B6 mAbs but did not bind $GS1b_4$, SBA and PNA lectins. In contrast all 12 clones that were $H\text{-}2K^{b+}$ lost the MAA, but gained the expression of the lectin binding sites. These findings paralleled those described for the H-2K transfected clones.

Tumor cells are a heterogeneous population consisting of individual clones with various phenotypic characteristics. It was of interest to determine whether individual clones isolated from the original BL6 melanoma differ in $H\text{-}2K^b$ antigen expression and other studied cell properties. We investigated 16 individual clones from the BL6 melanoma of which 15 were found to be $H\text{-}2K^{b-}$. These 15 clones expressed MAA, but did not bind the SBA, PNA and $GS1B_4$ lectins. One clone

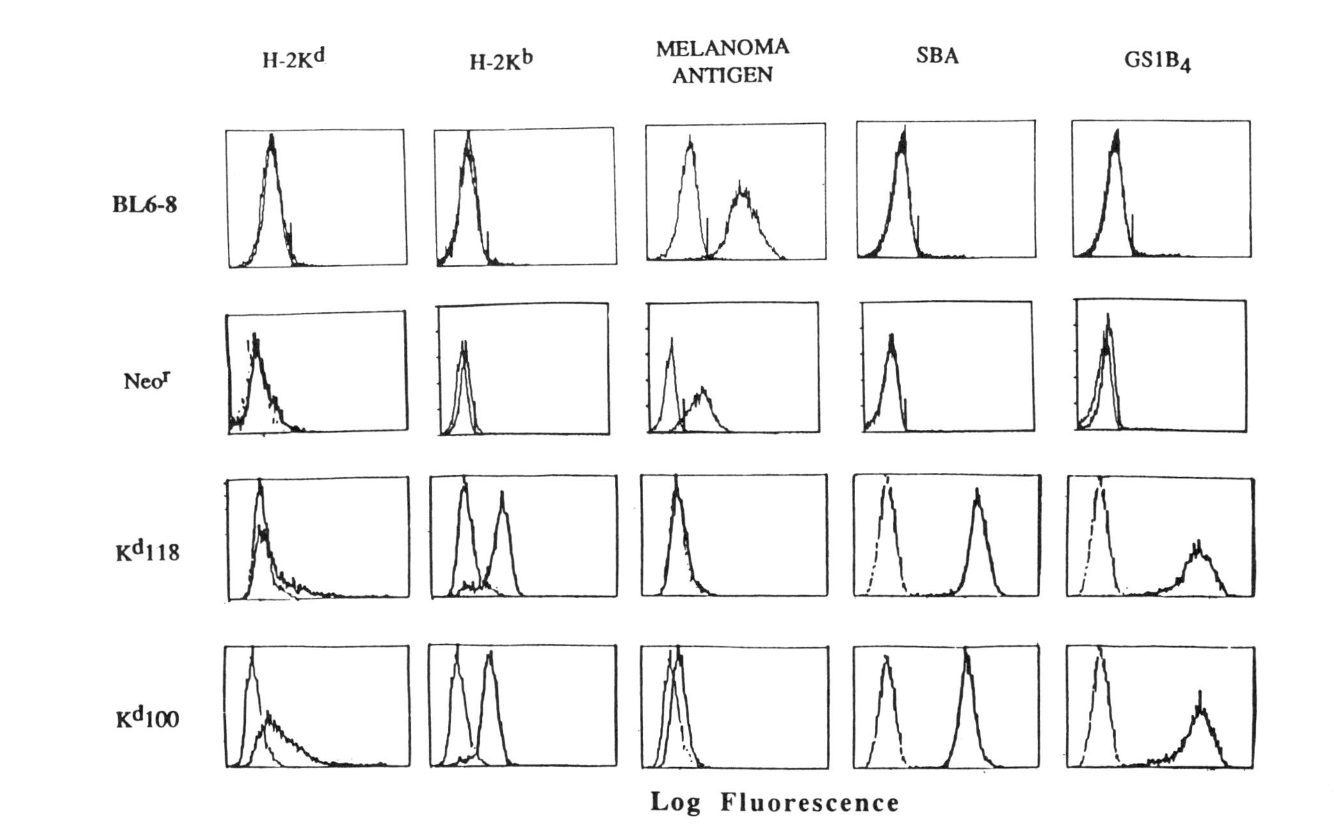

Figure 5. Phenotypic characterization of BL6-8 melanoma cells transfected with *neo*^{r} or H-2K^d gene. Expression of H-2K^d (mAb 31-3-4), H-2K^b (mAb 28-13-3), melanoma associated antigen (mAb MM2-9B6) was analyzed using flow cytometric analysis. Expression of lectin binding carbohydrates was tested using biotinylated SBA and GS1B$_4$ lectins.

(BL6-29) was found to be positive for H-2K^b antigen and reacted strongly with GS1B$_4$, SBA and PNA lectins, but did not express MAA (Gorelik et al., 1991).

Thus, all these data demonstrate that expression of the H-2K^b genes is associated with the specific alteration of cell membrane properties of the BL6 melanoma cells. These changes are identical in all clones expressing transfected or endogenous H-2K^b gene, and closely associated with an increased sensitivity of these clones to NK/NC, as well as TNF-mediated lysis.

These changes appear to be associated with H-2K^b gene expression and are not induced by the transfection procedure *per se*, since transfection with class II H-2IA^k and *neo^r* gene did not affect tumor cell characteristics. Furthermore, induction of H-2K^b antigen expression by a procedure unrelated to transfection (i.e. MNNG treatment) results in identical changes caused by H-2K^b gene transfection. Finally, a clone from the original BL6 melanoma displaying spontaneous expression of H-2K^b gene reveals the same phenotypic changes noted in the H-2K^{b+} transfected clones.

Since activation of the endogenous H-2K^b gene and transfection of the exogenous H-2K^b gene have similar effects, the observed alterations of tumor cell characteristics do not appear to depend on the plasmid constructions of the transfected H-2K^b gene or on the place of its insertion in the genome of the BL6 melanoma cell.

V. EFFECT OF H-2K^b MOLECULES ON THE METASTATIC PROPERTIES OF BL6 MELANOMA CELLS

B16 melanoma and its high and low metastatic lines have been intensively investigated in order to identify the cell properties responsible for their metastatic behavior. The results of these studies indicate that to generate distant metastases tumor cells have to possess a complex of particular properties which will allow them to invade blood vessels and disseminate, extravasate, and proliferate in various organs and tissues (Fidler et al., 1981; Nicolson, 1982).

Although these metastatic properties are still far from being completely identified, it is generally considered that lack of one of these properties could prevent the fulfillment of the whole metastatic cascade (Fidler et al., 1981; Nicolson, 1982). Tumor cells might also be nonmetastatic not because they lack one or several metastatic properties, but because they possess one or more characteristics which do not permit them to complete the metastatic process. This antimetastatic tendency might result from complex interactions between tumor cells and host defense mechanisms. It remains unclear, however, which particular host mechanisms participate in the control of metastatic spread and growth. They might conceivably be either immunological or nonimmunological in nature.

Immunogenicity of tumor cells could render the cells nonmetastatic even if they possessed all of the necessary metastatic properties. Indeed, some immunogenic tumors fail to form distant metastases in the immunocompetent host, but in immunosuppressed mice the metastatic potentials of tumor cells can be fully realized (Eccles and Alexander, 1975; Fidler and Kripke, 1980). Thus, metastatic cells should be nonimmunogenic or weakly immunogenic to avoid destruction by T cell-mediated immunity. Metastatic tumor cells might escape immune destruction by losing their ability to express tumor-associated transplantation antigens (TATA) or by expressing TATA that are distinctive from those of tumor cells growing at the primary site (Gorelik et al., 1982; Gorelik, 1987). Because T cells recognize TATA in association with MHC antigens (Doherty et al., 1984), reduction or loss of MHC antigens might be an additional mechanism by which tumor cells escape recognition and destruction by T lymphocytes, and have a chance to survive and metastasize.

Data accumulated in the past 20 years demonstrate that various experimental tumors, as well as tumors obtained from cancer patients, manifest alterations in the expression of MHC antigens (see for review Gorelik, 1987). Furthermore, analysis of H-2 antigen expression on cells derived from primary and metastatic murine tumors revealed substantial differences in the expression of MHC antigens (Gorelik, 1987). Although metastatic cells might display reduced levels of MHC antigen expression, the association between metastatic ability and MHC antigen expression is not always a negative correlation. Tumor cells which express low levels of MHC gene products are not always highly metastatic. On the contrary, some tumor cells with higher levels of H-2 antigen expression are more metastatic than cells expressing low levels of MHC determinants (Haywood and McKhann, 1971; De Baetselier et al., 1980; Katzav et al., 1983). In addition, it might be the ratio between expression of H-2K and H-2D gene products rather than absolute levels of MHC gene products that is important to the metastatic ability of tumor cells (Eisenbach et al., 1983). Therefore, although quantitative alterations in H-2 antigen expression by primary and metastatic tumor cells have been demonstrated, the significance of these findings is far from fully understood.

Using class I H-2 gene transfection, it was demonstrated that the appearance of MHC antigens made transfected 3LL and F10 tumor cells highly immunogenic and nonmetastatic. Decrease in metastatic potential of these cells was attributed to increase in tumor cell immunogenicity. Indeed, in immunosuppressed mice the metastatic potential of H-2 gene transfected tumor cells was virtually restored (Wallich et al., 1985; Plaskin et al., 1988; Gelber et al., 1989). However, in some tumor models a reduction of metastatic ability with increased MHC antigen expression was found, and it was suggested that this reduction in metastasis formation may be mediated by immunologically independent mechanisms (Gattoni-Celli et al., 1989, 1990).

We also investigated whether expression of the endogenous or transfected H-2K^b gene might alter metastatic ability of BL6 melanoma cells. To analyze the association between class I H-2K^b antigen expression and the metastatic properties of these tumor cells, we first compared the abilities of MNNG-treated H-2K^b BL6 melanoma cells to form experimental metastases in C57BL/6 mice. As we described above BL6T2 melanoma cells are highly immunogenic. Therefore, it is expected that the differences in metastasis formation by BL6 and BL6T2 melanoma cells observed in immunocompetent mice could be diminished or disappear in a host with deficient T cell-mediated immunity. Metastasis formation was investigated in C57BL/6 and athymic nude mice. After i.v. inoculation of 1×10^5 tumor cells BL6T2 melanoma cells did not develop metastatic foci in the lungs, whereas all mice inoculated with BL6 cells formed numerous pulmonary metastases. The generation of BL6T2 metastases did not increase in athymic nude mice, indicating that T cell-mediated immunity was not primarily responsible for the weak lung colonizing ability of BL6T2 melanoma cells.

Postulating that NK cells might be of primary importance in controlling BL6T2 metastasis formation, we next studied whether differences in metastatic potentials between BL6 and BL6T2 cells were reduced or disappeared in mice that lack functionally active NK cells. With suppression of NK cell activity by anti-asialo GM1 serum a significant increase in the number of lung colonies were found in C57BL/6 and nude mice inoculated with BL6 or BL6T2 cells. Although an increase in BL6T2 metastases in NK-depressed mice was observed, numbers of BL6T2 metastases were 20 times lower than those formed by BL6 cells in NK-deprived C57BL/6 as well as nude mice. All BL6 metastatic nodules were black and BL6T2 metastases were white.

Collectively, the data obtained with immunocompetent C57BL/6 and athymic nude mice indicate that differences in the lung colonizing abilities of H-2K^b low BL6 and H-2K^{b}high BL6T2 cells cannot be entirely attributed to either the function of T cell- or NK cell-mediated immunity.

To obtain further confirmation of these findings, the metastatic ability of H-2K^{b-} clone BL6-8 transfected with *neo^r* and/or H-2K^b gene was tested in the immunocompetent C57BL/6 and T-cell deficient nude mice with normal or depleted NK cell activity. Transfection of *neo^r* gene did not significantly change the metastatic ability of the BL6-8 clone. Numerous black metastatic foci were found in the lungs (Figure 6). In contrast, clones CL8-1 and CL8-2 with high expression of the transfected H-2K^b gene produced few metastases in the lungs of C57BL/6 mice. As was previously observed with BL6T2 melanoma cells, all metastatic tumors which developed after inoculation of H-2K^{b+} clones were amelanotic and were larger in size than metastasis from H-2K^{b-} clones. If low numbers of metastatic foci were due to the high immunogenicity of the tested clones, we would expect substantially more metastatic foci in T-cell deficient mice. However, numbers of metastatic foci did not increase when H-2K^b positive clones were inoculated into nude mice, suggesting that their low metastatic ability is not dependent on T cell-mediated immunity. Suppression of NK cell activity by anti-asialo GM1 treatment significantly increased numbers of metastatic foci in the lungs of

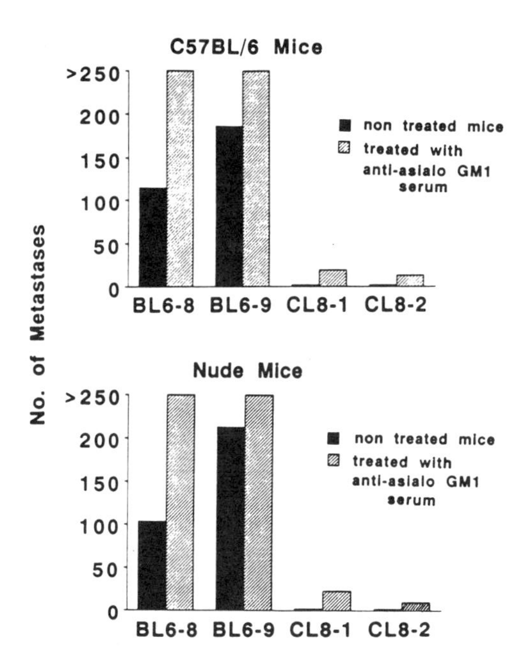

Figure 6. Effect of H-2K^b gene transfection on the metastatic properties of BL6-8 melanoma cells. BL6-8 cells were transfected with *neo^r* gene (clone BL6-9) or H-2K^b gene (clones CL8-1 and CL8-2). Tumor cells (2×10^5) were inoculated i.v. into C57BL/6 or nude mice treated or nontreated with anti-asialo GM1 antibodies, and metastatic foci in the lungs were determined 18 days after tumor cell inoculation.

C57BL/6 and nude mice inoculated with either H-2K^{b+} or H-2K^{b-} clones (Figure 6). However, even in the absence of T and NK cells, H-2K^{b+} clones formed significantly lower numbers of metastases than H-2K^{b-} melanoma cells. These data indicate that low metastatic ability of H-2K^{b+} melanoma clones is not solely dependent on T and NK cell function.

Our preliminary data indicate that in triple-immunodeficient mice that carry nu, bg and xid mutations and are thus deficient in T, NK and B cell function, H-2K^{b+} BL6 melanoma cells generated much fewer metastatic nodules in the lungs than did H-2K^{b-} melanoma cells. Similarly, no restoration of metastatic potentials was observed when H-2K^b transfected BL6 melanoma cells were inoculated i.v. into x-irradiated (500r) and anti-asialo GM1 treated C57BL/6 mice. These data indicate that low metastatic ability of H-2K^{b+} BL6 melanoma does not depend on immune system function, but rather is attributable to some other properties of these cells that make them poorly metastatic.

It remains unknown what cell properties could be affected following expression of H-2K^b gene which render tumor cells low metastatic. It might be a result of alteration of some cell surface properties that are important for completion of the metastatic process after tumor cells enter into the blood. As described above, H-2K^b antigen expression after MNNG treatment or H-2K^b gene transfection was associated with a loss of MAA recognizable by MM2-9B6 or MM2-3C6 mAbs and alteration of cell membrane carbohydrates sites reacting with GS1B, SBA and PNA lectins. In the H-2K^{b-} BL6 melanoma clones SBA and PNA binding sites are actually present but they are heavily sialylated. After neuraminidase treatment these clones were able to bind SBA and PNA lectins.

Sialylation of SBA and PNA binding sites in various Bl6 melanoma lines was previously demonstrated (Raz et al., 1980; Irrimura and Nicolson, 1984). It is of interest that several other tumor systems showed a negative correlation between metastatic capacity and ability to bind SBA lectins (Dennis et al., 1981; Altevogt et al., 1983). Weakly metastatic Eb lymphoma cells had more binding sites for SBA lectin than the highly metastatic Esb line (Altevogt et al., 1983). When Esb-M cells were selected from the highly metastatic Esb line on their ability to adhere to plastic, this subline was found to be weakly metastatic and to have increased binding sites for SBA (Altevogt et al., 1983). Similarly, when lectin-binding sites of high and low metastatic variants of MDAY-D2 were investigated, it was also found that weakly metastatic MDW40 cells had strong binding properties for SBA lectin (Dennis et al., 1981).

The involvement of GS1B$_4$ binding protein in metastasis formation was previously suggested (Grimstad et al., 1984). GS1B$_4$ binding carbohydrates were found in the murine endothelial and epithelial basement membranes, and laminin (Shibata et al., 1982).

Using flow cytometric analysis, we tested the expression of laminin and GS1B$_4$ binding carbohydrates on cell surface of B16, B16F1, B16F10, BL6, MNNG-treated H-2K^{b+} BL6T2 lines and H-2K^b gene transfected CL8-1 melanoma cells. Using rabbit anti-laminin antibodies, it was found that all tested melanoma cell lines express cell surface laminin. However, no expression of GS1B$_4$ binding sites was found on B16, B16F1, B16F10 melanoma cells, suggesting that laminin on these cells is differently glycosylated and missing α-galactosyl epitopes capable of binding GS1B$_4$ lectin. In contrast H-2K^{b+} BL6T2 and CL8-1 cells expressed both GS1B$_4$ binding carbohydrates and laminin.

It is possible that alterations in cell surface properties in H-2K^{b+} BL6 melanoma impair their interactions with the endothelial cells and extracellular basement membrane. Indeed, our data demonstrate that H-2K^{b+} BL6 melanoma cells had inferior ability in comparison to H-2K^{b-} cells to bind murine pulmonary endothelial cell monolayers, as well as laminin or collagen IV. Thus, immunological and nonimmunological alterations in BL6 melanoma cell characteristics induced by H-2K^b gene expression are responsible for observed reduction in metastatic ability of these cells.

VI. MECHANISMS OF THE PLEIOTROPIC EFFECTS OF H-2K^b GENE IN BL6 MELANOMA CELLS

It is unclear how and by what mechanisms the H-2K^b gene could affect various cell properties in BL6 melanoma, that are not directly coded by this gene. The possible explanation for these effects is based on the assumption that the H-2K^b gene interacts and affects function of some non-MHC genes.

It was demonstrated that the expression of MHC gene can be affected by some oncogenes or viruses (Van Der Eb et al., 1983; Versteeg et al., 1988). It was also described that this interaction can be in the opposite direction and the H-2 gene could regulate the expression of some murine viruses (Yetter et al., 1983) or oncogenes (Alon et al., 1987).

Our data indicate that the expression of MAA is inhibited by H-2K^b gene products. It was demonstrated that expression of MAA recognized by MM2-9B6 and MM2-3C6 mAbs is associated with the replication and budding of a B-tropic ecotropic retrovirus that was found to be tissue and strain specific (Leong et al., 1988). This was confirmed by experiments in which B-tropic virus, isolated from B16F10 melanoma cells, induced in SC-1 fibroblasts the cell membrane antigen that reacted with MM2-9B6 mAb (Leong et al., 1988). In contrast, infection with LP-BM5 or 1909 B-tropic viruses, isolated from normal tissue of C57BL\6 mice and N-tropic viruses from different strains of mice, failed to induce cell surface MAA. Infection of SC-1 fibroblasts with NB-tropic Moloney virus also did not induce antigen reacting with MM2-9B6 mAb (Leong et al., 1988). It is considered that virus isolated from B16F10 melanoma is a recombinant form of the endogenous, ecotropic virus of C57BL\6 mice (Leong et al., 1988). Furthermore, in the H-2K^{b+}, MNNG-treated, T2-39 cells these MAA were not detected and immunoelectron microscopy did not find budding virus in these cells (Leong et al., 1988). Preliminary data indicate that in all BL6 melanoma clones expressing the transfected or endogenous H-2K^b gene loss of MAA was always associated with loss of the budding retrovirus particles (Jacqueline Muller, unpublished observations). Thus, it is possible to assume that H-2K^b gene could affect the function of B-tropic retrovirus that seem to be responsible for expression of MAA in BL6 melanoma cells.

Expression of H-2K^b antigen in BL6 melanoma cells was also associated with alterations in the lectin-binding carbohydrate sites on the cell membrane. Although the appearance of SBA, PNA and GS1B$_4$ binding sugars on the cell membrane coincided with the appearance of H-2K^b molecules, it seems unlikely that these sugars are the carbohydrate component of the H-2K^b glycoprotein. Western blot analysis of GS1B$_4$ binding cell surface glycoproteins of H-2K^{b+} clones CL8-1, CL8-2, T2-39 detected several glycoproteins, with a predominant species at Mr 95 and 160–170 kDa, that obviously differ from 45 kDa class I MHC products. In confirmation of the flow cytometric analysis, no GS1B$_4$ binding glycoproteins were found by Western blotting in the cell membrane preparations from the *neo*^{r}

and class II H-21A^k transfected melanoma clones. These alterations in the cell surface carbohydrates of BL6 melanoma cells expressing H-2K^b gene occurred as a result of reexpressing of α1,3 galactosyltransferase gene and appearance of α-galactosyl epitopes. It is possible to assume that the ecotropic retrovirus is responsible for down-regulation of α1,3 galactosyltransferase gene expression in BL6 melanoma, since it was up-regulated with disappearance of this retrovirus.

Similarly, this retrovirus seems to be responsible for the regulation of BL6 melanoma cells resistance to TNF and NCMC. This assumption is based on the findings that sensitivity of BL6 melanoma cells to TNF and NCMC increased with loss of the B-tropic ecotropic retrovirus. Thus, the observed pleiotropic effects of H-2K^b gene in BL6 melanoma cells are mediated via interaction between H-2K^b gene and B-tropic ecotropic retrovirus specific for melanomas of C57BL/6 origin.

Mechanisms responsible for this retrovirus elimination remain unknown. Southern blot analysis of DNA extracted from H-2K^{b+} and H-2K^{b-} BL6 melanoma clones was performed to assess whether elimination of B-tropic ecotropic retrovirus in H-2K^{b+} cells is associated with loss or alteration of the DNA proviral sequences. PstI restriction enzyme cleaved DNA was hybridized with the envelope coding fragment specific for the murine ecotropic retroviruses. DNA of H-2K^{b-} clones contained a single strong band about 8.8 kb. In all H-2K^{b+} clones this band was substantially reduced and a new strong band of 5.4 kb appeared (Arifa Kahn, unpublished observations). These alterations were found in all tested clones that expressed transfected or endogenous H-2K^b genes. The observed changes in the proviral DNA sequences are probably due to recombination or deletion and could be responsible for disappearance of the budding retroviral particles. It should be determined how the H-2K^b gene can induce such changes in the retroviral DNA, as well as how this retrovirus is capable of controlling many phenotypic characteristics of BL6 melanoma cells.

Thus, our data demonstrate that class I H-2K^b gene has broad biological effects in BL6 melanoma cells beyond regulation of the determinants interacting with the immune system.

REFERENCES

Allison, J., Campbell, I., Morahan, G., Mandel, T., Harrison, L., & Miller, J. (1988). Diabetes in transgenic mice resulting from over-expression of class I histocompatibility molecules in pancreatic cells. Nature 333, 529–533.

Alon, Y., Hammerling, G., Segal, S., & Bar-Eli, M. (1987). Association in the expression of Kirstein-ras oncogen and the major histocompatibility complex class I antigen in fibrosarcoma tumor cell variants exhibiting different metastatic capabilities. Canc. Res. 47, 2553–2557.

Altevogt, P., Fogel, M., Chengsong-Popov, R., Dennis, J., Robinson, P., & Schirrmacher, V. (1983). Different patterns of lectin binding and cell surface sialylation detected on related high and low metastatic tumor lines. Canc. Res. 43, 5138–5144.

Bonner, J. & Tyan, M. (1983). Glucocorticoid-induced cleft palate in the mouse: two major histocompatibility complex, H-2 loci with different mechanisms. Genetics 103, 263–276.

Boon, T. (1983). Antigenic tumor cell variants obtained with mutagens. Adv. Canc. Res. 39, 121–151.

Chvatchko, Y., Van Obbengham, E., Kiger, N., & Fehlman, M. (1983). Immunoprecipitation of insulin receptor by antibodies against class I antigens of the murine H-2 major histocompatibility complex. FEBS Lett. 263, 207–211.

Class, F. & Van Rood, J. (1985). The interaction of drugs and endogenous substances with HLA-class I antigens. Prog. Allergy 36, 136–153.

Collins, L., Lapierre, W., Fiers, J., Strominger, M., & Pober, J. (1986). Recombinant tumor necrosis factor increases mRNA levels and surface expression of HLA-A, B antigens in vascular endothelial cells and dermal fibroblasts *in vitro*. Proc. Natl. Acad. Sci. USA 83, 446–452.

DeBaetselier, P., Katzav, S., Gorelik, E., Feldman, M., & Segal, S. (1980). Differential expression of H-2 gene products in tumor cells is associated with their metastogenic properties. Nature 288, 179–181.

Dennis, J., Donaghue, T., & Kerbel, R. (1981). Membrane associated alterations detected in poorly tumorigenic lectin-resistant variant sublines of a highly malignant and metastatic murine tumor. J. Natl. Canc. Inst. 66, 129–139.

Doherty, P., Knowles, B. & Wettstein, P. (1984). Immunological surveillance of tumors in the context of major histocompatibility complex restriction of T-cell function. Adv. Canc. Res. 42, 1–65.

Due, C., Simmonsen, M., & Olsson, L. (1986). The major histocompatibility complex I heavy chain as a structural subunit of the human cell membrane insulin receptor: Implications for the range of biological functions of histocompatibility antigens. Proc. Natl. Acad. Sci. USA 83, 6007–6011.

Eccles, S. & Alexander, P. (1975). Immunologically mediated restraint of latent tumor metastasis. Nature 257, 52–54.

Eisenbach, L., Segal, S., & Feldman, M. (1983). MHC imbalance and metastatic spread in Lewis lung carcinoma clones. Int. J. Canc. 32, 143–120.

Faustman, D., Li, X., Lin, H., Fu, Y., Eienbarth, G., Arruch, J., & Guo, J. (1991). Linkage of faulty MHC class I to autoimmune diabetes. Science 254, 1756–1761.

Fidler, I. & Kripke, M. (1980). Tumor cell antigenicity, host immunity and cancer metastasis. Cancer Immunol. Immunother. 7, 201:205.

Fidler, I., Gerstein, D., & Hart, I. (1981). The biology of cancer invasion and metastasis. Adv. Cancer Res. 28, 149–250.

Gattoni-Celli, S., Marozzi, A., Timpane, R., Kirsch, K., & Isselbacher, K. (1990). Partial suppression of metastatic potential of malignant cells in immunodeficient mice caused by transfection of H-2K^b gene. J. Natl. Canc. Inst. 82, 960–962.

Gattoni-Celli, S., Strauss, R., Willett, C., Pozzatti, R., & Isselbacher, K. (1989). Modulation of the transformed and neoplastic phenotype of rat fibroblasts by MHC-1 gene expression. Canc. Res. 49, 3392–3395.

Gelber, C., Plaksin, D., Vadai, E., Feldman, M., & Eisenbach, L. (1989). Abolishment of metastatic formation by murine tumor cells transfected with "foreign" H-2K genes. Canc. Res. 49, 2366–2373.

Goldman, A. & Katsumata, M. (1986). Murine glucocorticoid receptors: New evidence for a discrete receptor influenced by H-2. Arch. Biochem. Biophys. 249, 316–325.

Goldstein, I., Blake, D., Ebisu, S., Williams, T., & Murphy, L. (1981). Carbohydrate binding studies on the Bandeiraea simplicifolia I isolectins. J. Biol. Chem. 256, 3890–3893.

Gorelik, E. (1987). MHC antigen expression and metastatic properties of tumor cells. In: Immune Response to Metastases (Herberman, R., Wiltrout, R., & Gorelik, E., eds.), pp. 35–56. CRC Press, Boca Raton, FL.

Gorelik, E., Fogel, M., De Baetselier, P., Katzav, S., Feldman, M., & Segal, S. (1982). Immunobiological diversity of metastatic cells. In: Cancer Invasion and Metastasis (Liotta, L. & Hart, I., eds.), pp. 134–146. Martinus Nijhoff, Boston.

Gorelik, E., Gunji, Y & Herberman, R. (1988). H-2 antigen expression and NK sensitivity of BL6 melanoma cells. J. Immunol. 140, 2096–2102.

Gorelik, E., Jay, G., Kim, M., Hearing, V., DeLeo, A., & McCoy, J.P. (1991). Effect of H-2K^b gene on expression of melanoma associated antigen and lectin binding sites on BL6 melanoma cells. Canc. Res. 51, 5212–5218.

Gorelik, E., Jay, G., Kwiatkowski, B., & Herberman, R. (1990). Increased sensitivity to MHC-nonrestricted lysis of BL6 melanoma cells by transfection with class I H-2K^b gene. J. Immunol.145, 1621–1632.

Gorelik, E., Peppoloni, S., Overton, R., & Herberman, R. (1985). Increase in H-2 antigen expression and immunogeneicity of BL6 melanoma cells treated with N-methyl-N′-nitro-nitrosoguanidine. Canc. Res. 45, 5341–5347.

Grimstad, I., Varani, J., & McCoy, J.P. (1984). Contribution of α-D-galactopyranosyl end groups to attachment of highly and low metastatic murine fibrosarcoma cells to various substrates. Exp. Cell Res. 155, 345–358.

Harel-Bellan, A., Quillet, A., Marchiol, C., DeMars, R., Turz, T., & Fradelizi, D. (1986). Natural killer susceptibility of human cells may be regulated by genes in the HLA region on chromosome 6. Proc. Natl. Acad. Sci. USA 83, 5688–5692.

Hart, I. (1979). The selection and characterization of an invasive variant of the BL6 melanoma. Am. J. Pathol. 97, 587–600.

Haywood, G., & McKhann, C. (1971). Antigenic specificities of murine sarcoma cells. Reciprocal relationship between normal transplantation antigens (H-2) and tumor-specific immunogenicity. J. Exp. Med. 133, 1171–1187.

Helenius, A. Morein, B., Fries, E., Simons, K., Robinson, P., Schirrmacher, V., Terhorst, C., & Strominger, J. (1978). Human (HLA-A and HLA-B) and murine (H-2K and H-2D) histocompatibility antigens are cell surface receptors for Semliki forest virus. Proc. Natl. Acad. Sci. USA 75, 3846–3852.

Herberman, R., Ortaldo, C., & Reynolds, J. (1986). Mechanism of cytotoxicity by natural killer (NK) cells. Annu. Rev. Immunol. 4, 651–680.

Irimura, T., & Nicolson, G. (1984). Carbohydrate chain analysis by lectin binding to electrophoretically separated glycoproteins from muring B16 melanoma sublines of various metastatic properties. Canc. Res. 44, 791–798.

Karre, K., Lunggren, H., Pionted, G., & Kiessling, R. (1986). Selective rejection of H-2 deficient lymphoma variants. Analysis of mechanisms. J. Exp. Med. 162, 1745–1759.

Katzav, S., DeBaetselier, P., Tartakovsky, B., Feldman, M., & Segal, S. (1983). Alterations in major histocompatibility complex phenotypes of mouse cloned T10 sarcoma cells: association with shifts from nonmetastatic to metastatic cells. J. Natl. Canc. Inst. 71, 317–324.

Kenyon, D. & Raska, K. (1986). Region of E1A of highly oncogenic adenovirus 12 in transformed cells protects against NK but not LAK cytolysis. Virology 155, 644–651.

Kittur, D., Shimizu, Y., DeMars, R., & Edidin, M. (1987). Insulin binding to human B lymphoblasts is a function of HLA haplotype. Proc. Natl. Acad. Sci. USA 84, 1351–1355.

Koller, B., Marrack, P., Kappler, J. and Smithies, O. (1990). Normal development of mice deficient in β_2M, MHC class I proteins, and CD8$^+$ T cells. Science 248, 1227–1230.

Lafuse, W., & Edidin, M. (1980). Influence of the mouse major histocompatibility complex, H-2 on liver adenylate cyclase activity and on glucagon binding to liver cell membrane. Biochemistry 14, 49–54.

Lattime, E., Pecoraro, G., & Stutman, O. (1981). Natural cytotoxic cells against solid tumors in mice. III. A comparison of effector cell antigenic phenotype and target cell recognition structure with those of NK cells. J. Immunol. 126, 2011–2019.

Lattime, E. Pecoraro, G., & Stutman, O. (1982). Natural cytotoxic cells against solid tumors in mice IV. Natural cytotoxic (NC) cells are not activated natural killer (NK) cells. Int. J. Canc. 30, 471–480.

Leiden, J., Karpinski, B., Gottschalk, L., & Kornbluth, J. (1989). Susceptibility to natural killer cell-mediated cytolysis is independent of the level of target cell class 1HLA-A expression. J. Immunol. 142, 2140–2148.

Leong, S., Muller, J., Yetter, R., Gorelik, E., Takami, T., & Hearing V. (1988). Expression and modulation of a retrovirus-associated antigen by murine melanoma cells. Canc. Res. 48, 4954–4958.

Ljunggren, H. & Karre, K. (1990). In search of the "missing- self" MHC molecules and NK recognition. Immunol. Today 11, 237–241.

Muller, U., Jongeneel, V., Nedospasov, S., Lindahl, K., & Steinmetz, M. (1987). Tumor necrosis factor and lymphotoxin genes map close to H-2D in the mouse major histocompatibility complex. Nature 325, 265–267.

Nicolson, G. (1982). Cancer metastasis: organ colonization and cell-surface properties of malignant cells. Biochim. Biophys. Acta 695, 113–176.

Nishmura, M., Stroynowski, I., Hood, L., & Ostrand-Rosenberg, S. (1988). H-2K^b antigen expression has no effect on natural killer susceptibility and tumorigenicity of murine hepatoma. J. Immunol. 141, 4403–4409.

Ortaldo, J., Mason, L., Mathieson, B., Liang, S., Flick, D. & Herberman, R. (1986). Mediation of mouse natural cytotoxic activity by tumor necrosis factor. Nature 321, 700–705.

Palumbo, D. & Vladutiu, A. (1979). Estrogen receptor in uteri of mice of different H-2 genotypes. Experientia 35, 1103–1104.

Plaksin, D., Gelber, C., Feldman, M., & Eisenbach, L. (1988). Reversal of the metastatic phenotype in the Lewis lung carcinoma cells after transfection with syngeneic H-2K^b gene. Proc. Natl. Acad. USA 85, 4463–4467.

Raz, A., McLellan, W., Hart, I. Bucana, C., Hoyer, L., Sela, B-A., Dragsten, P., & Fidler, I. (1980). Cell surface properties of B16 melanoma variants with differing metastatic potential. Canc. Res. 40, 1645–1651.

Roth, J. (1987). Subcellular organization of glycosylation in mammalian cells. Biochem. Biophys. Acta 906, 405–416.

Sawada, Y., Fohring, B., Shenk, T., & Raska, R. (1985). Tumorigenicity of adenovirus-transformed cells: Region E1A of adenovirus 12 confers resistance to natural killer cells. Virology 147–413.

Schreiber, A., Schlessenter, A., & Edidin, M. (1984). Interaction between major histocompatibility complex antigen and epidermal factor receptor on human cells. J. Mol. Biol. 98, 725–731.

Schrier, P., Bernards, R., Vaessen, R., Houweling, A., & Van Der Eb, A. (1983). Expression of Class I major histocompatibility antigens switched off by highly oncogenic adenovirus 12 in transformed rat cells. Nature 305, 771–775.

Shibata, S., Petters, B., Roberts, D., Goldstein, I., & Liotta, L. (1982). Isolation of laminin by affinity chromatography on immobilized Griffonia simlicifolia I lectin. FASEB Letters 142, 194–199.

Smith, D., Larsen, R., Mattox, S., Low, J., & Cumminges, R. (1990). Transfer and expression of murine UDP-Gal: α-D-Gal- α1,3 galactosyltransferase gene in transfected Chinese hamster ovary cells: competition reaction between α1,3 galactosyltransferase and endogenous α2,3 sialyltransferase. J. Biol. Chem. 265, 6225–6231.

Solano, A., Cremaschi, G., Sanchez, M., Borda, E., Sterin-Borda, L., & Posesta, E. (1980). Molecular and biological interaction between major histocompatibility complex class I antigens and luteinizing hormone receptors or beta-adrenergic receptors triggers cellular response in mice. Proc. Nat. Acad. Sci. USA 85, 5087–5091.

Storkus, W., Alexander, J., Payne, A., Cresswell, P., & Dawson, J. (1989). The alpha 1/alpha 2 domains of class I HLA molecules confer resistance to natural killer. J. Immunol. 143, 3853–3861.

Storkus, W., Howell, D., Salter, R., Dawson, J., & Cresswell, P. (1987). NK susceptibility varies inversely with target cell class I HLA antigen expression. J. Immunol. 138, 1657–1659.

Tanaka, K., Gorelik, E., Nozumi, N., & Jay, G. (1988). Rejection of BL6 melanoma induced by the expression of a transfected MHC class I gene. Mol. Cell. Biology 8, 1857–1861.

Versteeg, R., Noordermer, I., Kruse-Wolter, M., Ruiter, D., & Schrier, P. (1988). c-Myc down-regulates Class I HLA expression in human melanomas. EMBO J. 7, 1023–1029.

Vilcek, J. & Lee, T. (1991). Tumor necrosis factor. New insights into the molecular mechanisms of its multiple actions. J. Biol. Chem. 266, 7313–7319.

Wallich, R., Bulbuc, N., Hammerling, G., Katzav, S., Segal, S. & Feldman, M. (1985). Abrogation of metastatic properties of tumor cells by *de novo* expression of H-2K antigens following H-2 gene transfection. Nature 315, 301–305.

Yamazaki, K., Beauchamp, G., Bard, J., Thomas, L., & Boyse, E. (1982). Chemosensory recognition of phenotypes determined by the TLA and H-2K regions of chromosome 17 of the mouse. Proc. Natl. Acad. Sci. USA 79, 7828–7833.

Yetter, R., Hartley, J., & Morse, H. (1983). H-2-linked regulation of xenotropic murine leukemia virus expression. Proc. Natl. Acad. Sci. USA 80, 505–509.

THE ROLE OF ANGIOGENESIS IN TUMOR PROGRESSION AND METASTASIS

Janusz W. Rak, Erik J. Hegmann, and Robert S. Kerbel

Advances in Molecular and Cell Biology
Volume 7, pages 205–251

ISBN: 1-55938-624-X

"The classification of the constituents of a chaos, nothing less here is essayed."

Herman Melville, Moby-Dick

I. INTRODUCTION

"No tumor is an island. . ."

Advances in molecular technology within the last two decades have made possible the detailed analysis of some of the basic mechanisms operating during cell proliferation, transformation and subsequent tumor progression. As a result of the wealth of information which has accumulated in the field of tumor biology, cancer has been labelled as a "molecular" disease mainly if not exclusively related to abnormal (i.e. faster and/or unlimited) cell proliferation. The underlying assumption in many studies has been that the whole process can be understood by analyzing the intrinsic properties of the cancer cell related to growth control mechanisms. Consequently, clonally-derived tumor cells grown in culture were often perceived as a satisfactory representation of tumors growing *in vivo*. This belief appeared justified by a series of major discoveries leading to the identification of a large number of dominant genes (oncogenes) capable of conferring a transformed phenotype to otherwise normal cells (Lewin, 1991). Many of these genes encode proteins responsible for the abnormal expression of growth factors, growth factor receptors or molecules involved in the intracellular transduction and regulation of growth factor signals. Acrine, autocrine or paracrine growth stimulation was thus thought to induce unlimited, autonomous proliferation of cancer cells (Aaronson, 1991; Bishop, 1991; Cantley et al., 1991; Cross and Dexter, 1991; Hunter, 1991).

This simple causality was challenged by the discovery of the transforming properties of genes encoding transcription factors and splicing factors, which can exert a rather broad spectrum of gene expression control (Hunter, 1991; Lewin, 1991). In addition, the finding of a series of recessive genes (tumor suppressor genes, Aaronson, 1991), which must be inactivated in order to allow for development of neoplasias or cancers, suggested that such growth *in vivo* may be brought about by a more complex multifactorial change (Liotta et al., 1991). "Metastasis suppressor" genes have also been recently detected (Hart et al., 1991; Liotta et al., 1991). Ironically, it turned out that some of the genes which seem to fulfill the criteria of tumor-, metastasis- or growth-suppressors did not interfere directly with intracellular growth regulatory machinery. Rather, lack of expression is essential for the operation of certain tumor-host or tumor-tumor interactions promoting expansion of solid tumors. A number of genes involved in the regulation of cell immunogenicity, proteolytic activity, adhesiveness or other properties can be put into this category since their expression leads to so called non-tumorigenic and/or non-metastatic phenotypes without necessarily having a direct effect on cell proliferation *in vitro*. This mode of action has been postulated for example, in the case of TIMP-1 (Khokha et al., 1989), and E-cadherin (Schipper et al., 1991), DCC (Fearon et al., 1990a), connexin 32 (Eghbali et al., 1991) and certain other molecules (Moroco et al., 1990; Mahoney et al., 1991).

It has also been shown that components of an "invasive phenotype" displayed by advanced breast carcinomas, such as secretion of the extracellular matrix degrading enzyme stromelysin III, can be provided in a "trojan-horse" manner by normal host stromal cells within the tumor, rather than by tumor cells themselves (Basset et al., 1990). Somewhat similar "complementarity" between tumor and stromal cells with respect to expression of plasminogen activator (uPA) and its receptor (uPAR) has been observed in human colon carcinoma (Pyke et al., 1991a; Pyke et al., 1991b). Tumor cell immunogenicity has also been found to be dependent on features of accompanying stroma (Singh et al., 1992). Cooperation in tumor cell growth and metastasis involving different tumor cell populations has also been postulated to occur (Poste et al., 1981; Miller, 1983; Miller et al., 1988, 1990; Lyons et al., 1989; Heppner and Miller, 1989b). For example, it has been shown that co-injection of poorly tumorigenic (CR) or non-tumorigenic cell populations (U et al., 1987; Lippman et al., 1991) may endow a cell mixture with a highly tumorigenic phenotype. These findings suggest that the malignant phenotype can be attributed to functional "units" comprising different tumor and host cell types (Rak, 1989; Miller and Heppner, 1990).

Clinical and experimental observations indicate that cancer may in fact be perceived as a systemic disease capable of a widespread dissemination of tumor cells, and the induction of pathological changes indirectly related to tumor growth (paraneoplastic syndromes), a systemic immune response to a growing tumor and a multifocal response of host tissue to tumor expansion (angiogenesis, inflammatory reaction, fibrosis, etc.) (Van den Hoff, 1988; Paweletz and Knierim, 1989). There have also been intriguing reports indicating diffuse biochemical changes in unaffected tissues of cancer patients (Jensen et al., 1982; Benedetto et al., 1990). It is apparent therefore that the complexity of changes seen in cancer at the cell, tissue and systemic levels cannot be understood solely on the basis of a detailed analysis of genes involved in the regulation of (tumor) cell proliferation, even though there seems to be growing agreement about the contribution of these genes to tumor development and progression (Fearon and Vogelstein, 1990b; Bishop, 1991; Hunter, 1991; Weinberg, 1991).

Likewise, the results of gene transfer or gene "knockout" experiments using cancer cells must be interpreted with caution when assessing the role of a particular gene in cancer development or progression. For example, it is becoming increasingly clear that an alteration in expression of a single gene frequently may lead to coordinate direct or indirect change in the expression of many other genes responsible for maintaining the cell phenotype (Liotta et al., 1991). This can be attributed to the interactive nature of the molecular machinery controlling various functions of normal and transformed cells. Limited availability and/or arbitrary choice of relevant assays used in "gene function" studies may therefore lead to an oversimplified picture. For example, it is now widely recognized that the molecular signals delivered by a particular "growth factor" evoke a variety of responses in a given cell including not only cell proliferation, but also differentiation, production of other cytokines, enzymes and extracellular matrix proteins and expression of adhesion molecules, ultimately changing the potential nature of cell-cell interactions (Mignatti et al., 1989; Roberts and Sporn, 1989; Madri et al., 1991; Auerbach, 1991a; D'Amore, 1992). In addition, a particular molecular mechanism may appear to be rather versatile in terms of its mode of action, function and contribution to cell phenotype. There are examples of plasma membrane-bound cytokines (e.g. TGF-α) which both elicit a signal and mediate cell attachment through a specific receptor ("juxtacrine" mechanism) (Massague et al., 1990). Similarly, there is evidence suggesting that integrins and perhaps certain adhesion molecules may be involved in signal transduction (Dedhar, 1989; Hynes, 1992). Furthermore, the interdependence between the mechanisms regulating cell attachment, spreading, cell shape, and cytokine-mediated responses has been recently demonstrated (Ingber, 1991). In this context, the traditional paradigm ascribing growth factors the role of "cell targeted molecular orders" is being replaced by the view that a relevant target for such regulatory molecules is the multicellular structure of a particular tissue—rather than a single cell (Sporn, 1992). Consistent with this view, it has been proposed that tumor-derived signals may switch tissue homeostasis to a permanent "wound healing" mode (Dvorak, 1986).

With all of these considerations in mind, a tumor cell seems to be a component rather than an equivalent of a tumor. Thus, cancer can be perceived as a situation in which abnormalities in molecular signal exchange may compromise tissue homeostasis in a "progressive" way, giving rise to an outgrowth of a complex, dynamic multicellular structure. Two very different cellular "components" of this structure seem to be essential for progressive macroscopic tumor growth: transformed, genetically altered neoplastic or cancer cells being the driving force of the process, and non-transformed (but frequently "activated") host cells which have been recruited to the tumor site. Some of the latter cells are essential for creating permissive conditions for tumor growth and metastasis, that is they may be necessary but not sufficient for metastasis.

Evidence for the coordinated regulation of intracellular and extracellular growth-permissive signals in transformed cells has been described recently. It has been shown that loss of a tumor suppressor gene in BHK cells was closely accompanied by downregulation of a 140-kDa fragment of the extracellular matrix molecule thrombospondin (Rastinejad et al., 1989; Good et al., 1990). This glycoprotein (which most likely is not the suppressor gene product) provides a negative control for the angiogenic activity of those cells and also possibly for surrounding cells. In the absence of the 140-kDa glycoprotein, transformed BHK cells undergo a "switch to the angiogenic state" and become capable of tumor formation in animals (Rastinejad et al., 1989; Bouch, 1990; Good et al., 1990; Bouch et al., 1991). Similar coordinate changes in the angiogenic state and expression of a tumor suppressor gene were reported in human osteosarcoma, Wilms' tumor, retinoblastoma and glioblastoma (Bouch, 1990; Bouch et al., 1991) as well as breast cancer (Zajchowski et al., 1990). It is not clear how this coordinate action of conceivably two separate genes is brought about and whether similar mechanisms operate in normal tissues during regeneration or embryogenesis. Nevertheless, tilting the balance between angiogenesis inhibitors and stimulators toward the latter appears to be a consistent host tissue response to solid tumor growth.

In summary, the progressive expansion of a three-dimensional tumor mass and the spread of cancer cells from one site to another is intimately dependent on a number of critical tumor cell-host cell interactions. Perhaps the most important of these is the contribution of endothelial cells. In this regard several lines of evidence indicate that tumor growth beyond 1–2 mm diameter tumors is dependent on angiogenesis (Folkman, 1990). The process of new blood vessel formation apparently is triggered by interactions between tumor cells and vascular endothelium in the surrounding mesenchyme. Some other host cells, especially macrophages (Polverini et al., 1977; Leibovich et al., 1987), mast cells (Blood and Zetter, 1990) and neutrophils, can also participate in this cell-cell "dialogue" (Blood and Zetter, 1990). Consequently, tumor neovascularization allows for tumor expansion which in turn is accompanied by further angiogenesis, thus generating a potentially deadly self-perpetuating system. This review is devoted to a discussion of some of the possible consequences of the interactions between tumor cells and the local vasculature which ultimately influence and facilitate tumor progression and the formation of metastases.

II. TUMOR PROGRESSION

A tumor cell population not only expands, but more importantly, gradually acquires new properties over time. This multistep process was noted long ago and the term "tumor progression" was coined by Foulds to describe it (Foulds, 1954). As a result of this process, tumor cells are believed to become "autonomous" i.e. refractory to physiological regulatory mechanisms operating within the tissue of origin (Foulds, 1954). Alternatively, it was postulated that, rather than autonomy, they acquire an aberrant pattern of "social" behavior within the tissue (Heppner, 1989a). For example, they cross anatomical barriers, migrate to and expand within "ectopic" tissue compartments and in distant organs, compromising the function of these organs and thus leading to the death of the affected individual. The dynamics of progression varies among different tumor types and among individual tumors. Recent studies strongly implicate gradual accumulation of different genetic abnormalities during progression of colon carcinoma (and most likely other tumors as well) (Fearon and Vogelstein, 1990b). Analysis of colon tumors has indicated that certain genetic abnormalities are more likely to be found at defined stages of progression (Fearon and Vogelstein, 1990b). Numerous exceptions however suggest that the order of the changes is not inviolate: it is their accumulation that is crucial (Fearon and Vogelstein, 1990b). It is not known whether the earlier events somehow facilitate the subsequent ones or if all of them are a result of the genetic instability of tumor cells causing secondary aberrations with different frequencies in different genetic loci (Cifone and Fidler, 1981; Nicolson, 1991a).

Through a series of clonal expansions the tumor cell population as a whole gradually becomes "malignant," that is competent for local invasion and metastasis. It is unclear whether all the tumor cells within that population have the same potential to acquire the malignant phenotype. Even if this is the case, different tumor cell subpopulations progress asynchronously and possibly along somewhat different pathways, since upon isolation from a single lesion the cells frequently display a remarkable diversity with respect to metastatic capacity, organ preference and many other properties (Heppner, 1984; Heppner, 1989a; Aslakson et al., 1991). These multiple pathways of progression can sometimes be linked to initial genetic events during tumor development (Brown et al., 1990). Although these processes are still poorly understood, it seems conceivable that they may be, to some extent, cell type or tissue specific, since the repertoire of genetic and phenotypic alterations of metastatic cells derived from various types of tumors are usually somewhat different.

Continuous selective pressure from the local microenvironment in concert with genetic instability of tumor cells are both thought to drive clonal evolution of tumor cell subpopulations toward expression of greater malignancy (Nowell, 1986). During this process, metastatically competent tumor cells seem to emerge, perhaps at relatively early stages (Fidler, 1973b; Fidler, 1990). Initially they are present as a "cryptic" minority within the primary tumor population (Fidler, 1973b). In a series of classical experiments Fidler and his colleagues demonstrated the presence of metastatically-competent cell subpopulations by isolating highly metastatic clones from the poorly metastatic parental tumor (mouse melanoma) population (Fidler, 1973b). Similar procedures based on repetitive rounds of isolation and re-inoculation of putatively rare tumor cells obtained from different "target" organs yielded metastatic variants with greater degrees of metastatic aggressiveness, having also different organ preferences (Nicolson, 1988; Fidler, 1990; Aslakson et al., 1991). It has been postulated that even as a cryptic minority subpopulation, metastatic cells disseminate forming micrometastases which can later form deadly secondary outgrowths. According to the "decathlon champion"model of metastasis, these highly specialized cells have an intrinsic capability to successfully achieve ("win") all the steps of the metastatic cascade, for example, invade blood vessels, survive in the circulation, home and extravasate in a distant organ, and expand therein. Their "multidisciplinary fitness" was postulated to compensate for their low numbers, that is "cryptic" nature (Fidler, 1990).

Although possible in some cases, this scenario was inconsistent with the finding that even after intravenous injection of a large number of highly metastatic tumor cells, the efficiency of metastasis formation was found to be in the range of 0.1%

or less (Weiss, 1990). In this context, a better understanding of random events in the metastatic "decathlon" or "cascade" (Fidler, 1990; Aslakson and Miller, 1992) would seem necessary (Weiss, 1990) even though it had been shown that the whole process, or at least some of its steps (Aslakson and Miller, 1992), are highly selective in as much as its outcome can be predicted based on the characteristics of tumor cells. At the same time, experiments on metastatic inefficiency (Weiss, 1990) demonstrated that the fate of a single cell expressing the metastatic phenotype is largely unpredictable due to stochastic cell death or survival events involved in the metastatic process. Furthermore, the expected differences in metastatic competence among sublines isolated from primary tumor sites and from secondary deposits, was in fact rarely found, especially in the case of advanced human tumors (Kerbel, 1990a). For example, in the case of human colon carcinoma, cells isolated from advanced primary lesions (e.g. Duke's D) and those from distant metastases were similar with respect to their capacity for liver metastasis formation in nude mice, whereas cells obtained from earlier stage lesions, for example Duke's A or B, were essentially devoid of liver metastasizing ability. (Morikawa et al., 1988a; Morikawa et al., 1988b).

Studies of the progression of human cutaneous melanoma show a similar trend. This type of tumor has been fairly well characterized because of its accessibility to visual observation and histological examination, even at relatively early stages of development (Clark et al., 1984; Kerbel, 1990a). Again, rather than finding differences between primary and secondary outgrowths, dramatic changes of cell phenotype and behavior were found during progressive growth of the primary lesion. Initially, transformed melanocytes residing in the dermal-epidermal interface form dysplastic nevi in the microenvironment occupied exclusively by keratinocytes. Occasionally the nevi undergo further transformation and expand horizontally forming the radial growth phase (RGP) melanoma. At a certain point tumor cells penetrate the basement membrane and enter the dermal mesenchyme. It has been shown that melanoma cells are initially unable to grow rapidly in this microenvironment and some of the infiltrates regress (Clark, 1991). It remains a subject of speculation whether the mechanism of this regression is related to an immune response or to cell differentiation, or necrosis (or apoptosis) somehow induced by the surrounding foreign tissue microenvironment. Eventually some of the transformed cells develop resistance to growth inhibitory factors of the dermis and expand vertically (Clark, 1991). Clinical data indicate that the "thin" lesions remain curable for some time even after entering the dermal mesenchyme. The prognosis becomes much worse, that is curability drops from 93.2% to 59.8% and 33.3% when the thickness of VGP melanoma extends beyond 0.76 mm and then beyond 3.6 mm, respectively (Clark et al., 1989). Even though the local tumor can be easily removed at that time, the probability of regional and distant metastases increases dramatically. At this stage the tumor invades the highly vascularized papillary dermis (Folkman, 1987a; Clark et al., 1989).

If one assumes that stable, qualitative changes occur in the phenotypes of cells within the progressing primary melanoma lesions, and that at a certain point such changes are manifested by a majority of the cell population, it follows that consistent differences between cells isolated from tumors at different stages of progression within the primary tumor and in metastases should be detectable. Indeed the differences between cell lines representative of early and advanced melanoma have been reported with respect to multiple properties (Cornil et al., 1989; Herlyn, 1990a; Cornil et al., 1991; Lu et al., 1992). For example, cell lines derived from advanced VGP and metastatic melanomas appear to have reduced growth requirements *in vitro*, increased anchorage-independent growth capacity, abnormal cell adhesion molecule expression and are able to readily form tumors in nude mice (Cornil et al., 1989; Herlyn, 1990a). In addition, differential growth stimulation in coculture with human keratinocytes has been observed (Rak et al., 1991b). The cell lines representative for RGP and early VGP are sensitive to the growth inhibitory effect of IL-6 secreted by dermal fibroblasts into conditioned media (Cornil et al., 1991; Lu et al., 1992) and to factor(s) present in dermal microvascular endothelial cell conditioned media (Rak et al., 1991b; Hegmann et al., 1992). Conversely, cells isolated from both advanced VGP and metastatic melanomas develop a "pleiotropic resistance to multiple inhibitory cytokines" such as IL-6, IL-1, TNFα, and TGF-β as well as others (Lu et al., 1992; Kerbel et al., 1992). This pattern of response to growth regulatory factors may be relevant for their ability to withstand the selective pressure in the orthotopic as well as ectopic mesenchyme where melanocytes are not normally found. It is also highly suggestive that negative selection of the less malignant (i.e. metastatically-incompetent) cellular variants, and positive selection of potentially metastatic clones may take place and result in dominance of the latter within the primary lesion; this could occur in part as the result of the interaction with host cells in the three different local tissue compartments as depicted in Figure 1.

The selective growth advantage of a metastatic subpopulation leading to dominance of these cells in the primary tumor was first demonstrated in the murine mammary carcinoma SP1 (Waghorne et al., 1987, 1988a, 1988b) and then in human mammary carcinoma (Price et al., 1990) by using "genetic tagging" and cell lineage analysis techniques (Kerbel et al., 1987; Talmadge and Zbar, 1987; Korczak et al., 1988; Waghorne et al., 1988b). Moreover, the degree of dominance of the metastatic subpopulation (within the mouse SP1 tumor) correlated with yield of metastatic deposits found in lungs of the tumor-bearing animals. Tumor removal prior to completion of the dominance process reduced, or prevented, distant metastasis formation from tumors initially composed of a mixture of metastatic and non-metastatic tumor cells, but not from similar tumors comprised of metastatic cells only (Theodorescu et al., 1991). Thus the concept of metastatic cell "clonal dominance" in primary tumors was proposed, whereby the cell subpopulation composition within the primary tumor undergoes a progressive change through the gradual enrichment of metastatically competent subline(s). Such enrichment could

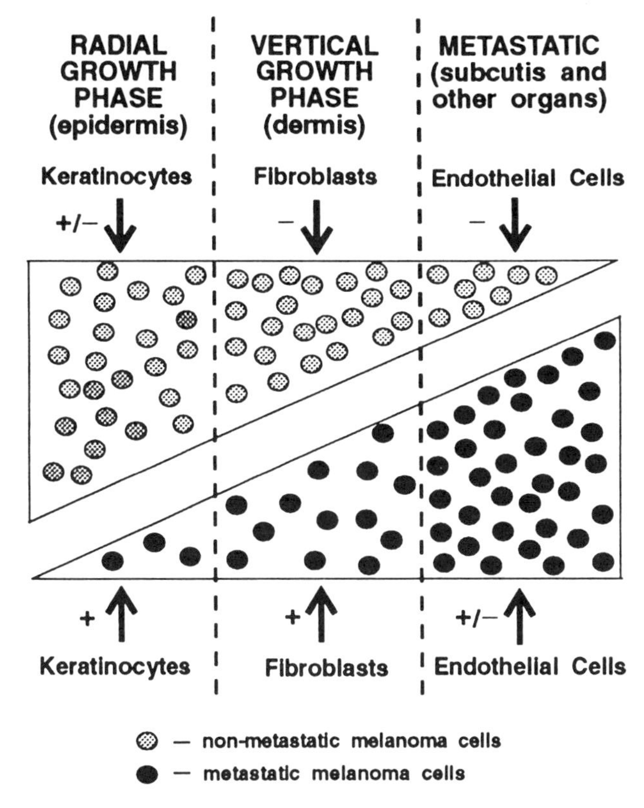

Figure 1. A model of melanoma progression under the influence of the local microenvironment. Stimulatory and/or inhibitory cytokines released from keratinocytes, fibroblasts, or endothelial cells exert differential effects on growth of melanoma cells isolated from lesions at different stages of progression. These interactions in situ may change the composition of the primary tumor leading to dominance of metastatically competent cellular variants.

help compensate for the inefficiency of the metastatic process by generating a sufficient output of malignant cells capable of entering the circulation (Kerbel, 1990a). This type of phenomenon can also explain intriguing paradoxes such as clonality of initially multifocal tumors (Woodruff, 1988) or clonality of metastases arising from heterogeneous cell aggregates (Rak, 1989). Miller et al. working with another experimental system also demonstrated that a mixture of two sister subpopulations, 4TO7 (highly organ colonizing, poorly metastatic) and 168 (non-metastatic), both being derived from the same mouse mammary tumor, could lead to rapid dominance of 4TO7 cells (Miller et al., 1987; Miller et al., 1988). Co-dominance of other mixed tumor cell populations was also observed (Miller and Heppner, 1990).

It should be kept in mind that the "metastatic equation" has at least two variables, one being the selective growth advantage of metastatically competent cells and the other the degree of metastatic competence itself. Therefore one can speculate that if the degree of metastatic competence is extremely high due to unusually rapid progression of some tumor cell subpopulations, the "threshold" level of their cellular dominance in the primary tumor necessary for distant metastasis formation may be less than in the case of other tumors undergoing slower progression. In fact, the unusually rapid progression ("type II" progression) (Bell et al., 1989) may explain the pathogenesis of the relatively rare syndrome called "unknown primary tumor" (UPT) (Bell et al., 1989) which is manifested by full blown metastatic disease in the absence of a known primary lesion (Bell et al., 1989). It is unknown whether the primary lesion is simply not detectable, or undergoes regression due to some kind of "concomitant immunity" mechanism which spares the disseminated metastatic cells.

Experiments demonstrating changes in cell subpopulation composition during tumor growth come mostly from "whole" tumor analysis (Kerbel, 1990a, Kerbel and Theodorescu, 1990b; Miller and Heppner, 1990). However these processes probably begin to occur at the microscopic level in the nests of tumor cell parenchyma which are surrounded by stromal components. It is conceivable that as a result of local interactions between tumor and host cells, only local or zonal changes initially take place (Miller et al., 1989). Multiple foci of morphologically different tumor cell populations were frequently observed in histological specimens from the same tumor (Heppner and Miller, 1983). Growth interactions, differential vulnerability to microenvironmental factors, competition for nutrients, differential invasion of the host tissue or mutual invasion between various tumor cell populations may eventually lead to the local dominance of the most malignant clones.

The molecular mechanisms(s) of clonal dominance are unknown. In some cases mutation of the p53 gene seems to be involved (Sidransky et al., 1992). Paracrine secretion of stimulatory or inhibitory growth factors such as TGF-β1 by tumor cell subpopulations seems conceivable (Kerbel and Theodorescu, 1990b). Interestingly, host factors appear to be an indispensable component of these interactions. For example, secretion of an active form of TGF-β by the non-metastatic SP1 cells *in vitro* either did not lead to change in composition of a mixture of non-metastatic SP1 cells containing a minority subpopulation of metastatically-competent variant SP1 cells (Samiei and Waghorne, 1991) or did so, but at an efficiency 20–40 times lower than upon orthotopic or subcutaneous co-injection of the cells into the animals (Rak, 1991a). These observations indicate that the host microenvironment may be crucial for enrichment of a more aggressive variant in the tumor population. This is consistent with the model of melanoma progression discussed earlier (Figure 1).

Stromal (Van den Hoff, 1988, 1991; Clarke et al., 1992) and inflammatory (Heppner, 1988) cellular components have been shown to actively participate in these processes. They not only exert their selective growth modulatory, cytotoxic, supportive, or mutagenic effects upon tumor cell subpopulations but also undergo a profound functional change often referred to as "activation" (Van den Hoff, 1988). Coexistence of tumor cell foci expressing different degrees of malignancy together with "activated" or phenotypically "transformed" fibroblasts were demonstrated in prostate carcinoma (Chung et al., 1989; Russel et al., 1990; Fletcher et al., 1991; Gleave et al., 1991; Hayashi and Cuhna, 1991). Some of these outgrowths form fibrosarcomas instead of carcinomas upon implantation to nude mice (Russel et al., 1990). A number of studies, some of which were cited earlier, indicate that stromal cells may contribute to invasion or metastasis of tumor cells by virtue of growth interactions or release of proteolytic enzymes at the tumor site (Pritchett et al., 1989; Basset et al., 1990; Hayashi and Cuhna, 1991). Somewhat analogous alterations in appearance and function of tumor-associated vascular endothelial cells have been reported, and are summarized in a later section of this review.

In summary, tumor progression appears to be associated with a stepwise accumulation of various genetic and phenotypic alterations, resulting in the generation of metastatically competent tumor cell variants (clones) endowed with a selective growth advantage at the primary site. This is followed by quantitative enrichment of the malignant variant(s) within the primary tumor cell population. The selective growth advantage of the metastatic (sub)clone may be achieved by growth promoting interactions of this clone with the rest of the tumor parenchyma and/or between tumor and adjacent stromal cells, not just by an intrinsic growth advantage, for example secretion of greater levels of autocrine growth factors. Dominance of the metastatically competent clone may help compensate for the overall inefficiency of the metastatic process leading to formation of secondary deposits. Changes of metastatically-competent tumor cell composition may initially occur locally within the primary tumor. Conceivably, in some cases, partial dominance of a metastatic clone(s) in highly vascularized areas of tumor masses may be sufficient for initiation of the metastatic process.

Further discussion of these issues may be found in any of the numerous review articles published on tumor progression (Nowell, 1986; Fearon and Vogelstein, 1990b; Nicolson, 1991a), metastasis (Weiss, 1985; Nicolson, 1988; Hart et al., 1989; Hart et al., 1991), tumor cell heterogeneity (Heppner, 1984, 1989a; Heppner and Miller, 1983, 1989b), tumor cell interactions (Miller and Heppner, 1990), tumor stromal interactions (Van den Hoff, 1988) and clonal dominance (Kerbel, 1990a). For the purposes of this review we wish to stress this information simply to highlight that, regardless of the accumulation of genetic alterations which occur within a tumor cell population over time, their ability to confer a series of growth advantages to the cells is in part explained by an enhanced ability of such cells to exploit more effectively and efficiently their surrounding tissue microenvironments. An example of this is induction of angiogenesis.

III. TUMOR ANGIOGENESIS AND MECHANISMS OF BLOOD VESSEL FORMATION

The development of tumor vasculature seems to be a necessary (but not sufficient) precondition for continuous primary tumor growth and metastasis. The confidence that solid tumor growth is in fact "angiogenesis-dependent" rests mostly on correlative evidence, as summarized recently by Folkman and collaborators (Folkman, 1985, 1990; Ingber et al., 1990; Kandel et al., 1991). This includes the following:

1. The rapid, exponential growth of tumors implanted in subcutaneous transparent chambers or chorioallantoic membranes begins upon vascularization.
2. When tumors are grown under avascular conditions, that is in isolated, perfused organs, cornea, or the anterior chamber of the eye, their growth is restricted to a size of approximately 1–2 mm or less. However, when they become vascularized their growth is dramatically accelerated by a factor of many orders of magnitude. Similar observations have been made in the case of metastatic foci of human retinoblastoma in the vitreous or anterior chamber of the eye. Ovarian carcinoma metastasizes to the peritoneal cavity as avascular seed which expands only after vascularization.
3. When liver metastases in the rabbit model exceed the size of approximately 1 mm, they become vascularized. The growth and vascularization of a subcutaneous transplantable tumor is accompanied by a four-fold increase in blood vessel content in that region.
4. The growth acceleration and metastasis in malignant melanoma can be correlated with the appearance of neovascularization.
5. The thymidine labelling index of solid tumor cells decreases with increasing distance from the nearest open capillary.
6. In several systems the mitotic potential of the endothelial cells in the tumor bed was shown to correlate with the growth rate of the tumor (chick chorioallantoic membrane, aging mice).
7. It has been reported that systemic administration of potentially angiogenic basic FGF facilitated growth of the human colon carcinoma in nude mice, which could be abrogated by specific neutralizing antibodies to basic FGF. Tumor cells themselves did not respond to basic FGF or express high affinity receptors for this factor in tissue sections. Instead strong binding of basic FGF to tumor vasculature was detected.
8. The natural angiogenesis inhibitor, thrombospondin, was found to be down-regulated simultaneously with the acquisition of the transformed phenotype by BHK cells (Bouch, 1990) or in human breast cancer cells (Zajchowski et al., 1990) due to the loss of expression of a putative tumor suppressor gene.

9. Exogenous angiogenesis inhibitors decrease the growth rate of tumors and may cause tumor regression.
10. Several studies based on the epidemiology and natural history of human tumors as well as tumors developing in mice bearing oncogenic transgenes indicate that the switch to the angiogenic state may contribute to the transition towards a higher state of malignancy. This issue shall be addressed in more detail in a subsequent section.

It should be noted that some forms of tumor growth may be independent of angiogenesis. These include ascites formation, mesothelioma and superficial expansion along nerve sheaths (Folkman, 1985, 1990; Mahadevan and Hart, 1990). Several excellent review articles published recently offer detailed descriptions of angiogenesis in relation to tumor growth and metastasis (Folkman, 1987a; Paweletz and Knierim, 1989; Blood and Zetter, 1990; Bouch, 1990; Mahadevan and Hart, 1990; Liotta et al., 1991). Therefore, we would like to give only a brief update of the general knowledge on this subject. We do so in the hope of creating a context for discussion of the issues which we feel may be important for our understanding of the consequences of certain aspects of tumor angiogenesis which have not yet been considered by others in any significant detail.

All tissues in the body, with only a few exceptions, are highly vascularized. This is a result of the *de novo* formation of blood vessels by differentiating angioblasts during embryogenesis (vasculogenesis), vessel enlargement (vascular expansion), and angiogenesis (Auerbach, 1991a). The latter term was coined after Hertig in 1935 (Steiner, 1992a) and occurs in either transcapillary (intussusceptive) formation of tissue pillars (Burri, 1991) or as the multistep process of sprouting of a new capillary from pre-existing postcapillary venules or capillaries (Folkman, 1985; Paweletz and Knierim, 1989).

The angiogenic process begins when the endothelial cells in the vessel wall are exposed to an angiogenic factor(s) released from surrounding tissue. The endothelial cells closest to the stimulus respond to it by a series of characteristic morphological and functional changes. They become metabolically active and increase the number of mitochondria as well as the volume of the endoplasmic reticulum and Golgi apparatus (Paweletz and Knierim, 1989). The protrusions are formed on the abluminal side of the cells and normally very tight intercellular contacts become weak and gaps are formed. Degradation and penetration of the surrounding basement membrane is followed by bulging of the abluminal side of the vessel, due both to the migration of the sprouting cells out of the vessel wall and neighboring endothelial cells towards the base of the sprout. The elongation of the sprout in the form of a solid cord of endothelial cells aligned in a bipolar fashion occurs as a result of proliferation of the cells at the base of the sprout, but can continue for a distance of 1–2 mm solely as a result of migration of the cells at the leading tip, and cell elongation. This was demonstrated in animals in which cell proliferation has been prevented by sublethal irradiation (Sholley et al., 1984).

Lumen formation follows shortly after sprouting endothelial cells form a ring-like shape and connect with adjacent cells in the sprout. Alternatively, lumen formation may proceed simultaneously with cord elongation initially in the form of intercellular space at the site of initial bulging of the vessel wall (Konerding et al., 1991). During embryogenesis a lumen can be formed by the fusion of intracellular vacuoles. Blood flow begins after the neighboring sprout tips approach each other and anastomose; that is fuse with one another, forming a capillary loop which may then further elongate toward the source of the angiogenic stimulus (Paweletz and Knierim, 1989). All the essential functions necessary for the generation of a network of capillary tubules seem to be preprogrammed in endothelial cells since network formation can be reproduced by pure populations of endothelial cells cultured on appropriate biomatrices or on plastic (Folkman and Haudenschild, 1980; Folkman, 1985; Montesano et al., 1990; Montesano et al., 1991). However, full functional maturity may require arrival of and interaction with mural cells (e.g. pericytes) (D'Amore, 1992). Non-endothelial cells (e.g. fibroblasts) can also sometimes form networks of cords while growing in matrices but they are unable to form a lumen (Sage, 1992). *In vitro* studies suggest that endothelial cells may be heterogeneous with respect to this capability (Iruela-Arispe and Sage, 1992; Sage, 1992) as they are for some other properties (Auerbach, 1991a), raising questions concerning the influence of such cell heterogeneity on angiogenesis *in vivo*.

A new vessel is rather leaky and normally the process of maturation begins soon after its formation has been completed (Paweletz and Knierim, 1989). Endothelial cell proliferation ceases soon after the arrival of pericytes to the abluminal surface of the capillary (D'Amore, 1992). This is accompanied by flattening of the endothelial lining and deposition of the basement membrane components. Later, the basement membrane forms a continuous layer. Although endothelial cells are generally quiescent in mature vessels, they may proliferate and/or migrate locally thus participating in a continuous remodelling of the vessel wall (Paweletz and Knierim, 1989).

This series of events is under strict molecular control exerted by various products of endothelial and other accessory cells (pericytes, smooth muscle cells, mast cells, macrophages, lymphocytes, neutrophils, platelets) (Sidky and Auerbach, 1975; Polverini et al., 1977; Blood and Zetter, 1990; Ambrus et al., 1991). Different regulators of angiogenesis have been described and for simplicity they can be categorized as either stimulators (angiogenic factors) or inhibitors. These factors are summarized in Table 1.

Most of these factors have been tested *in vitro* for mitogenic, migratory or endothelial cell tubule-forming activity and in various angiogenic assays *in vivo* (Auerbach et al., 1991b). However, these assays may not reflect the complexity and "fine tuning" of the process in the natural microenvironment (Gullino, 1991). The redundancy of angiogenic factors seems to support this notion (Blood and Zetter, 1990). Another indication is the existence of several paradoxical findings

Table 1. Some Regulators of Angiogenesis*

Stimulators (in vivo)	*Inhibitors (in vivo or in vitro)*
aFGF and bFGF	Interferon α
EGF/TGF α	Interferon β
PD-ECGF	Platelet factor-4
VEGF/VPF	TIMP-1
Substance P	TIMP-2
PDGF	CDI
Angiogenin	Thrombospondin
Angiotensin II	Penicillamine
TGF-β	Vitamin D3 analogues
IL-1	Minocycline
IL-6	Herbimycin A
TNFα	Fumagillin
Prostaglandins	Angiostatic steroids
Fibrin	Interferon-γ
Angiotropin	Tissue extracts from aorta, vitreous, lens

Note: *(Bickell et al., 1991; D'Amore et al., 1988; Folkman et al., 1987c; Blood et al., 1990a; Bouch, 1990; Klagsbrun, 1992)

in this field. For example, although TNFα is known to be a potent stimulator of angiogenesis and endothelial cell migration, certain studies have demonstrated angiotoxicity of this cytokine. This paradox may be resolved if one takes local concentration of the factor into consideration. For example, injection of low doses of TNFα can promote angiogenesis *in vivo* whereas much higher doses can inhibit this process in mice (Fajardo et al., 1992). Similarly, TGF-β exerts an angiogenic effect *in vivo*, although its effect on endothelial cell proliferation and migration *in vitro* is inhibitory. Perhaps this too is related to dose of the factor, although there are other plausible possibilities. For example, it has been shown that TGF-β induces an inflammatory response and thus may in turn indirectly stimulate an angiogenic response (Knighton et al., 1991). On the other hand, while endothelial cells grown in conventional monolayer cultures are indeed inhibited by TGF-β, the same cells in the phase of tubule formation are growth stimulated by exposure to TGF-β (Iruela-Arispe and Sage, 1992). In this regard the responsiveness of endothelial cells to angiogenic cytokines is regulated by the interaction of these cells with extracellular matrix proteins (Ingber and Folkman, 1989; Ingber, 1991). For example, the mitogenic response to bFGF appears to be dependent on the attachment and spreading of cells on fibronectin (Ingber, 1991). Tubule formation appears to be stimulated by the interaction of the luminal surface of the endothelium with collagen I fibers acting as a template and a stimulatory signal transduced by the $\alpha_2\beta_1$ integrin (Jackson et al., 1991). In addition, several angiogenic factors are sequestered in the extracellular matrix, where they are protected from degradation. During an inflammatory or angiogenic process, they can be released from the matrix to express their activity. Indeed, this appears to be the way the most potent known angiogenic factor, bFGF, is mobilized from extracellular storage sites (Vlodavsky et al., 1990, 1991).

There are also important enzymatic activities involved in angiogenesis. The endothelial cells of sprouting capillary buds elaborate metalloproteinases including type IV collagenase, serine proteases, cathepsins and various other enzymes responsible for the initial dissolution of the basement membrane (Liotta et al., 1991). Proteolysis by itself seems to be an important factor in vascular morphogenesis. Montesano and colleagues (Montesano et al., 1990) demonstrated that the transformed phenotype of an endothelial cell line transfected with polyoma middle T antigen can be abrogated, and tubule forming capacity restored, by the addition of exogenous serine protease inhibitors to the culture media. A natural cartilage derived inhibitor (CDI) of angiogenesis appears to be a potent inhibitor of metalloproteinases (Moses et al., 1990). Interestingly, although some inhibition of vessel formation by tissue inhibitors of metalloproteinases (TIMP) has been reported (Bouch, 1990; Langer, 1992), a recent finding indicates that the two activities ascribed to both TIMP and CDI are dissociable (Langer, 1992). The fine balance between cytokines such as bFGF, and TGF-β on the one hand, and proteases and matrix proteins on the other, seems to represent a check point of angiogenesis (Pepper et al., 1990, 1991). Of even greater importance this molecular "cross-talk" appears to take place at the cell surface rather than in solution, and in some cases, may require a heterologous cell population (Rifkin, 1991, 1992). For example, during the maturation of new capillaries, which is associated with the emergence of pericytes, TGF-β is activated. It had been shown that direct interaction between pericytes and endothelial cells is required to cleave the inactive TGF-β, liberating the active form (Antonelli-Orlidge et al., 1989). This process, in which plasminogen activator (uPA), plasmin and other components are involved, probably occurs after binding of those components to their receptors on the surfaces of endothelial cells and pericytes in the appropriate configuration (Rifkin, 1991). The resulting activation of TGF-β may then contribute to the down-regulation of endothelial cell growth, changes in production of extracellular matrix proteins and vessel maturation (Antonelli-Orlidge et al., 1989; Folkman et al., 1989b).

In summary, angiogenesis appears to be a highly organized and tightly regulated multistep process. Although an understanding of the molecular mechanisms is still in its infancy, a number of angiogenesis regulators have been characterized and their relevant genes cloned. Various cytokines, extracellular matrix proteins and enzymes capable of degrading them, as well as enzyme inhibitors, all seem to be involved in angiogenesis. The question remains as to whether there is a single common mechanism of angiogenesis, or several different pathways, and what actually happens when this tissue reaction is triggered by tumor cells as opposed to inflammatory or repair processes.

A. Features of Tumor Neovascularization

There are similarities between the formation of new blood vessels during physiological angiogenesis (wound healing, ovulation) and tumor-associated angiogenesis. The latter process, in fact, was a model for some of the studies discussed above (Folkman, 1986). The possibility exists, however, that these generic similarities are superficial and that angiogenesis in the tumor microenvironment has some distinct features. This belief is supported by morphological studies indicating that the maturation process of new capillaries induced by a tumor is incomplete (Paweletz and Knierim, 1989; Blood and Zetter, 1990). Presumably, the continuous presence of various angiogenic stimuli, proteases, as well as abnormalities in extracellular matrix and cytokine levels, explains the obvious dysfunction of the process as a whole. Unlike normal capillaries, where endothelial cells usually become quiescent, the mitotic rate of these cells in tumor vessels can be elevated up to 50-fold (Denekamp, 1990). Though not as high as in placenta or granulation tissue, it is apparently more persistent. The tumor-associated microvessel wall is thin and development of the basement membrane is retarded. An absence of pericytes has also been noted. Unlike those in normal tissues, tumor blood vessels have no collateral potential, lack innervation, and their responsiveness to pharmacological stimulation by vasoconstrictors and vasodilators is compromised (Paweletz and Knierim, 1989; Blood and Zetter, 1990; Denekamp, 1990). Only rare lymphatics are found in the tumor tissue (Folkman, 1990). Within the tumor mass, the blood vessels are tortuous, saccular, dilated and multiple arterio-venous shunts are present (Jain, 1990). Several different categories of tumor blood vessels can be distinguished on the basis of ultrastructural studies (Jain, 1990). Sinusoids, fenestrated capillaries and blood channels without apparent endothelial lining have been found. It has been postulated that even tumor cells themselves may contribute to the vessel wall as well as vascular sprouts (Paweletz and Knierim, 1989). Some of these features are so consistent that they can be used diagnostically, for example, in renal cell carcinoma (Prout and Garnick, 1982).

The architecture of tumor blood vessels may undergo a significant change over time (Jain, 1990), and intra-tumor microvessel heterogeneity often exists. In the tumor periphery, the rapidly developing blood vessels have been found to be leaky for macromolecular tracers (Nagy et al., 1988; Dvorak et al., 1991a). This leakiness may be a consequence of secretion of vascular permeability factor (VPF) (Keck et al., 1989; Senger et al., 1990; Dvorak et al., 1991a) . The activity of the specific vascular endothelial cell growth factor (VEGF) has been recently ascribed to a molecule closely related to VPF (Gospodarowicz et al., 1989; Leung et al., 1989; Conn et al., 1990; Dvorak et al., 1991b). This angiogenic cytokine which can be secreted by tumor cells, accumulates in the vicinity of the blood vessels and may be responsible for leakage of fibrinogen from the vascular bed and formation of extracellular fibrin matrices—which by themselves facilitate further angiogenesis. This mechanism may play a pivotal role in tumor-induced angiogenesis (Dvorak et al., 1991b). Structure and function of tumor blood vessels can also be affected by hypoxia, acidification and nutrient deprivation in the tumor mass (Ogawa et al., 1991). Organ-specific features of microcirculation and/or local characteristics of endothelial cells (Auerbach, 1991a; Auerbach et al., 1987) may also bear on development of vasculature in various tumors (Jain, 1990). Interestingly, the vasculature within the same tumor appears to be highly heterogeneous. Not only are there differences in blood vessel appearance and density between the necrotic center, the periphery and the zone dividing these two regions, but also differences within these zones are frequently observed (Jain, 1990). It is tempting to speculate that these areas may co-localize with domains occupied by different locally expanding tumor cell subpopulations expressing different angiogenic potentials.

The vascular space within tumors varies between 1–20% (Jain, 1990). The number of endothelial cells present in the tumor mass is usually estimated as being up to 10% (Paweletz and Knierim, 1989), but in some hemangiomas host endothelium recruited to the tumor site may reach up to 90% of the total cell number (Williams et al., 1989). This rather high ratio was noticed in early studies by Algire and others, and became the conceptual foundation for the notion that tumors receive an excessive blood supply (Blood and Zetter, 1990). However, this was difficult to reconcile with the fact that tumors often develop large necrotic areas. More accurate measurements revealed, in fact, a decrease in the capillary blood flow accompanying tumor growth (Blood and Zetter, 1990). Various abnormalities of the tumor microcirculation can account for this effect. The formation of bypassing shunts, blood vessel compression and/or outgrowth by the tumor mass, stasis and frequent changes in direction of blood flow in tumor microvessels as well as high interstitial pressure obstructing penetration of nutrients (and drugs) all create a situation in which a large proportion of an "excessive" tumor vasculature may not be functionally normal (Jain, 1990).

B. Properties of Tumor-Associated Endothelium

The central role of endothelial cells in the blood vessel formation process implies that the numerous abnormalities of tumor microcirculation are likely to be due to tumor-induced abnormalities in the phenotype and function of these cells. One obvious experimental approach to study the ways in which tumor-associated endothelial cells are different from those in healthy tissue would be to isolate and compare them to each other. Successful isolation of endothelial cells from tumors by using specific endothelial markers, selective media and cell-sorting techniques, has been reported recently by some investigators (Costello and Del Maestro, 1990; Neville, 1991). It should be noted, however, that there are several inherent technical and conceptual difficulties associated with this strategy. For example, it is not known whether potential differences between "normal" and tumor-associated endothelium are stable, or inducible by the presence of the tumor microenvironment. There is a tacit assumption that tumor vasculature will express endothelial markers in the same fashion as it does in the healthy tissue. However, to cite one example, it is known that hypoxia is likely to alter von Willebrand Factor (vWF) antigen expression by endothelial cells (Ogawa et al., 1991). Since isolation of tumor endothelial cells usually requires different kinds of experimental manipulations than those used in the case of normal tissue, there is a certain degree of uncertainty as to the influence of such manipulations on the cell phenotype. Finally, the endothelial cells isolated from the tumor mass may be representative of a portion of the vasculature which has been encompassed by the tumor rather than of newly formed vessels.

Another difficulty in characterizing tumor-associated endothelial cells seems to be related to the fact that tumor angiogenesis is a result of the interactions between populations of tumor cells and endothelial cells, both of which are heterogeneous (Heppner and Miller, 1983; Zetter, 1988; Auerbach, 1991a). Endothelial cell heterogeneity had been discussed primarily in terms of the origin of these cells. Initially it was postulated that they were the progeny of angioblasts from the yolk sac which migrate to, and differentiate within organs during embryogenesis. Alternatively, it was thought that they may be generated locally (Auerbach, 1991a). There are also differences between microvascular endothelial cells and those isolated from large vessels (Blood and Zetter, 1990; Beitz et al., 1991). Organ-specific characteristics have been found in endothelial cells derived from various murine organs (Auerbach, 1991a; Belloni et al., 1992). This is illustrated for example by the differential expression of P-glycoprotein by endothelial cells obtained from various anatomical locations (Cordon-Cardo et al., 1990b). Another example is the presence of site-specific adhesion molecules on the surface of endothelial cells whereby they participate in the selective homing of lymphocytes in lymphatic organs (Sher-Taylor et al., 1988). In addition, different endothelial adhesion molecules and growth factors have been incriminated in the phenomenon of organ-specific patterns of metastasis (Auerbach et al., 1987; Nicolson, 1988; Sher-Taylor et al., 1988). These organ-specific differences may not be the only source of differences between endothelial cells. It has been shown that the lining of the vessel wall may contain "domains" with high and low turnover rates, with larger and smaller sizes of endothelial cells (Heimark and Schwartz, 1988). Distinct subpopulations of highly and poorly "angiogenic" cells have been isolated from cloned rat endothelial cells grown in culture (Iruela-Arispe and Sage, 1992). This suggests that certain functional heterogeneity (stem cell-like hierarchy?) of endothelial cells may exist even within the same anatomical location. It remains unknown whether tumors or repair reactions mobilize the whole population of endothelial cells in the local capillary bed or whether they recruit certain subsets of those cells to form new blood vessels.

Whichever is the case, histological techniques using specific antibodies have revealed very interesting differences between tumor-associated and "normal" endothelial cells. The monoclonal antibody EN 7/44, for example, was raised by immunizing mice with highly vascular fragments of human breast carcinoma. This antibody specifically recognizes budding blood vessels in tumorous and hyperplastic breast tissue (Hagemeier et al., 1986). Neither the nature of the relevant antigen, nor what role, if any, it plays in the angiogenic process is known (Hagemeier et al., 1986). An elevated expression of the c-ets1 oncogene encoded transcription factor has been observed in endothelia found in malignant and benign tumors, granulation tissue, embryonal murine endothelium and exponentially growing brain or aortic endothelial cells in tissue culture (Wernert et al., 1992). Also, tumor-associated fibrocytes and Kaposi sarcoma cells are positive for this marker. In marked contrast, quiescent endothelia from healthy tissue were virtually negative. Similarly, confluent endothelial cells in culture, unless treated with TNFα, expressed several-fold less of the c-ets1 transcript than corresponding cell cultures in log phase (Wernert et al., 1992). Interestingly, c-ets1 is thought to activate transcription of various proteases such as uPA, stromelysin and collagenase genes through the "PEA3" DNA binding motif (Wernert et al., 1992). Gene expression of c-ets1 during tumor angiogenesis seems to be a feature reminiscent of embryonal endothelium. By analogy, this observation raises the possibility that endothelial cell-specific expression of certain other genes during embryogenesis may also have some relevance for tumor angiogenesis. Perhaps tumor-induced rapid blood vessel formation is associated with expression of the "embryonal phenotype" of endothelial cells. Possible markers for such a study include, for example, newly discovered flk-1 and tek thyrosine kinases (T. Yamaguchi, J. Rak, D. Dumont, unpublished observation) or fli-1 (B. Motro, personal communication) transcripts of which have been found to be specifically expressed by murine embryonal endothelium. Are these genes also expressed by tumor-associated endothelium? If so, what mechanisms activate their expression, and what role do they play in endothelial cell function? These questions are the subject of on-going research. Finally, two other recent interesting examples of phenotypic alterations in activated proliferating endothelium are worth noting. First, the high affinity receptor for bFGF is selectively overexpressed by endothelium in murine tumors (Janet Gross-Dzubow, personal communication). Second, the unique positioning of the fibronectin receptor on the luminal surface of rapidly proliferating endothelial cells has also been reported (Thorpe et al., 1991).

Although many published studies stress differences between tumor-associated versus "normal" quiescent endothelial cells, it is unclear whether the phenotypic abnormalities observed are unique for tumor blood vessels or whether they accompany any situation where rapid angiogenesis is induced, regardless of the stimulus. There are a few scattered reports suggesting that a permanent change in behavior of tumor-associated endothelial cells during interaction with tumor cells may occur. The SK HEP-1 cell line isolated from the ascites fluid of a liver adenocarcinoma patient was recently found to express multiple phenotypic markers of endothelial cells (Heffelfinger and Darlington, 1991). These cells are immortal and extremely heterogeneous with respect to their capacity to interact with extracellular matrix, invade different substrata and form tubular networks *in vitro* (Heffelfinger and

Darlington, 1991). It has been suggested that these apparently transformed cells are of endothelial rather than hepatocyte origin. Obviously the initial diagnosis of the donor would have to be confirmed and laboratory artefact ruled out before this finding can be considered as evidence of hepatoma-associated transformation of endothelium. However, a precedent exists since concomitant transformation of stromal fibroblasts in the course of prostate cancers had been reported (Russel et al., 1990). Despite extensive studies in this area, the mechanism of endothelial cell recruitment and the nature of the phenotypic change they undergo at the tumor site are still poorly understood.

C. Tumor Cell–Endothelial Cell Interactions

Various types of interactions between tumor cells and endothelium repeatedly occur during the natural history of cancer. The rearrangements of tissue architecture during tumor expansion result in invasion of the local blood vessels by the tumor cells. These vessels "reciprocate" forming sprouts which approach the tumor site, leading to a gradual bidirectional interposition (Paweletz and Knierim, 1989; Liotta et al., 1991). Metastatic tumor cells penetrate the vessel wall, first toward the lumen in the primary site (intravasation), and then toward the extravascular space of the secondary organ (extravasation) where, at least in some instances, they again trigger angiogenesis (Liotta et al., 1991). Failure to do so would preclude development of microscopic metastases into macroscopic tumor deposits. Thus, organ-specific patterns of metastases may in some cases be accounted for by competence to induce angiogenesis in a particular foreign organ environment but not in others. Behind these superficially simple "mechanics" of intravasation and extravasation there is a series of cellular events comprising cell proliferation, cell death, adhesion, migration, invasion, and matrix remodelling. These events are driven by several possible combinations of molecular mechanisms which are currently being analyzed in many laboratories; some of them have already been reported and reviewed by several authors (Blood and Zetter, 1990; Liotta et al., 1991). At this point it is difficult to generalize about the possible ways tumor and vascular endothelial cells communicate with each other. What seems to be clear, however, is that the potential to interact may differ depending upon the characteristics of a particular tumor or endothelial cell subpopulation. And conversely, various subsets of tumor cells appear to have the potential to respond in different ways to signals presented by nearby endothelial cells.

Endothelial cells are capable of responding to a wide array of mitogens and growth inhibitors (as summarized in Table 1). This issue has been the focus of numerous studies on growth interactions between tumor and endothelial cells since several of those factors may be elaborated by various tumor cell types, and may serve as an equivalent of the "tumor angiogenic factor" activity (TAF) (Folkman and Klagsbrun, 1987b). It should be noted however that angiogenic and mitogenic activities are separable, although largely overlapping (Folkman and Klagsbrun,

Table 2. Growth Factors Produced by Endothelial Cells Which Could Affect the Behavior of Adjacent Tumor Cells

Growth Factor	*Abbreviation*	*Reference*
Platelet Derived Growth Factor	PDGF	(Borsum, 1991)
Connective Tissue Growth Factor	CTGF	(Bradham et al., 1991)
Basic Fibroblast Growth Factor	bFGF	(Borsum, 1991)
Endothelial Derived Growth Factor	EDGF	(Borsum, 1991)
Heparin-like inhibitor		(Borsum, 1991)
Peptide inhibitor of SMC growth		(Borsum, 1991)
Granulocyte Colony Stimulating Factor	G-CSF	(Borsum, 1991)
Macrophage Colony Stimulating Factor	M-CSF	(Cross and Dexter, 1991)
Granulocyte-Macrophage Col. Stim. Factor	GM-CSF	(Borsum, 1991)
Transforming Growth Factor β	TGF-β	(Antonelli-Orlidge et al., 1989)
Insulin-like Growth Factor I	IGF-I	(Boes et al., 1991)
Interleukin-1	IL-1	(Borsum, 1991)
Interleukin-6	IL-6	(Bussolino et al., 1991)
Interleukin-8	IL-8	(Bussolino et al., 1991)
Leukemia Inhibitory Factor	LIF	(Lubbert et al., 1991)
Transferrin		(Nicolson et al., 1991b)
Melanoma Growth Stimulatory Activity	MGSA	(Wen et al., 1989)

1987b; D'Amore and Braunhut, 1988; Bickell and Harris, 1991; Klagsbrun, 1992). The other side of this "coin," that is regulation of tumor cell proliferation by products of endothelial cells (Table 2), has not attracted nearly as much experimental attention. This is presumably because the nutrients and cytokines found in the blood are thought to be a more powerful source of stimuli than endothelial cells lining the vessels *per se*. However, this view seems to be somewhat inconsistent with common knowledge concerning interactions between endothelial cells and platelets, leukocytes and mural cells, which apparently can take place in the presence of blood-derived factors (Honn et al., 1987; D'Amore, 1992). Since these local intravascular or perivascular interactions are often mediated by membrane- or matrix-bound growth factors (Rifkin, 1991), one may postulate that similar growth regulatory cross-talk between tumor cells and endothelial cells is likely, for example in situations where small numbers of tumor cells liberated by a primary tumor mass are confronted with an excess of endothelial cells. This would normally occur when a single tumor cell or small emboli of tumor cells have arrested in a distant capillary bed. In addition, the perfusion of the tumor capillary bed is rather poor and therefore there might be excessive vascularity locally accompanied by insufficient blood flow and nutrient (e.g. cytokine) supply. For these reasons, it may be worthwhile to consider the potential consequences of the fact that endothelial cells elaborate various extracellular matrix proteins, chemoattractants and cytokines (some of them listed in Table 2) some of which may modulate tumor cell growth, both at the primary site and at distant metastatic sites.

In this context, it has been shown that tumor cells may selectively respond to growth and motility factors secreted by endothelial cells isolated from organs to which those tumor cells preferentially metastasize (Nicolson et al., 1991b). If this selective interaction is important for a "seed and soil" effect in metastasis, one can speculate that perhaps similar selective interactions may operate within the primary tumor.

Not only can the proliferation of tumor cells be modulated by endothelial cells but so might their very survival. For example, it has been shown that tumor cells attached to endothelial monolayers become resistant to NK-mediated lysis (Kaminski and Auerbach, 1988). However, microvascular endothelial cells themselves, when activated with appropriate cytokines (for example, TNFβ and IFNγ), have been shown to acquire the capability to lyse murine melanoma cells *in vitro* (Li et al., 1991). Conversely, at least for Walker 256 carcinoma cells and some human tumor cell lines, it has been demonstrated that killing of endothelial cells by tumor cells may occur (Al-Mondhiry and McGarvey, 1987; Shaughnessy et al., 1991). Besides stimulation of angiogenesis, one of the most prominent areas of research in the field of tumor-endothelial interactions seems to be tumor cell attachment to, and penetration through the vascular wall in metastasis (Blood and Zetter, 1990). Several mechanisms of tumor cell entrapment in the distant organ may be operational. Mechanical embolization or arrest of multicellular aggregates seems to be an efficient way of tumor cell dissemination (Fidler, 1973a; Lione and Bosmann, 1978; Miller and Heppner, 1990). However, there is also evidence that tumor cells may specifically bind to endothelial cell adhesion molecules in a way somewhat similar to the "homing" process of leukocytes (Sher-Taylor et al., 1988). It is not currently clear whether tumor cells utilize exactly the same pathways of interaction, including "rolling" on selectins followed by stable attachment via cellular adhesion molecules (Springer, 1992). Nevertheless, the binding of human and murine melanoma cells to the INCAM-110 (VCAM-1) receptor or human colon carcinoma cells to ELAM-1 selectin both expressed on cytokine-activated endothelial cells has been recently reported (Rice and Bevilacqua, 1989; Lauri et al., 1991). Also the gp IIb/IIIa integrin was shown to mediate IL-1 stimulated adhesion between melanoma and endothelial cells in culture (Burrows et al., 1991). *In vivo* treatment of animals with IL-1 increased lung colony formation of various tumors probably through a similar mechanism (Giavazzi et al., 1990).

Following attachment, tumor cells extravasate, that is leave the lumen of the blood vessel. They can then penetrate the surrounding basement membrane after induction of endothelial cell retraction (El-Sabban and Pauli, 1991). Prior to that however, some other forms of selective endothelial-tumor cell signal exchange may occur. It has been recently reported that metastatic tumor cells seem capable of establishing gap junctions with endothelium (El-Sabban and Pauli, 1991). Through these tight contact sites, dye transfer and presumably exchange of regulatory molecules can occur (El-Sabban and Pauli, 1991). Elevated levels of connexin mRNA was observed in metastatic (but not in non-metastatic) tumor cells which were also unable to transfer intracellular dye to endothelial cells (El-Sabban and Pauli, 1991). It is intriguing to speculate about what kind of material may be delivered or transferred to endothelial cells from adjacent metastatic tumor cells along with the dye, how endothelial cells would react, and whether similar interactions take place in the primary tumor. It has been shown that metastatic tumor cells sometimes have a propensity to form somatic cell hybrids spontaneously with each other *in vivo* and with non-metastatic counterparts, the latter becoming more aggressive as a result of the fusion (Miller et al., 1989). Another question which emerges from these studies is whether various endothelial cell populations establish junctions with different subpopulations of metastatic tumor cells to the same degree.

Tumor cell invasion and angiogenesis are both associated with expression of the invasive phenotype (Liotta et al., 1991). One of the essential components of this phenotype is the capability by invading cells to break down the extracellular matrix through the release of various proteolytic enzymes (Liotta et al., 1991). Several classes of proteolytic enzymes have been shown to be involved, including metalloproteinases, serine proteases and cathepsins (Liotta et al., 1991). These enzymes are capable of digestion of collagens I, IV, and V, fibronectin, laminin and other matrix components (Liotta et al., 1991; Basset et al., 1990). The obvious implication of this is that the local net activity of these enzymes is likely to exceed the levels achievable by endothelial or tumor cells alone. Therefore, an additive or even synergistic effect on invasiveness of both endothelial cells and tumor cells may be expected. For example, breaking of the basement membrane, which is necessary for intravasation, may be achieved by proteolysis and invasion of tumor cells inside the vessel or invasion of sprouting endothelial cells outside the vessel (Mahadevan and Hart, 1990). It seems important to keep in mind that proteases may have an influence on endothelial morphology and behavior since protease inhibitors reverse the transformed phenotype of the endothelial cell line (Montesano et al., 1990). Also, it is possible that stromal cells may contribute to cancer progression and metastasis by secreting stromelysin III as is the case in breast cancer (Basset et al., 1990). In this regard endothelial cells may have considerable potential to contribute to this process.

The description of the numerous types of tumor cell-endothelial cell interactions and their possible significance probably deserves a separate monograph. Our goal is simply to point out that those interactions are reciprocal and seem to extend beyond considerations of angiogenesis in the primary site, and tumor cell arrest in the microcirculation at the secondary site. In this sense, one may ask to what extent does the vasculature contribute to the phenomena associated with, or referred to as tumor progression?

IV. THE RELATIONSHIP BETWEEN TUMOR VASCULARIZATION, PROGRESSION AND METASTASIS

Despite extensive work being done in this area, there seems to be little agreement as to the exact time frame and the content of the term "tumor progression." Besides sequential genetic changes and instability in general (Cifone and Fidler, 1981; Nowell, 1986; Fearon and Vogelstein, 1990b; Nicolson, 1991a), it is thought that the acquisition of invasiveness and metastatic competence, increased growth rate, growth autonomy, loss of a differentiated phenotype, loss of antigenicity and expression of a drug-resistant phenotype are the relevant criteria to assess tumor progression (Nowell, 1986). It seems reasonable (for the purpose of this article) to establish the beginning of this process at, or prior to, the development of distinguishable premalignant lesions; at this stage the process can be observed and predictions can be made as to the frequency and direction of further steps or stages of progression. This is admittedly arbitrary and one can argue that point mutations, initial disruption of tissue homeostasis, or "progressive state" selection in the cell population which otherwise remains visually normal, may precede, and contribute to causality of the later developments (Benedetto et al., 1990; Rubin, 1990; Farber and Rubin, 1991; Hunter, 1991). These exceptionally early stages—which may precede the development of overt cancer by decades in humans—are poorly understood, and most likely unrelated to the state of the local vasculature.

The transition from precancerous neoplastic lesion to cancerous growth seems to be strongly correlated with and often preceded by the induction of angiogenesis. Early studies demonstrated that in the murine mammary gland, various classes of hyperplastic alveolar nodules (HAN) progress to tumors with a frequency correlating with their angiogenic potential (Folkman, 1985 and references therein). Human hyperplastic breast epithelium was also shown to be able to induce angiogenesis upon implantation in the rabbit iris (Brem et al., 1978). In contrast, normal epithelium, fibroadenoma, fibrous, adipose or fibrocystic tissue, lipoma and gynecomastia failed to generate a similar significant vascular response (Brem et al., 1978). Morphologically "normal" epithelium isolated from breast cancer patients was shown to be twice as angiogenic as the material isolated from non-cancerous breast tissue (Jensen et al., 1982). Mouse fibroblasts maintained in serial passage *in vitro* first become angiogenic and sometimes afterwards acquire a tumorigenic phenotype (Ziche and Gullino, 1982). An elegant study performed by Folkman and colleagues demonstrated that the acquisition of angiogenic potential indeed may be a precondition for progression from hyperplasia to neoplasia *in vivo* (Folkman et al., 1989a). A large proportion of transgenic mice expressing the large T SV40 oncogene driven by the rat insulin promoter develop hyperplasias of beta pancreatic islets. In many cases these give rise to pancreatic tumors. These islets can be analyzed *in situ* or isolated and tested for angiogenic potential *in vitro*. It turns out that the angiogenic potential developed in islets destined to become cancerous, prior to their overt tumorigenic conversion (Folkman et al., 1989a). In this regard it is known that in cases of histologically malignant "carcinoma *in situ*" of the cervix, bladder and breast, the lesions may remain silent and avascular for several years (Weidner et al., 1991 and references therein).

At present little is known of the nature of this "switch" to the angiogenic state during tumor progression. However, significant progress has been made recently as a result of collaboration between groups led by Folkman and Hanahan through application of transgenic mouse technology. These investigators chose the model of spontaneously developing fibromatosis which gradually progresses to fibrosarcoma in transgenic mice expressing the bovine papilloma virus gene. They noticed that neovascularization rapidly increases upon transition from benign to an aggressive form of the lesion. Subsequently, it was demonstrated that the change is associated with a "switch" to a massive extracellular export of highly angiogenic bFGF which remained intracellular in normal cells and benign lesions (Kandel et al., 1991).

The obvious requirement for access to blood vessels by metastases has prompted several investigators to examine more carefully the impact of tumor vascularity on tumor dissemination (Mahadevan and Hart, 1990; Weidner et al., 1991). The early experiments by Liotta demonstrated that the appearance of tumor cells in the blood stream corresponds to the ingrowth time of the blood vessels in the transplantable murine fibrosarcoma while the number of lung metastases correlated with tumor vascularity (Liotta et al., 1974). Furthermore, blood vessel invasion was found to be a prognostic indicator in a study comprising 175 breast cancer patients (Weigand et al., 1982). The recent work by Weidner and colleagues demonstrated a strong correlation between incidence of metastasis during the clinical course of the initial, invasive breast carcinomas and blood vessel density in the primary tumor tissue specimens (Weidner et al., 1991). Realizing that very small or compressed capillaries, or venules, in conventional tissue sections may be extremely difficult or impossible to visualize, these investigators, unlike many others, highlighted endothelial cells (and thereby vessels) by immuno-staining for factor VIII related antigen (which stains endothelial cells specifically). They concluded that the blood vessel count may be an independent and accurate predictor of the regional and distant dissemination of the disease. Interestingly, since the blood vessel count was performed in the most intense vascular areas ("hotspots") of each tissue specimen (Weidner et al., 1991), one can reason that even localized increases in vascularity increase the risk for development of metastases. Furthermore, these hypervascular regions may represent localized regions of clonal outgrowth of malignant cell subpopulations, that is dominance of both highly metastatic and angiogenic subpopulations of the breast carcinoma.

Human malignant melanoma also provides an interesting possible tumor model to study the relationship of angiogenesis and metastasis. Radial growth phase (RGP) melanomas represent a prevascular, curable disease (Folkman, 1987a). In the vertical growth phase (VGP), thickness of the lesion begins to increase, and this is associated with the rapid acceleration of tumor growth and metastasis. In

1970 Breslow introduced a threshold "thickness" of 0.76 mm or greater in which the prognosis (as to recurrence and metastasis of VGP melanomas) worsens dramatically (Breslow, 1970). Although it is not known why this particular thickness delineates two different phases of the disease, the vertical expansion of melanoma is often associated with the entry of the cells into an ectopic tissue compartment, that is the dermal mesenchyme (Clark, 1991). This microenvironment was shown to be highly growth inhibitory for the proliferation of early-phase melanoma cells (Clark, 1991; Cornil et al., 1991; Lu et al., 1992) and frequent regression of the tumor infiltrates is often observed (Clark, 1991). Stromal cells such as fibroblasts (Cornil et al., 1991; Lu et al., 1992) and endothelial cells are capable of secreting several possible inhibitory cytokines for early-stage melanoma, including IL-6, which probably contribute to these phenomena (Kerbel et al., 1992). At the same time "thin" VGP lesions become thicker, their demand for blood vessels (i.e. nutrients, etc.) is likely to increase. Under such pressures the melanoma cell population undergoes an adaptation due to selective clonal expansion. Cellular variants resistant to the inhibitory cytokines may emerge, thus enabling tumor expansion in the presence of fibroblasts (Cornil et al., 1991; Lu et al., 1992; Kerbel et al., 1992) and endothelium (Rak et al., 1991b; Hegmann et al., 1992). The endothelial cells begin to respond to the angiogenic stimuli released by tumor cells thus initiating neovascularization of the lesion.

The exact time frame of the switch to the angiogenic state in melanomas is not known. Srivastava et al. reported that there are prognostically significant differences in vascularity after transition from "thin" to intermediate thickness primary melanomas (0.76–4.0 mm) (Srivastava et al., 1986, 1988, 1989; Folkman, 1987a). This would suggest that the angiogenic state is still developing at the time when the metastatic spread is likely to begin. Thus, the eventual manifestation of the metastases in melanoma may be dependent in some way upon the degree of vascularization of the primary lesions when first detected (Folkman, 1987a). The vascularization may simply facilitate the dissemination of pre-existing metastatically competent cells or perhaps stimulate their growth and clonal expansion. Smolle et al. (1989) reported similar findings. A more recent study, however, based on analysis of 107 melanoma cases, does not confirm the results obtained by Srivastava but rather suggests that, at least with the range of thickness of 0.85–1.25 mm, tumor vascularity is of no prognostic value (Carnochan et al., 1991). The authors postulate that the breakpoint in tumor vascularization may occur at the level of 0.76 mm thickness, in agreement with the "all or nothing" results obtained by Doppler ultrasound analysis of tumor blood flow (Carnochan et al., 1991). In all these studies an extensive vasculature at the tumor base was observed. However, unlike the experiments of Weidner et al., these investigators did not preselect the most intensely vascular areas ("hotspots") of tumors for quantitation (Weidner et al., 1991).

The diverse capacity to stimulate angiogenesis has been observed among human melanomas in clinical studies (cited above), as well as in experiments with melanoma xenografts grown in nude mice (Solesvik et al., 1982). Some preliminary *in vitro* studies indicate that highly aggressive malignant melanoma cell lines elaborate greater amounts of endothelial cell growth stimulatory factors than their less aggressive counterparts (Wesseling et al., 1990; and Vacca, 1991). Although angiogenic potential is thought to develop independently of other characteristics expressed by metastatically-competent melanoma cells, the latter result suggests that at certain stages in progression these two features "couple" with each other in a way which still needs to be clarified. In this respect some light may be shed on this subject by the work of several authors who found a change in cytokine expression pattern during melanocyte transformation (Halaban et al., 1988; Albino et al., 1991) and melanoma progression (Herlyn et al., 1990b). These studies suggest that expression of bFGF, TGF-α and TGF-β may be particularly important in melanomagenesis (Albino et al., 1991). bFGF was proposed to be an autocrine growth stimulator for melanoma cells (Halaban et al., 1988). It is striking that most of these growth factors are potent angiogenic factors and thus may simultaneously drive tumor cell proliferation in an autocrine manner as well as tumor neovascularization.

Although the search continues for evidence as to whether tumor progression is associated with, or supported by, a continuous stepwise increase in angiogenic capacity (Allen and Maher, 1991; Boukamp et al., 1991; Christofori et al., 1991; Maxwell et al., 1991; Schweigerer et al., 1991), it is apparent that this characteristic by itself does not induce or explain the malignant behavior and metastasis of tumor cells. Thus several tumor types including adrenal adenomas and hemangiomas remain benign despite having an excessive vasculature (Folkman, 1985). This has been interpreted as an indication that development of the malignant phenotype is independent from angiogenesis (Folkman, 1985). In this context it is difficult to understand why selection of more angiogenic and metastatic variants seem to continue beyond initial phases of tumor progression. We provide a possible resolution to this paradox in a later section. Before doing so we discuss one other aspect of angiogenesis which may be of possible relevance to cancer progression, and cancer treatment, and which has not received any significant experimental attention.

V. THE PARADOX OF THE RESISTANCE OF TUMOR VASCULATURE TO CANCER CHEMOTHERAPY

As discussed above, tumor vasculature is considered to be a vital component of the tumor tissue as a whole, necessary for tumor growth and subsequent progression. Many cases have been reported illustrating the detrimental effects on tumor growth of a compromised blood supply. For example, anti-angiogenic compounds such as heparin-cortisone acetate (Folkman et al., 1989b), flavone acetic acid (FAA) (Denekamp, 1990) and fumagillin (Ingber et al., 1990) were shown to reduce tumor growth in mice even though these agents are not necessarily directly toxic to tumor cells. Research into the discovery of new anti-angiogenic compounds has flourished in the last several years, and indeed it has been proposed that the targeting of normal tissues considered essential for tumor growth (i.e. tumor vasculature) is a logical and fruitful area for experimental therapeutics (Denekamp, 1990). This approach may help circumvent the problem of the development of resistance by tumor cells to therapeutic agents due to the genetic instability and phenotypic flexibility of their genome (Kerbel, 1991).

In addition to the compounds "designed" to have anti-angiogenic properties in tumor tissue, other compounds initially believed to have direct cytotoxic effects on tumor cells may actually mediate tumor regression indirectly by vascular occlusion or destruction (Denekamp, 1990). Agents such as TNF, various interleukins, interferon and endotoxin, the radiation sensitizer and protector misonidazole and ethiophos WR2721, respectively, as well as the treatment regimens of photodynamic therapy, radiation and hyperthermia all may have anti-tumor effects in mice mediated in part, or in full, by damaging the tumor vasculature (Denekamp, 1990). Whether this is also true in humans is highly questionable and remains a subject of considerable experimental and clinical importance.

Although tumor vessels may be sensitive to certain anticancer agents, the cytotoxic effects of cancer chemotherapy agents on tumor angiogenesis is less clear. As discussed earlier, one of the most compelling differences between normal and tumor-associated blood vessels is the rapid cycling of a significant proportion of tumor vessel endothelial cells (Denekamp, 1990). Given that many chemotherapeutic drugs are often toxic to normal rapidly cycling cells in the bone marrow and gastrointestinal tract, for example, one might assume that the cycling endothelial cells in tumor vessels would also be susceptible to such chemotherapy-induced cytotoxicity. Thus, even if tumor cells were resistant to a given chemotherapy drug, one would expect indirect tumor kill via reduced blood supply, as has been observed after interferon therapy *in vivo* of interferon-resistant mouse leukemias (Dvorak and Gresser, 1989) in which vascular collapse mediated by interferon was found to precede tumor regression. By analogy, a significant tumor regression might be expected after treatment of drug-resistant tumors with classical cell-cycle specific chemotherapeutic compounds because of induced vascular collapse. Unfortunately this is infrequently observed following chemotherapy in humans for most types of solid tumors. We postulate that one possible reason for this may be the expression of a drug-resistant phenotype by tumor-associated endothelial cells themselves (Kerbel, 1991).

Evidence exists for the presence of drug resistance mechanisms in endothelial cells in both normal and tumor tissues. The 170-kDa conserved membrane protein P-glycoprotein is associated with a multidrug resistance phenotype *in vitro* (Bradley et al., 1988; Roninson, 1992) and with clinical drug resistance (Chan et al., 1990, 1991; Roninson, 1992) and tumor progression (Weinstein et al., 1991). P-glycoprotein has been localized by using immunohistochemistry in normal tissues as well, including human endothelials of the blood-brain and blood-testes barriers and papillary dermis (Cordon-Cardo et al., 1990a), endometrium (Axiotis et al., 1991), ovaries and female genital tract (Finstad et al., 1990), esophagus and bronchi (Cordon-Cardo et al., 1990a). In addition, P-glycoprotein has been observed in endothelial cells within non-Hodgkin's lymphoma, leiomyosarcoma and lymphoma (Schlaifer et al., 1990). Since blood-brain barrier endothelial cells express P-glycoprotein, brain tumors may be shielded from chemotherapeutic drugs in the blood, as proposed by Cordon-Cardo and colleagues (Cordon-Cardo et al., 1990a) and Sugawara (Sugawara, 1990). Also, the vessel endothelial cells within brain tumors express P-glycoprotein (Becker et al., 1991; Nabors et al., 1991). Chemotherapy resistance observed in primary brain tumors may therefore be due in part to the ability of P-glycoprotein to confer resistance to tumor-associated endothelial cells, thereby protecting the tumor from hemorrhagic necrosis followed by tumor regression (Nabors et al., 1991).

In addition to P-glycoprotein, endothelial cells also express other markers associated with a drug-resistant phenotype. Human endothelial cells of capillaries and arteries of the digestive, respiratory and genital tracts as well as the meninges, brain and skeletal muscle were found to be positive *in situ* for glutathione S-transferase, one member of an important family of isoenzymes involved in cellular detoxification (Terrier et al., 1990). Depletion of glutathione levels in mouse sinusoidal endothelial cells resulted in increased sensitivity to the alkylating agent dacarbazine (DeLeve and Kaplowitz, 1992). MRK20, a 85-kDa protein first described in Adriamycin-resistant cancer cell lines, is also expressed in endothelial cells of the liver and tonsils (Sugawara et al., 1988). Induction of a drug resistance phenotype may also be a mechanism by which tumor-associated endothelial cells could gain resistance to anticancer agents. Possible induction of P-glycoprotein expression in tumor stroma has been observed in human breast carcinoma (Wishart et al., 1991). As mentioned previously, gap junction mediated dye transfer to endothelial cell monolayers was more effective from metastatic B16-F10 melanoma cells compared to less aggressively metastatic B16 melanoma cells (El-Sabban and Pauli, 1991). In this regard, human ovarian carcinoma cells initially sensitive to L-phenylalanine mustard gained resistance after contact-dependent intercellular transfer of glutathione from a subline resistant to this agent (Frankfurt et al., 1991). Conceivably such a transfer could also occur between tumor cells and stromal cells. Thus it seems possible that tumor vessel endothelial cells may be capable of expressing several defence mechanisms against potentially toxic chemotherapy agents.

Only a handful of studies have directly examined the anti-angiogenic activity of chemotherapy drugs, but these seem to indicate that tumor vasculature is sensitive to certain chemotherapy agents. Baguley et al. (1991) reported that the vinca alkaloids vinblastine, vincristine, and vindesine, as well as colchicine, induce growth delays in colon 38 adenocarcinoma in mice, in part due to vascular necrosis. The authors also show that two multidrug-resistant P388 leukemia cell lines were sensitive to vinblastine and vincristine when grown subcutaneously, but not when grown as intraperitoneal ascites (avascular), suggesting that when blood vessel capillaries of solid tumors are compromised by certain chemotherapy agents, tumor growth is retarded. Chick embryo chorioallantoic membrane angiogenesis was inhibited by vinca alkaloids and anthracyclines, while certain agents such as cyclophosphamide, melphalan, 5-fluorouracil, methotrexate and *cis*-platinum had little effect on developing vessels (Steiner, 1992b). Repair of partially denuded endothelial cell monolayers can be inhibited by Mitomycin C (Coomber, 1991) and such *in vitro* monolayers retract upon addition of bleomycin, Adriamycin, 1,3-bis(2-chloroethyl)-1 nitrosourea and vincristine (Nicolson and Custead, 1985), and also *in vivo* with agents such as bleomycin (Orr et al., 1986).

Such studies would suggest that specific chemotherapy agents may inhibit angiogenesis while others have little effect, indicating that endothelial cells may indeed be relatively resistant to a variety of chemotherapeutic agents. With vessel chemoresistance, reduction of the tumor burden would be possible only through direct killing of tumor cells which may be chemoresistant themselves. Thus tumor progression toward multidrug resistance may be in part due to development of chemoresistance in tumor vessel endothelial cells. If this is so, it would seem reasonable to suggest that agents which can reverse the drug-resistant phenotype should be directed at tumor-associated endothelial cells—not just the tumor cell population *per se*. Such a strategy could render "drug-resistant tumors" sensitive to the drugs they are resistant to by an indirect mechanism, as discussed earlier.

VI. THE CONTRIBUTION OF ANGIOGENESIS TO THE DOMINANCE OF METASTATIC CLONES: A HYPOTHESIS

Endothelial cells are not only capable of responding to a variety of intercellular signals but also of elaborating many potent cytokines which are able to affect tumor cell growth in perivascular zones. The excess of blood vessels and poor blood flow in tumors may create better conditions for such interactions. Tumor cells are heterogeneous with respect to many properties, including metastatic potential, responsiveness to various cytokines and angiogenic capacity, and it is likely that those subpopulations expand within certain territories of the tumor. Nests of different tumor cell types have been found even within the same tumor. Poorly and highly vascularized territories have also been identified within the tumor mass. The latter seems to be of prognostic significance in breast cancer (Weidner et al., 1991). It seems conceivable that these intratumoral domains may appear as a result of local expansion of tumor cell clones endowed with different potentials to elicit a vascular reaction or responsiveness to endothelium-derived cytokines. In this sense, interactions between tumor cells and endothelium are most likely local, mutual and selective. We postulate that the tumor cell-endothelial cell interactions may contribute to tumor progression by influencing tumor cell composition in

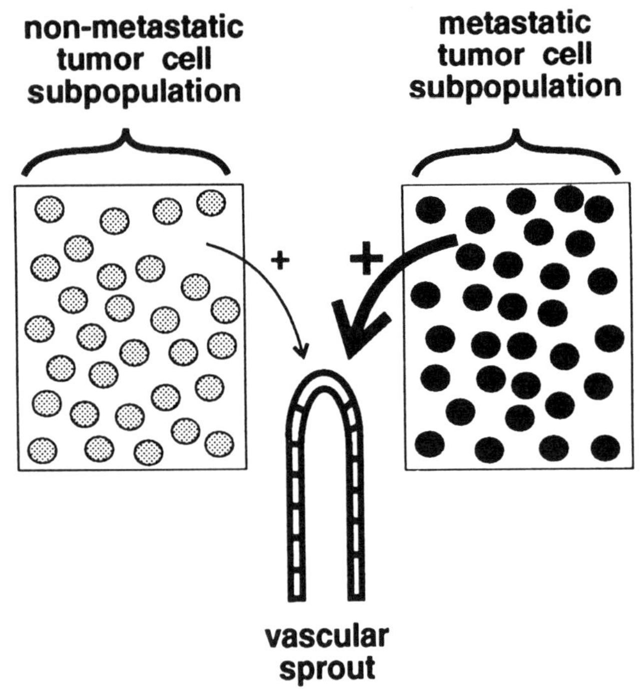

Figure 2. Possible role of differential angiogenesis in clonal dominance of metastatic cells in perivascular areas. The competition for blood vessels between tumor cell subpopulations occupying distinct territories within the tumor may endow more angiogenic (metastatic) cell subpopulations with a selective growth advantage.

perivascular areas, thus leading to initially local, and then later, general dominance of metastatically competent cells in these tumors.

There are several potential mechanisms which may contribute to this "perivascular dominance":

1. Tumor cell subpopulations may secrete various amounts of angiogenic stimulators (and inhibitors) thus competing for neovascularization. Differential angiogenesis would lead to greater vascularity in the areas occupied by hyper-angiogenic tumor subpopulations (Figure 2). This is in agreement with results demonstrating heterogeneity of tumor cells with respect to their angiogenic potential. Some growth factors (e.g. bFGF) elaborated by aggressive tumor cells are known to have angiogenic activity.
2. Growth inhibitory cytokines secreted locally by endothelial cells may contribute to the selective (relative) growth advantage of metastatic tumor cells

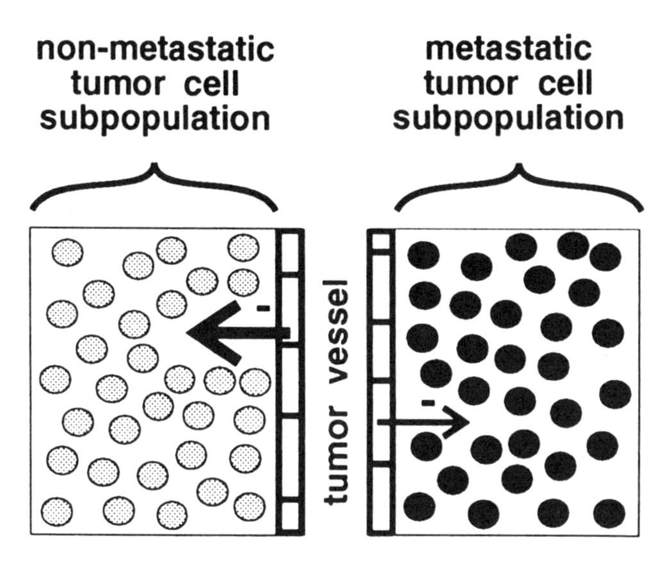

Figure 3. Possible role of differential tumor cell endothelial cell interactions in clonal dominance of metastatic cells in perivascular areas. Metastatic tumor cells frequently express a pleiotropic resistance to growth inhibitory cytokines released by stromal cells including endothelium. This may lead to a growth advantage of these cells in the perivascular space.

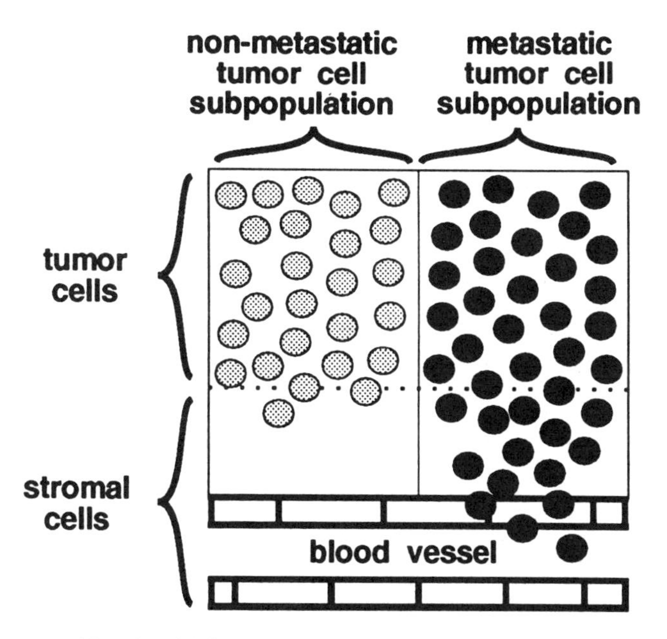

Figure 4. Possible role of differential invasiveness in clonal dominance of metastatically competent tumor cell variants. More invasive tumor cells may take over the perivascular space more readily than their non-invasive counterparts.

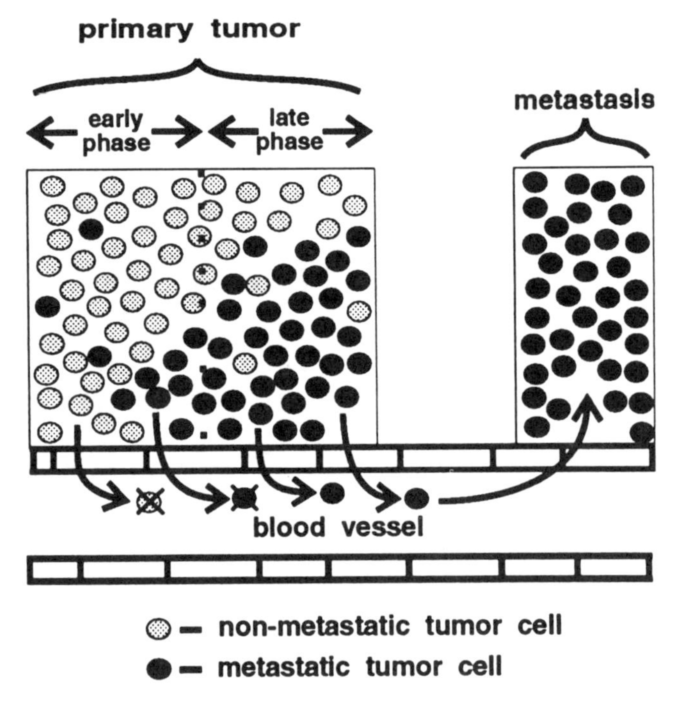

Figure 5. Consequences of dominance of the metastatically competent tumor cell subpopulation in the perivascular area. Local enrichment in more malignant cells in close proximity to blood vessels may contribute to a further growth advantage of these cells. This may also create permissive conditions for initiation of the metastatic process by increasing the frequency of intravasation of metastatically competent tumor cells.

which can manifest pleiotropic resistance to multiple growth inhibitors (Kerbel et al., 1992). In coculture of human endothelial cells with human melanoma cells, the melanoma cell lines derived from advanced (metastatic) tumors were found to be more resistant to growth inhibition mediated by several cytokines than the lines representative of RGP or early VGP lesions (Rak et al., 1991b; Hegmann et al., 1992) (Figure 3).

3. Differential invasiveness may lead to a faster and greater access to blood vessels by the more aggressive tumor cell variants (Figure 4).
4. The proximity of more aggressively metastatic tumor cell variants to the blood vessels may endow these tumor cells with even greater growth advantage, as well as an enhanced opportunity to enter the circulation and initiate the metastatic process (Figure 5).

VII. SUMMARY

Several aspects of cancer growth and metastasis can be better understood when certain tumor cell-host cell interactions are taken into account. The capacity of tumor cells to elicit various host tissue responses seems to be as important for development of solid tumors as tumor cell proliferation itself. In this respect, the process of tumor angiogenesis apparently plays a vital role. There is growing evidence to suggest that the consequences of tumor blood vessel formation can be extended far beyond nutrient supply. Tumor blood vessels also contribute to metastasis. In addition, vascular endothelium may engage in mutual interactions with tumor cells, whereby both populations undergo progressive changes which ultimately serve to facilitate the growth of nascent metastases. We have tried to highlight several questions concerning tumor cell-endothelial interactions and the tumor angiogenic process which we feel merit much more detailed experimental analysis. These are as follows:

1. Does a quantitative relationship between capacity to metastasize and induce angiogenesis exist in certain types of cancers?
2. Can endothelial cells help determine the outcome of metastasis and the pattern of organ-specific metastases by the liberation of locally acting growth factors, especially inhibitory polypeptides.
3. Are proliferating vascular endothelial cells sensitive—or highly resistant—to the toxic effects mediated by certain anti-cancer therapeutic drugs? If so, can this resistance be reversed?
4. Is there a "geographical" relationship between "clonal dominance" of metastatically-competent tumor cells in primary tumors and angiogenic vascular hotspots?

Answers to these questions should contribute to a better understanding of how angiogenesis contributes to the later stages of tumor progression and metastasis, and how its manipulation can be used as a new clinical strategy for cancer treatment.

REFERENCES

Aaronson, S.A. (1991). Growth factors and cancer. Science 254, 1146–1167.

Al-Mondhiry, H. & McGarvey, V. (1987). Tumor interaction with vascular endothelium. Haemostasis. 17, 245–253.

Albino, A.P., Davis, B.M., & Nanus, D.M. (1991). Induction of growth factor RNA expression in human malignant melanoma: markers of transformation. Canc. Res. 51, 4815–4820.

Allen, L.E. & Maher, P.A. (1991). Basic FGF and FGF receptor expression in human bladder carcinoma cell lines with different invasive potentials. J. Cell. Biochem. Suppl. 15F, Abs., CF 201.

Ambrus, J.L., Ambrus, C.M., Forgach, P., Stadler, S., Halpern, J., Sayyid, S., & Toumbis, C. (1991). Role of platelets in tumor induced angiogenesis. Effect of platelet derived growth factor (PDGF), pentoxifylline (PX). and sodium diethyldithiocarbamate (DDTC). Angiogen., Int. Symp., St. Gallen, Mar 13–15, Abs. 58.

Antonelli-Orlidge, A., Saunders, K.B., Smith, S.R., & D'Amore, P.A. (1989). An activated form of transforming growth factor β is produced by cocultures of endothelial cells and pericytes. Proc. Natl. Acad. Sci. USA 86, 4544–4548.

Aslakson, C.J., Rak, J.W., Miller, B.E., & Miller, F.R. (1991). Differential influence of organ site on three subpopulations of a single mouse mammary tumor at two distinct steps in metastasis. Int. J. Canc. 47, 466–472.

Aslakson, C.J. & Miller, F.R. (1992). Selective events in the metastatic process defined by analysis of the sequential dissemination of subpopulations of a mouse mammary tumor. Canc. Res. 52, 1399–1405.

Auerbach, R., Lu, W.C., Pardon, E., Gumkowski, F., Kaminska, G., & Kaminski, M. (1987). Specificity of adhesion between murine tumor cells and capillary endothelium: an *in vitro* correlate of preferential metastasis *in vivo*. Canc. Res. 47, 1492–1496.

Auerbach, R. (1991a). Vascular endothelial cell differentiation: Angiogenesis in tumor progression organ-specificiynd selective affinities as the basis for developing anti-cancer strategies. Int. J. Radiat. Biol. 60, 1–10.

Auerbach, R., Auerbach, W., & Polakowski, I. (1991b). Assays for angiogenesis. Pharmac. Ther. 51, 1–11.

Axiotis, C.A., Monteagudo, C., Merino, M.J., Lapore, N., & Neumann, R.D. (1991). Immunohistochemical detection of p-glycoprotein in endometrial adenocarcinoma. Am. J. Path. 138, 799–806.

Baguely, B.C., Holdaway, K.M., Thomsen, L.L., Zhuang, L., & Zwi, L.J. (1991). Inhibition of growth of colon 38 adenocarcinoma by vinblastine and colchicine: evidence for a vascular mechanism. Eur. J. Canc. 27, 482–487.

Basset, P., Bellocq, J.P., Wolf, C., Stoll, I., Hutin, I., Limacher, J.M., Podhajeer, O.L., Chenard, M.P., Rio, M.C., & Chambon, P. (1990). A novel proteinase gene specifically expressed in stromal cells of breast carcinoma. Nature 348, 689–704.

Becker, I., Becker, K.-F., Meyermann, R., & Holt, V. (1991). The multidrug-resistance gene MDR1 is expressed in human glial tumors. Acta Neuropathol. 82, 516–519.

Beitz, J., Kim, I.-S., Calabresi, P., & Frakelton, R.A. (1991). Human microvascular endothelial cells express receptors for platelet-derived growth factor. Proc. Natl. Acad. Sci. USA 88, 2021–2025.

Bell, C.W., Pathak, S., & Frost, P. (1989). Unknown primary tumors: establishment of cell lines, identification of chromosomal abnormalities, and implications for a second type of tumor progression. Canc. Res. 49, 4311–4315.

Belloni, P.N., Carney, D.H., & Nicolson, G.L. (1992). Organ-derived microvessel endothelial cells exhibit differential responsiveness to thrombin and other growth factors. Microvas. Res. 43, 20–45.

Benedetto, C., Bajardi, F., Ghiringhello, B., Nohammer, G., Phitakpraiwan, P., & Rojanapo, W. (1990). Quantitative measurments of the changes in protein thiols in cervical intraepithelial neoplasia and in carcinoma of the human uterine cervix provide evidence for the existence of a biochemical field effect. Canc. Res. 50, 6663–6667.

Bickell, R. & Harris, A.L. (1991). Novel growth regulatory factors and tumor angiogenesis. Eur. J. Canc. 27, 781–785.

Bishop, J.M. (1991). Molecular themes in oncogenesis. Cell 64, 235–248.

Blood, C.H. & Zetter, B.R. (1990). Tumor interactions with the vasculature: angiogenesis and tumor metastasis. Biochim. Biophys. Acta 1032, 89–118.

Boes, M., Dake, B.L., & Bar, R.S. (1991). Interactions of cultured endothelial cells with TGFβ, bFGF, PDGF, and IGF-I. Life Sciences 48, 811–821.

Borsum, T. (1991). Biochemical properties of vascular endothelial cells. Virchows Arch. B. Cell Pathol. 60, 279–286.

Bouch, N. (1990). Tumor angiogenesis: the role of oncogenes and tumor suppressor genes. Canc. Cells 2, 179–185.

Bouch, N., Polverini, P.J., Tolsma, S.S., Frazier, W.A., & Good, D. (1991). Tumor suppressor gene control of angiogenesis. J. Cell Biochem. Suppl. 15F, 216 Abstract, CF 026.

Boukamp, P., Tomakidi, P., & Fusenig, N.E. (1991). Angiogenesis, an early and essential step of epidermaltumor cell invasion: an *in vivo* model to study preinvasive stages. J. Cell. Biochem. Suppl. 15F, Abstract, CF 403.

Bradham, D.M., Igarashi, A., Potter, R.L., & Grotendorst, G.R. (1991). Connective tissue growth factor: a cysteine-rich mitogen secreted by human vascular endothelial cells is related to the SRC-induced immediate early gene product CEF-10. J. Cell Biol. 114, 1285–1294.

Bradley, G., Juranka, P.F., & Ling, V. (1988). Mechanism of multidrug resistance. Biochim. Biophys. Acta 948, 87–128.

Brem, S.S., Jensen, H.M., & Gullino, P.M. (1978). Angiogenesis as a marker of preneoplastic lesions of the human breast. Canc. 41, 239–244.

Breslow, A. (1970). Thickness, cross-sectional areas and depth of invasion in the prognosis of cutaneous melanoma. Ann. Surg. 172, 902–908.

Brown, K., Buchmann, A., & Balmain, A. (1990). Carcinogen-induced mutations in the mouse c-Ha-ras gene provide evidence of multiple pathways of tumor progression. Proc. Natl. Acad. Sci. USA 87, 538–542.

Burri, P.H. (1991). Intussusceptive microvascular growth, a new mechanism of capillary network expansion. Angiogenesis, Int. Symp., St. Gallen, Mar 13–15, Abstract 88.

Burrows, F.J., Haskard, D.O., Hart, I.R., Marshall, J.F., Selkirk, S., Poole, S., & Thorpe, P.E. (1991). Influence of tumor derived interleukin 1 on melanoma-endothelial cell interactions. Canc. Res. 51, 4768–4775.

Bussolino, F., De Rossi, M., Sica, A., Colotta, F., Wang, J.M., Bocchietto, E., Padura, I.M., Bosia, A., Dejana, E., & Mantovani, A. (1991). Murine endothelioma cell lines transformed by polyoma middle T oncogene as target for and producers of cytokines. J. Immununol. 147, 2122–2129.

Cantley, L.C., Auger, K.R., Graziani, A., Kapeller, R., & Soltoff, S. (1991). Oncogenes and signal transduction. Cell 64, 281–302.

Carnochan, P., Briggs, J.C., Westbury, G., & Davies, A.J.S. (1991). The vascularity of cutaneous melanoma: a quantitative histological study of lesions 0.85–125 mm thickness. Br. J. Canc. 64, 102–107.

Chan, H.L., Haddad, G., Thorner, P.S., DeBoer, G., Lin, Y.P., Ondrusek, N., Yeger, H., & Ling, V. (1991). P-glycoprotein expression as a predictor of the outcome of therapy for neuroblastoma. New Engl. J. Med. 325, 1608–1614.

Chan, H.S., Thorner, P.S., Haddad, G., & Ling, V. (1990). Immunohistochemical detection of p-glycoprotein: prognostic correlation in soft tissue sarcoma of childhood. J. Clin. Oncol. 8, 689–704.

Christofori, G., Radvanyi, F., Lacey, M., & Hanahan, D. (1991). Transgenic mouse model for tumor progression and angiogenesis. J. Cell Biochem. Suppl. 15F, 248 Abstract, CF 407.

Chung, L.W.K., Chang, S.M., Bell, C., Zhau, H.E., Ro, J.Y., & von Eschenbach, A.C. (1989). Co-inoculation of tumorigenic rat prostate mesenchymal cells with non-tumorigenic epithelial cells results in the development of carcinosarcoma in syngeneic and athymic animals. Int. J. Canc. 43, 1179–1187.

Cifone, M.A. & Fidler, I.J. (1981). Increasing metastatic potential is associated with increasing genetic instability of clones isolated from murine neoplasms. Proc. Natl. Acad. Sci. USA 78, 6949.

Clark, W. (1991). Tumor progression and the nature of cancer. 1991 Gordon Conference on Cancer, Aug 11–16 Newport, R.I., Lecture.

Clark, W.H., Elder, D.E., Guerry, D., Epstein, M.N., Greene, M.H., & Van Horn, M. (1984). A study of tumor progression: The precursor lesions of superficial spreading and nodular melanoma. Hum. Pathol. 15, 1147–1165.

Clark, W.H., Elder, D.E., Guerry IV, D., Braitman, L.E., Trock, B.J., Schultz, D., Synnestvedt, M., & Halpern, A.C. (1989). Model predicting survival in stage I melanoma based on tumor progression. J. Natl. Canc. Inst. 81, 1893–1904.

Clarke, R., Dickson, R.B., & Lippman, M.E. (1992). Hormonal aspects of breast cancer. Growth factors, drugs and stromal interactions. Critical Reviews in Oncology/Hematology 12, 1–23.

Conn, G., Bayne, M.L., Soderman, D.D., Kwok, P.W., Sullivan, K.A., Palisi, T.M., Hope, D.A., & Thomas, K.A. (1990). Amino acid and cDNA sequence of a vascular endothelial cell mitogen that is homologous to platelet-derived growth factor. Proc. Natl. Acad. Sci. USA 87, 2628–2632.

Coomber, B.L. (1991). Mitomycin C inhibits endothelial monolayer regeneration *in vitro*. J. Cell Biol. 115, 2128 Abstract.

Cordon-Cardo, C., O'Brien, J.P., Boccia, J., Casals, D., Bertino, J.R., & Melamed, M.R. (1990a). Expression of the multidrug resistance gene product (p-glycoprotein) in human normal and tumor tissues. J. Histochem. Cytochem. 38, 1277–1287.

Cordon-Cardo, C., O'Bren, J.P., & Casals, D. (1990b). Multidrug resistance gene (P-glycoprotein). is expressed by endothelial cells at blood-brain barrier sites. Proc. Natl. Acad. Sci. USA 86, 695–698.

Cornil, I., Man, S., Fernandez, B., & Kerbel, R.S. (1989). Enhanced tumorigenicity, melanogenesis, and metastases of a human malignant melanoma after subdermal implantation in nude mice. J. Natl. Canc. Inst. 81, 938–944.

Cornil, I., Theodorescu, D., Man, S., Herlyn, M., Jambrosic, J., & Kerbel, R.S. (1991). Fibroblast cell interactions with human melanoma cells affect tumor cell growth as a function of tumor progression. Proc. Natl. Acad. Sci. USA 88, 6028–6032.

Costello, P. & Del Maestro, R. (1990). Human cerebral endothelium: isolation and characterization of cells derived from microvessels of non-neoplastic and malignant glial tissue. J. Neuro-Onc. 8, 231–243.

Cross, M. & Dexter, M.T. (1991). Growth factors in development, transformation, and tumorigenesis. Cell 64, 271–280.

D'Amore, P. (1992). Endothelial cell mural cell interactions. J. Cell Biochem. Suppl. 16A, 36 Abs., CA 013.

D'Amore, P.A. & Braunhut, S.J. (1988). Stimulatory and inhibitory factors in vascular growth control. Endothel. Cells II, 13–60.

Dedhar, S. (1989). Signal transduction via the beta 1 integrins is a required intermediate in interleukin-1 beta induction of alkaline phosphatase activity in human osteosarcoma cells. Exp. Cell Res. 183, 207–214.

DeLeve, L.D. & Kaplowitz, N. (1992). Selective susceptibility of hepatic endothelial cells to dacarbazine toxicity: a model for hepatic eno-occlusive disease. J. Cell. Biochem. Suppl. 16A, 42. Abs. CA 104.

Denekamp, J. (1990). Vascular attack as a therapeutic strategy for cancer. Canc. Metastas. Rev. 9, 267–282.

Dvorak, H.F. (1986). Tumors: wounds that do not heal. New Engl. J. Med. 315, 1650–1659.

Dvorak, H.F. & Gresser, I. (1989). Microvascular injury in pathogenesis of interferon-induced necrosis of subcutaneous tumors in mice. J. Natl. Canc. Inst. 81, 497–502.

Dvorak, H.F., Nagy, J.A., & Dvorak, A.M. (1991a). Structure of solid tumors and their vasculature: implications for therapy with monoclonal antibodies. Canc. Cells, 3, 77–85.

Dvorak, H.F., Sioussat, T.M., Brown, L.F., Berse, B., Nagy, J.A., Sotrel, A., Manseau, E.J., Van De Water, L., & Senger, D.R. (1991b). Distribution of vascular permeability factor (vascular endothelial growth factor) in tumors: concentration in tumor blood vessels. J. Exp. Med. 174, 1275–1278.

Eghbali, B., Kessler, J.A., Reid, L.M., Roy, C., & Spray, D.C. (1991). Involvement of gap junctions in tumorigenesis: transfection of tumor cells with connexin 32 cDNA retards growth *in vivo*. Proc. Natl. Acad. Sci. USA 88, 10701–10705.

El-Sabban, M.E. & Pauli, B.U. (1991). Cytoplasmic dye transfer between metastatic tumor cells and vascular endothelium. J. Cell Biol. 115, 1375–1382.

Fajardo, L.F., Kwan, H.H., Kowalski, J., Prionas, S.D., & Allison, A.C. (1992). Dual role of tumor necrosis factor-α in angiogenesis. Am. J. Path. 140, 539.

Farber, E. & Rubin, H. (1991). Cellular adaptation in the origin and development of cancer. Canc. Res. 51, 2751–2761.

Fearon, E.R., Cho, K.R., Nigro, J.M., Kern, S.E., Simons, J.W., Ruppert, M.J., Hamilton, S.R., Presinger, A.C., Thomas, G., Kinzler, K.W., & Vogelstein, B. (1990a). Identification of a chromosome 18q gene that is altered in colorectal cancers. Science. 247, 49–56.

Fearon, E.R. & Vogelstein, B. (1990b). A genetic model for colorectal tumorigenesis. Cell 61, 759–767.

Fidler, I.J. (1973a). The relationship of embolic homogeneity, number, size, and viability to the incidence of experimental metastasis. Eur. J. Canc. 9, 223–227.

Fidler, I.J. (1973b). Selection of successive cell lines for metastasis. Nature 242, 148–149.

Fidler, I.J. (1990). Critical factors in the biology of human cancer metastasis: 28th G.H.A. Clowes Memorial Award Lecture. Canc. Res. 50, 6130–6138.

Finstad, C.L., Saigo, P.E., Rubin, S.C., Federici, M.G., Provencher, D.M., Hoskins, W.J., Lewis, J.L., Jr., & Lloyd, K.O. (1990). Immunohistochemical localization of p-glycoprotein in adult human ovary and female genital tract of patients with benign gynecological conditions. J. Histochem. and Cytochem. 38, 1677–1681.

Fletcher, J.A., Pinkus, G.S., Weidner, N., & Morton, C.C. (1991). Lineage-restricted clonality in biphasic solid tumors. Am. J. Pathol. 138, 1199–1207.

Folkman, J. & Haudenschild, C. (1980). Angiogenesis *in vitro*. Nature 288, 551–555.

Folkman, J. (1985). Tumor angiogenesis. Adv. Canc. Res. 43, 175–203.

Folkman, J. (1986). How is blood vessel growth regulated in normal and neoplastic tissue? G.H.A. Clowes Memorial Award Lecture. Canc. Res. 46, 467–473.

Folkman, J. (1987a). What is the role of angiogenesis in metastasis from cutaneous melanoma? Eur. J. Canc. Clin. Oncol. 23, 361–363.

Folkman, J. & Klagsbrun, M. (1987b). Angiogenic factors. Science 235, 442–447.

Folkman, J., Watson, K., Ingber, D., & Hanahan, D. (1989a). Induction of angiogenesis during transition from hyperplasia to neoplasia. Nature 339, 58–61.

Folkman, J., Weisz, P.B., Joullie, M.M., Li, W.W., & Ewing, W.R. (1989b). Control of angiogenesis with synthetic heparin substitutes. Science 243, 1490–1493.

Folkman, J. (1990). What is the evidence that tumors are angiogenesis dependent? J. Natl. Canc. Inst. 82, 4–6.

Foulds, L. (1954). Tumor progression. Canc. Res. 17, 337–339.

Frankfurt, O.S., Seckinger, D., & Sugarbaker, E.V. (1991). Intercellular transfer of drug resistance. Canc. Res. 51, 1190–1195.

Giavazzi, G., Garofalo, A., Bani, M.R., Abbate, M., Ghezzi, P., Boraschi, M., Mantovani, A., & Dejana, E. (1990). Interleukin 1-induced augmentation of experimental metastases from human melanoma in nude mice. Canc. Res. 50, 4771–4775.

Gleave, M., Hsieh, J.T., Gao, C., von Eschenbach, A.C., & Chung, L.W.K. (1991). Canc. Res. 51, 3753–3761.

Good, D.J., Polverini, P.J., Rastinejad, F., Le Beau, M.M., Lemons, R.S., Frazier, W.A., & Bouch, N.P. (1990). A tumor suppressor-dependent inhibitor of angiogenesis is immunologically and functionally indistinguishable from a fragment of thrombospondin. Proc. Natl. Acad. Sci. USA 87, 6624–6628.

Gospodarowicz, D., Abraham, J.A., & Schilling, J. (1989). Isolation and characterization of a vascular endothelial cell mitogen produced by pituitary-derived folliculo stellate cells. Proc. Natl. Acad. Sci. USA 86, 7311–7315.

Gullino, P.M. (1991). Microenvironment and angiogenic response. Angiogenesis, Int. Symp., St. Gallen, Mar 13–15 Abs. 136.

Hagemeier, H.-H., Vollmer, E., Goerdt, S., Schulze-Osthoff, K., & Sorg, C. (1986). Monoclonal antibody reacting with endothelial cells budding vessels in tumors and inflammatory tissues, and non-reactive with normal adult tissues. Int. J. Canc. 38, 481–488.

Halaban, R., Kwon, B., Ghosh, S., Delli Bovis, P., & Baird, A. (1988). bFGF as an autocrine growth factor for human melanocytes. Onco. Res. 3, 177–186.

Hart, I.R., Goode, N.T., & Wilson, R.E. (1989). Molecular aspects of the metastatic cascade. Biochim. Biophys. Acta 989, 65–84.

Hart, I.R. & Easty, D. (1991). Identification of genes controlling metastatic behavior. Br. J. Canc. 63, 9–12.

Hayashi, N. & Cuhna, G.R. (1991). Mesenchyme induced changes in the neoplastic characteristics of the Dunning prostatic adenocarcinoma. Canc. Res. 51, 4924–4930.

Heffelfinger, S. C. & Darlington, G. (1991). SK-HEP-1: a model for angiogenesis. J. Cell Biochem. Suppl. 15F, 250 Abs. CF 412.

Hegmann, E.J., Rak, J.W., & Kerbel, R.S. (1992). Differential growth inhibition of human melanoma cell lines from different stages of tumor progression by human endothelial cells. J. Cell. Biochem. Suppl. 16A, Keystone, Jan. 13–26 Abs. CA 405.

Heimark, R.L. & Schwartz, S.M. (1988). Endothelial morphogenesis. In: Endothelial Cell Biology in Health and Disease. (Simionescu, N. & Simionescu, M., eds.), pp. 123–143. Plenum Press, New York.

Heppner, G.H. & Miller, B.E. (1983). Tumor heterogeneity: biological implications and therapeutic consequences. Canc. Metastasis Rev. 2, 5–23.

Heppner, G.H. (1984). Tumor heterogeneity. Canc. Res. 44, 2259–2265.

Heppner, G.H. (1988). Role of inflammatory cells in tumor progression and metastasis. In: Proceedings on Tumor Progression and Metastasis UCLA Symposia, Vol. 78, pp. 151–155. Alan R. Liss, New York.

Heppner, G.H. (1989a). Tumor cell societies. J. Natl. Canc. Inst. 81, 648–649.

Heppner, G.H. & Miller, B.E. (1989b). Therapeutic implications of tumor heterogeneity. Semin. Oncol. 16, 91–105.

Herlyn, M. (1990a). Human melanoma: development and progression. Canc. Metastas. Rev. 9, 101–112.

Herlyn, M., Kath, R., Williams, N., Valyi-Nagy, I., & Rodeck, U. (1990b). Growth-regulatory factors for normal, premalignant and malignant human cells *in vitro*. Adv. Canc. Res. 54, 213–234.

Honn, K.V., Menter, D.G., Steinert, B.W., Taylor, J.D., Onoda, J.M. & Sloane, B.F. (1987). Analysis of platelet, tumor cell and endothelial cell interactions *in vivo* and *in vitro*. In: Prostaglandins in Cancer Research (Garaci, E., Paoletti, R., & Santoro, M.G., eds.), pp. 172–183. Springer-Verlag, Berlin.

Hunter, T. (1991). Cooperation between oncogenes. Cell, 64, 249–270.

Hynes, R.O. (1992). Integrins: Versatility, modulation, and signaling in cell adhesion. Cell 69, 11–25.

Ingber, D., Fujita, T., Kishimoto, S., Sudo, K., Kanamaru, T., Brem, H., & Folkman, J. (1990). Synthetic analogues of fumagillin that inhibit angiogenesis and suppress tumor growth. Nature 348, 555–557.

Ingber, D. (1991). Extracellular matrix and cell shape: potential control points for inhibition of angiogenesis. J. Cell Biol. 47, 236–241.

Ingber, D.E. & Folkman, J. (1989). How does extracellular matrix control capillary morphogenesis. Cell, 58, 803–805.

Iruela-Arispe, M.L. & Sage, E.H. (1992). Transforming growth factor-β promotes proliferation of endothelial cells that exhibit angiogenesis *in vitro*. J. Cell. Biochem., Symposia on Molecular & Cellular Biology, Keystone, Colorado, Abstract CA 407.

Jackson, C., Jenkins, K., & Schrieber, L. (1991). Mechanisms of type I collagen induced vascular tube formation. Angiogenesis, Int. Symp., St. Gallen, Mar 13–15 Abs. 46.

Jain, R.K. (1990). Vascular and interstitial barriers to delivery of therapeutic agents in tumors. Canc. Metastas. Rev. 9, 253–266.

Jensen, H.M., Chen, I., DeVault, M.R., & Lewis, A.E. (1982). Angiogenesis induced by "normal" human breast tissue: a probable marker for precancer. Science 218, 293–295.

Kaminski, M. & Auerbach, R. (1988). Tumor cells are protected from NK-cell-mediated lysis by adhesion to endothelial cells. Int. J. Canc. 41, 847–849.

Kandel, J., Bossy-Wetzel, E., Radvanyi, F., Klagsbrun, M., Folkman, J., & Hanahan, D. (1991). Neovascularization is associated with a switch to the export of bFGF in the multistep development of fibrosarcoma. Cell 66, 1095–1104.

Keck, P.J., Hauser, S.D., Krivi, G., Sanzo, K., Warren, T., Feder, J., & Connolly, D.T. (1989). Vascular Permeability Factor, an endothelial cell mitogen related to PDGF. Science 246, 1309–1312.

Kerbel, R.S., Waghorne, C., Man, S.M., Elliott, B., & Breitman, M.L. (1987). Alterations of the tumorigenic and metastatic properties of neoplastic cells is associated with the process of calcium phosphate-mediated DNA transfection. Proc. Natl. Acad. Sci. USA 84, 1263–1267.

Kerbel, R.S. (1990a). Growth dominance of the metastatic cancer cell: cellular and molecular aspects. Adv. Canc. Res. 55, 87–132.

Kerbel, R.S. & Theodorescu, D. (1990b). Tumor cell subpopulation interactions mediated by transforming growth factor β. In: Cell to Cell Interaction (Schaeffer, P. & Burger, M.M., eds.), pp. 100–113. Karger AG, Basel.

Kerbel, R.S. (1991). Inhibition of tumor angiogenesis as a strategy to circumvent acquired resistance to anti-cancer therapeutic agents. Bioessays 13, 31–36.

Kerbel, R.S., Cornil, I., & MacDougall, J. (1992). Transition in responsiveness to multiple inhibitory growth factors: possible contribution to tumor progression and metastasis. In: Metastasis: Basic Research and its Clinical Applications. Contrib. Oncol. (Rabes, H., Peters, P.E., & Munk, K., eds.), Vol. 44, pp. 185–202. Karger, Basel.

Khokha, R., Waterhouse, P., Yagiel, S., Lala, P.K., Overall, C.M., Norton, G., & Denhardt, D.T. (1989). Antisense RNA-induced reduction in murine TIMP levels confers oncogenicity on Swiss 3T3 cells. Science 243, 947–950.

Klagsbrun, M. (1992). Angiogenesis factors. In: Endothelial Cells (Ryan, U.S., ed.), 2nd edition, pp. 37–50. CRC Press, Boca Raton, FL.

Knighton, D.R., Fiegel, V.D., & Philips, G.D. (1991). Wound healing angiogenesis: the role of platelet derived growth factor in wound healing angiogenesis. J. Cell Biochem. Suppl. 15F, Abs. Q015.

Konerding, M.A., van Ackern, C., Hinz, S., Steinberg, F., & Streffer, C. (1991). Combined morphological approaches in the study of network formation in tumor angiogenesis. Angiogenesis, Int. Symp., St. Gallen, Mar 13–15, Abs.

Korczak, B., Robson, I.B., Lamarche, C., Bernstein, A., & Kerbel, R.S. (1988). Genetic tagging of tumor cells with retrovirus vectors: clonal analysis of tumor growth and metastasis *in vitro*. Mol. Cell. Biol. 8, 3143–3149.

Langer, R. (1992). Drug delivery systems for angiogenesis stimulators and inhibitors. The Molecular Biology of the Endothelial Cell. Keystone, Colorado, Jan 13–19, Abs. CA 022.

Lauri, D., Needham, L., Martin-Padura, I., & Dejana, E. (1991). Tumor cell adhesion to endothelial cells: endothelial leukocyte adhesion molecule-1 as an inducible adhesive receptor specific for colon carcinoma cells. J. Natl. Canc. Inst. 83, 1321–1324.

Leibovich, S.J., Polverini, P.J., Shepard, M.H., Wiseman, D.M., Shively, V., & Nuseir, N. (1987). Macrophage-induced angiogenesis is mediated by tumor necrosis factor-α. Nature 329, 630–632.

Leung, D.W., Cachianes, G., Kuang, W.-J., Goeddel, D.V., & Ferrara, N. (1989). Vascular Endothelial Growth Factor is a secreted angiogenic molecule. Science 246, 1306–1309.
Lewin, B. (1991). Oncogenic conversion by changes in transcription factors. Cell 64, 303-312.
Li, L., Nicolson, G.L., & Fidler, I.J. (1991). Direct *in vitro* lysis of metastatic tumor cells by cytoline-activated murine vascular endothelial cells. Canc. Res. 51, 245–254.
Lione, A. & Bosmann, H.B. (1978). Quantitative relationship between volume of tumor cell units and their intravascular survival. Br. J. Canc. 37, 248–253.
Liotta, L., Kleinerman, J., & Saidel, G. (1974). Quantitative relationships of intravascular tumor cells, tumor vessels, and pulmonary metastases following tumor implantation. Canc. Res. 34, 997–1004.
Liotta, L.A., Steeg, P.S., & Stetler-Stevenson, W.G. (1991). Cancer metastasis and angiogenesis: an imbalance of positive and negative regulation. Cell 64, 327–336.
Lippman, M.E., Wellstein, A., Lupu, R., & Dickson, R.B. (1991). Growth factor control of normal and malignant mammary cell growth. J. Cell Biochem. Suppl. 15F, 216 Abs. CF 025.
Lu, C., Vickers, M.F., & Kerbel, R.S. (1992). Interleukin-6: A fibroblast-derived growth inhibitor of human melanoma cells from early but not advanced stages of tumor progression. Proc. Natl. Acad. Sci. USA, submitted.
Lubbert, M., Mantovani, L., Lindemann, A., Mertelsmann, R., & Herrmann, F. (1991). Expression of leukemia inhibitory factor is regulated in human mesenhymal cells. Leukemia 5, 361–365.
Lyons, G.J., Siew, K., & O'Grady, R.L. (1989). Cellular interactions determining the production of collagenase by rat mammary carcinoma cell line. Int. J. Canc. 43, 119–125.
Madri, J.A., Bell, L., Marx, M., Merwin, J.R., Basson, C., & Prinz, C. (1991). Effects of soluble factors and extracellular matrix components on vascular cell behavior *in vitro* and *in vivo*: models of re-endothelialization and repair. J. Cell Biol. 45, 123–130.
Mahadevan, V. & Hart, I.R. (1990). Metastasis and angiogenesis. Acta Oncol. 29, 97–103.
Mahoney, P.A., Weber, U., Onofrechuk, P., Blessmann, H., Bryant, P.J., & Goodman, C.S. (1991). The fat tumor suppressor gene in Drosophila encodes a novel member of the cadherin gene superfamily. Cell 67, 853–868.
Massague, J., Pandiella, A., & Laiho, M. (1990). Growth stimulation by cell to cell contact and growth suppression: two aspects of the biology of transforming growth factors. In: Cell to Cell Interaction (Schaeffer, P. & Burger, M.M., eds.), pp. 122–142. S. Karger, Basel.
Maxwell, M., Naber, S.P., Wolfe, H.J., Hedley-Whyte, T., Galanopoulos, T., Neville-Golden, J., & Antoniades, H.N. (1991). Expression of angiogenic growth factor genes in primary human astrocytomas may contribute to their growth and progression. Canc. Res. 51, 1345–1351.
Mignatti, P., Tsuboi, R., & Rifkin, D.B. (1989). *In vitro* angiogenesis on the human amniotic membrane: requirement for basic fibroblast growth factor-induced proteases. J. Cell Biol. 108, 671–682.
Miller, B.E., Miller, F.R., Wilburn, D.J., & Heppner, G.H. (1987). Analysis of tumor cell composition in tumors composed of paired mixtures of mammary tumor cell lines. Br. J. Canc. 56, 561–569.
Miller, B.E., Miller, F.R., Wilburn, D., & Heppner, G.H. (1988). Dominance of a mammary tumor subpopulation in mixed heterogenous tumors. Canc. Res. 48, 5747–5753.
Miller, B.E., Miller, F.R., & Heppner, G.H. (1989). Therapeutic perturbation of the tumor ecosystem in reconstructed heterogeneous mouse mammary tumors. Canc. Res. 49, 3747–3753.
Miller, F.R. (1983). Tumor subpopulation interactions in metastasis. Inv. Metast. 3, 234–242.
Miller, F.R., Mohamed, A.N., & McEachern, D. (1989). Production of a more aggressive tumor cell variant by spontaneous fusion of two mouse tumor subpopulations. Canc. Res. 49, 4316–4321.
Miller, F.R. & Heppner, G.H. (1990). Cellular interactions in metastasis. Canc. Metast. Rev. 9, 21–34.
Montesano, R., Pepper, M.S., Mohle-Steinlein, U., Risau, W., Wagner, E.F., & Orci, L. (1990). Increased proteolytic activity is responsible for the aberrant morphogenetic behavior of endothelial cells expressing the middle T oncogene. Cell 62, 435–445.
Montesano, R., Pepper, M.S., Vassalli, J.-D., & Orci, L. (1991). Modulation of angiogenesis *in vitro*. Angiogenesis, Int. Symp., St. Gallen, Mar 13–15, Abs. 55.

Morikawa, K., Walker, S.M., Jessup, J.M., & Fidler, I.J. (1988a). *In vivo* selection of highly metastatic cells from surgical specimens of different primary human colon carcinomas implanted into nude mice. Canc. Res. 48, 1943–1948.

Morikawa, K., Walker, S.M., Nakajima, M., Pethak, S., Jessup, M., & Fidler, I.J. (1988b). Influence of organ enviornment on the growth, selection, and metastasis of human colon carcinoma cells in nude mice. Canc. Res. 48, 6863–6871.

Moroco, J.R., Solt, D.B., & Polverini, P.J. (1990). Sequential loss of suppressor genes for three specific functions during *in vivo* carcinogenesis. Lab. Invest. 63, 298–306.

Moses, M.A., Sudhalter, J., & Langer, R. (1990). Identification of an inhibitor of neovascularization from cartilage. Science 248, 1408–1410.

Nabors, M.W., Griffin, C.A., Zehnbauer, B.A., Hruban, R.H., Phillips, P.C., Grossman, S.A., Brem, H., & Colvin, O.M. (1991). Multidrug resistance gene (MDR1) expression in human brain tumors. J. Neurosurg. 75, 941–946.

Nagy, J.A., Brown, L.F., Senger, D.R., Lanir, N., Van De Water, L., Dvorak, A.M., & Dvorak, H.F. (1988). Pathogenesis of tumor stroma generation: a critical role for leaky blood vessels and fibrin deposition. Biochim. Biophys. Acta 305, 326.

Neville, M.E. (1991). Quantitation of endothelial cells within B16 melanomas *in vivo*. J. Cell Biochem. Suppl. 15F, Abs. CF 419.

Nicolson, G.L. & Custead, S.E. (1985). Effects of chemotherapeutic drugs on platelet and metastatic tumor cell-endothelial cell interactions as a model for assessing vascular endothelial integrity. Canc. Res. 45, 331–336.

Nicolson, G.L. (1988). Organ specificity of tumor metastasis: role of preferential adhesion invasion and growth of malignant cells at specific secondary sites. Canc. Metastas. Rev. 7, 143–188.

Nicolson, G.L. (1991a). Gene expression, cellular diversification and tumor progression to the metastatic phenotype. BioEssays. 13, 337–342.

Nicolson, G.L., Hamada, J-., & Cavanaugh, P.G. (1991b). Stimulation of growth and migration of liver-metastasizing lymphoma cells by molecules from murine liver endothelial cells. Angiogenesis, Int. Symp., St. Gallen, Mar 13–15, Abs. 125.

Nowell, P.C. (1986). Mechanisms of tumor progression. Canc. Res. 46, 2203–2207.

Ogawa, S., Leavy, J., Clauss, M., Koga, S., Shreeniwas, R., Joseph-Silverstein, J., Furie, M., & Stern, D. (1991). Modulation of endothelial cell (EC) function in hypoxia: alterations in cell growth and the response to monocyte-derived mitogenic factors. J. Cell Biochem. Suppl. 15F, Abs. CF 020.

Orr, F.W., Adamson, I.Y.R., & Young, L. (1986). Promotion of pulmonary metastasis in mice by bleomycin-induced endothelial injury. Canc. Res. 46, 891–897.

Paweletz, N. & Knierim, M. (1989). Tumor related angiogenesis. Crit. Rev. Oncol.-Hemat. 9, 197–242.

Pepper, M.S., Belin, D., Montesano, R., Orci, L., & Vassali, J.-D. (1990). Transforming growth factor-β 1 modulates basic fibroblast growth factor—induced proteolytic and angiogenic properties of endothelial cells *in vitro*. J. Cell Biol. 111, 743–755.

Pepper, M.S., Montesano, R., Vassali, J.-D., & Orci, L. (1991). Chondrocytes inhibit endothelial sprout formation *in vitro*: evidence for involvement of a transforming growth factor-β. J. Cell. Physiol. 146, 170–179.

Polverini, P.J., Cotran, R.S., Gimbrone, M.A., & Unanue, E.R. (1977). Activated macrophages induce vascular proliferation. Nature 269, 804–806.

Poste, G., Doll, J., & Fidler, I.J. (1981). Interactions among clonal subpopulations affect stability ofthe metastatic phenotype in polyclonal populations of B16 melanoma cells. Proc. Natl. Acad. Sci. USA 78, 6226–6230.

Price, J., Bell, C., & Frost, P. (1990). The use of a genotypic marker to demonstrate clonal dominance during the growth and metastasis of a human breast carcinoma in nude mice. Int. J. Canc. 45, 968–971.

Pritchett, T.R., Wang, J.K.M., & Jones, P.A. (1989). Mesenchymal-epithelial interactions between normal and transformed human bladder cells. Canc. Res. 49, 2750–2754.

Prout, G.R., Jr. & Garnick, M.B. (1982). The kidney and ureter. In: Cancer Medicine (Holland, J.F. & Frei, III, E., eds.), pp. 1880–1912, Lea & Febiger, Philadelphia.

Pyke, C., Kristensen, P., Ralfkiaer, E., Eriksen, J., & Dano, K. (1991a). The plasminogen activation system in human colon cancer: messenger RNA for the inhibitor PAI-I is located in endothelial cells in the tumor stroma. Canc. Res. 51, 4067–4071.

Pyke, C., Kristensen, P., Ralfkier, E., Grondahl-Hansen, J., Eriksen, J., Blasi, F., & Dano, K. (1991b). Urokinase-type plasminogen activator is expressed in stromal cells and its receptor in cancer cells at invasive foci in human colon adenocarcinomas. Am. J. Pathol. 138, 1059–1066.

Rak, J. (1989). Possible role of tumor stem-end cell interaction in metastasis. Med. Hypoth. 29, 17–19.

Rak, J. (1991a). Metastatic tumor cells maintain viability longer than non-metastatic counterparts in three dimensional culture. 1991 Gordon Conference on Cancer, Aug 11–16, Newport, R.I., Abs..

Rak, J., Hegmann, E., & Kerbel, R.S. (1991b). Differential *in vitro* effect of dermal keratinocytes and endothelial cells on the proliferation of metastatic and non-metastatic melanoma cell lines. 1991 Gordon Conference on Cancer, Aug 11–16, Newport, R.I., Abs.

Rastinejad, F., Polverini, P.J., & Bouch, N. (1989). Regulation of the activity of a new inhibitor of angiogenesis by a cancer suppressor gene. Cell, 56, 345–355.

Ribatti, D. & Vacca, A. (1991). Tumor progression and angiogenesis in human malignant melanoma. Angiogenesis, Int. Symp., St.Gallen, Mar 13–15, Abs. 44.

Rice, G.E. & Bevilacqua, M.P. (1989). An inducible endothelial cell surface glycoprotein mediates melanoma adhesion. Science. 246, 1303–1306.

Rifkin, D. (1992). Characterization of the activation of latent TGF-β. The Molecular Biology of the Endothelial Cell. Symposium on Molecular and Cellular Biology, Keystone, Colorado, Jan 13–19, Lecture.

Rifkin, D.B. (1991). Extracellular activities of bFGF and TGFβ. J. Cell Biochem. Suppl. 15F, Abs. CF 024.

Roberts, A.B. & Sporn, M.B. (1989). Regulation of endothelial cell growth, architecture and matrix synthesis by TGF-β. Am. Rev. Respir. Dis. 140, 1126–1128.

Roninson, I.B. (1992). The role of the mdr1 (p-glycoprotein) gene in multidrug resistance *in vitro* and *in vivo*. Biochem. Pharma. 43, 95–102.

Rubin, H. (1990). The significance of biological heterogeneity. Canc. Metast. Rev. 9, 1–20.

Russel, P.J., Brown, J., Grimmond, S., Stapleton, P., Russel, P., Raghavan, D., & Symonds, G. (1990). Tumor-induced stromal cell transformation: induction of mouse spindle-cell fibrosarcoma not mediated by gene transfer. Int. J. Canc. 46, 299–309.

Sage, H. (1992). The Molecular Biology of the Endothelial Cell. Symposium on Molecular and Cellular Biology, Keystone, Colorado, Jan 13–19, Lecture.

Samiei, M. & Waghorne, C.G. (1991). Clonal selection within metastatic SP1 mouse mammary tumors is independent of metastatic potential. Int. J. Canc. 47, 771–775.

Schipper, J.H., Frixen, U.H., Behrens, J., Unger, A., Jahnke, K., & Birchmeier, W. (1991). E-cadherin expression in squamous cell carcinomas of head and neck: inverse correlation with tumor differentiation and lymph node metastasis. Canc. Res. 51, 6328–6337.

Schlaifer, D., Laurent, G., Chittal, S., Tsuruo, T., Soues, S., Muller, C., Charcosset, J.Y., Alard, C., Brousset, P., Mazerrolles, C., & Delsol, G. (1990). Immunohistochemical detection of multidrug resistance associated p-glycoprotein in tumor and stromal cells of human cancers. Br. J. Canc. 62, 177–182.

Schweigerer, L., Schwab, M., & Fotsis, T. (1991). Endothelial cell growth factors in human neuroblastoma cells transfected with the human MYCN oncogene. Angiogenesis, Int. Symp., St. Gallen, Mar 13–15, Abs.

Senger, D.R., Connoly, D.T., Van De Water, L., Feder, J., & Dvorak, H.F. (1990). Purification and NH2-terminal amino acid sequence of guinea pig tumor-secreted vascular permeability factor. Canc. Res. 50, 1774–1778.

Shaughnessy, S., Lafrenie, R.M., Buchanan, M.R., Podor, T.J., & Orr, W.F. (1991). Endothelial cell damage by Walker carcinosarcoma cells is dependent on vitronectin receptor mediated tumor cell adhesion. Am. J. Pathol. 138, 1535–1543.

Sher-Taylor, B., Bargatze, R., Holzmann, B., Gallatin, W.m., Matthews, D., Wu, N., Picker, L., Butcher, E.C. & Weissman, I.L. (1988). Homing receptors and metastasis. Adv. Canc. Res. 51, 361–390.

Sholley, M.M., Ferguson, G.P., Seibel, H.R., Montour, J.L., & Wilson, J.K. (1984). Mechanisms of neovascularization. Vascular sprouting can occur without proliferation of endothelial cells. Lab. Invest. 51, 624–632.

Sidky, Y.A. & Auerbach, R. (1975). Lymphocyte-induced angiogenesis: a quantitative and sensitive assay of the graft-vs-host reaction. J. Exp. Med. 141, 1084–1100.

Sidransky, D., Mikkelsen, T., Schwechheimer, K., Rosenblum, M.L., Cavanee, W., & Vogelstein, B. (1992). Clonal expansion of p53 mutant cells is associated with brain tumor progression. Nature 355, 846.

Singh, S., Ross, S.R., Acena, M., Rowley, D.A., & Schreiber, H. (1992). Stroma is critical for preventing or permitting immunological destruction of antigenic cancer cells. J. Exp. Med. 175, 139–146.

Smolle, J., Soyer, H.-P., Hofmann-Wellenhof, R., Smolle-Juettner, F.-M., & Kerl, H. (1989). Vascular architecture of melanocytic skin tumors: a quantitative immunohistochemical study using automated image analysis. Path. Res. Pract. 185, 740–745.

Solesvik, O.V., Rofstad, E.K., & Brustad, T. (1982). Vascular structure of five human malignant melanomas grown in athymic nude mice. Br. J. Canc. 46, 557–567.

Sporn, M.B. (1992). Control and modulation of the local action of peptide growth factor. The Molecular Biology of the Endothelial Cell. Symposium on Molecular and Cellular Biology, Jan 13–19, Keystone, Colorado, Lecture.

Springer, T. (1992). The distinctive functions of selectins, integrins, and IG family molecules in regulation of leukocyte interaction with endothelium. The Molecular Biology of the Endothelial Cell. Symposium on Molecular and Cellular Biology. Keystone, Colorado, Jan 13–19, Lecture.

Srivastava, A., Laidler, P., Hughes, L.E., Woodcock, J., & Shedden, E.J. (1986). Neovascularization in human cutaneous melanoma: a quantitative morphological and Doppler ultrasound study. Eur. J. Canc. Clin. Oncol. 22, 1205–1209.

Srivastava, A., Laidler, P., Davies, R.P., Horgan, K., & Hughes, L.E. (1988). The prognostic significance of tumor vascularity in intermediate-thickness (0.76–4.0 mm thick) skin melanoma. Am. J. Pathol. 133, 419–423.

Srivastava, A., Hughes, L.E., Woodcock, J.P., & Laidler, P. (1989). Vascularity in cutaneous melanoma detected by Doppler sonography and histology: Correlation with tumor behavior. Br. J. Canc. 59, 89–91.

Steiner, R. (1992a). Angiogenesis: historical perspective. In: Angiogenesis: Key Principles, Science-Technology-Medicine. (Steiner, R. & Weis, P.B., eds.), pp. 449–454, Birkhauser Verlag, Basel.

Steiner, R. (1992b). Angiostatic activity of anticancer agents in the chick embryo chorioallantoic membrane (CHE-CAM). assay. In: Angiogenesis: Key Principles—Science, Technology, Medicine (Steiner, R., Weiss, P.S., & Langer, R., eds.), pp. 449–454. Birkhauser Verlag, Basel.

Sugawara, I., Ohkochi, E., Hamada, H., Tsuruo, T., & Mori, S. (1988). Cellular and tissue distribution of MRK20 murine monoclonal antibody-defined 85-kDa protein in adriamycin-resistant cancer cell lines. Jpn. J. Canc. Res. 79, 1101–1110.

Sugawara, I. (1990). Expression and functions of p-glycoprotein (mdrl gene product) in normal and malignant tissues. Acta Pathol. Jpn. 40, 545–553.

Talmadge, J.E. & Zbar, B. (1987). Clonality of pulmonary metastases from the bladder 6 subline of the B16 melanoma studied by Southern hybridization. J. Natl. Canc. Inst. 78, 315–320.

Terrier, P., Townend, A.J., Coindre, J.M., Triche, T.J., & Cowan, K.H. (1990). An immunohistochemical study of pi class glutathione s-transferase expression in normal human tissue. Am. J. Path. 137, 845–853.
Theodorescu, D., Cornil, I., Sheehan, C., & Kerbel, R.S. (1991). Dominance of metastatically competent cells in primary murine breast neoplasms is necessary for distant metastatic spread. Int. J. Canc. 47, 118–123.
Thorpe, P.E., Wallace, P.M., Knyba, R.E., Watson, G.J., Mahadevan, V.A., Land, H., Yerganian, G., & Brown, P.J. (1991). Targeting to proliferating vascular endothelium. Angiogenesis, Int. Symp., St. Gallen, Mar 13–15, Abs. 81.
U, H.S., Kelley, P., Ashbaugh, S., Tatsukawa, K., & Werner, R. (1987). Tumorigenicity in the nude mouse model of cocultures derived from two nontumorigenic cell types, human pituitary adenomas and mouse C3H 10T1/2 fibroblasts. Canc. Res. 47, 5678–5683.
Van den Hoff, A. (1988). Stromal involvement in malignant growth. Adv. Canc. Res. 50, 159–196.
Van den Hoff, A. (1991). The role of stromal cells in tumor metastasis: A new link. Canc. Cells, 3, 186–187.
Vlodavsky, I., Korner, G., Ishai-Michaeli, R., Bashkin, P., Bar-Sharit, R., & Fuks, Z. (1990). Extracellular matrix-resident growth factors and enzymes: possible involvement in tumor metastasis and angiogenesis. Canc. Metastas. Rev. 9, 203–226.
Vlodavsky, I., Fuks, Z., Ishai-Michaeli, R., Bashkin, P., Levi, E., Korner, G., Bar-Shavit, R., & Klagsbrun, M. (1991). Extracellular matrix-resident basic fibroblast growth factor: implication for the control of angiogenesis. J. Cell. Biochem. 45, 167–176.
Waghorne, C., Korczak, B., Breitman, M.L. & Kerbel, R.S. (1987). Analysis of growth, selection, and metastasis of tumor cell populations *in vivo* using random insertions of foreign DNA as genetic tags. In: Cancer Metastasis, Biological and Biochemical Mechanisms and Clinical Aspects: Advances in Exp. Medicine and Biology (Prodi, G., ed.). pp. 27–38.
Waghorne, C., Breitman, M.L. & Kerbel, R.S. (1988a). Application of gene transfer to the study of tumor progression and metastasis. In: UCLA Symposia on Molecular and Cellular Biology: Vol 78, Tumor Progression and Metastasis (Nicolson, G.L. & Fidler, I.J., eds.), pp. 135–141, Alan R. Liss, New York.
Waghorne, C., Thomas, M., Lagarde, A.E., Kerbel, R.S., & Breitman, M.L. (1988b). Genetic evidence for progressive selection and overgrowth of primary tumors by metastatic cell subpopulations. Canc. Res. 48, 6109–6114.
Weidner, N., Semple, J.P., Welch, W.R., & Folkman, J. (1991). Tumor angiogenesis and metastasis—correlation in invasive breast carcinoma. N. Eng. J. Med. 324, 1–8.
Weigand, R.A., Isenberg, W.M., Russo, J., Brennan, M.J., & Rich, M.A. (1982). Blood vessel invasion and axillary lymph node involvement as prognostic indicators for human breast cancer. Cancer 50, 962–969.
Weinberg, R.A. (1991). Tumor suppressor genes. Science 254, 1138–1146.
Weinstein, R.S., Jakate, S.M., Dominguez, J.M., Lebowitz, M.D., Koukoulis, G.K., Kuszak, J.R., Kluskens, L.F., Grogan, T.M., Saclarides, T.J., Roninson, I.B., & Coon, J.S. (1991). Relationship of the expression of the multidrug resistance gene product (p-glycoprotein) in human colon carcinoma to local tumor aggressiveness and lymph node metastasis. Canc. Res. 51, 2720–2726.
Weiss, L. (1985). Principles of Metastasis. Academic Press, New York.
Weiss, L. (1990). Metastatic inefficiency. Adv. Canc. Res. 54, 159–211.
Wen, D., Rowland, A., & Derynck, R. (1989). Expression and secretion of gro/MGSA by stimulated human endothelial cells. EMBO J. 8, 1761–1766.
Wernert, N., Raes, M.-B., Lassalle, P., Dehouck, M.-P., Gosselin, B., Vandenbunder, B., & Stehelin, D. (1992). c-ets1 proto-oncogene is a transcription factor expressed in endothelial cells during tumor vascularization and other forms of angiogenesis in humans. Am. J. Pathol. 140, 119–127.

Wesseling, P., Van Muijen, G.N.P., Duyvestijn, A.M., De Waal, R.M.W., & Rujter, D.J. (1990). Difference in angiogenic potential of human melanoma cell lines with different metastatic behavior in nude mice. Clin. Expl. Metastas. 8 Suppl., Abs. 75.

Williams, R.L., Risau, W., Zervwes, H.-G., Drexler, H., Aguzzi, A., & Wagner, E.F. (1989). Endothelioma cells expressing the polyoma middle T oncogene induce hemangiomas by host cell recruitment. Cell, 57, 1053–1063.

Wishart, G.C., Plumb, J.A., George, W.D., & Kaye, S.B. (1991). P-glycoprotein expression is found in stromal cells in breast cancer but not those of normal breast. Br. J. Canc. 64 (Suppl. 15), Abs. 8.

Woodruff, M.F.A. (1988). Tumor clonality and its biological significance. Adv. Canc. Res. 50, 197–230.

Zajchowski, D.A., Band, V., Trask, D.K., Kling, D., Connolly, J.L., & Sager, R. (1990). Suppression of tumor-forming ability and related traits in MCF-7 human breast cancer cells by fusion with immortal mammary epithelial cells. Proc. Natl. Acad. Sci. USA 87, 2314–2318.

Zetter, B.R. (1988). Endothelial heterogeneity: influence of vessel size, organ localization, and species specificity on the properties of cultured endothelial cells. In: Endothelial Cells (Ryan, U.S., ed.), 2nd edition, pp. 63. CRC Press, Boca Raton, FL.

Ziche, H. & Gullino, P.M. (1982). Angiogenesis and neoplastic progression *in vitro*. J. Natl. Canc. Inst. 63, 483–487.

INDEX

Lightner, Duke University Medical Center. **Cell-Substrate Adhesion: Induction of Cell Spreading and Apical/Basal Plasma Membrane Polarity,** *Bruce S. Jacobson, University of Massachusetts, Amherst.* **X-Ray Diffraction Studies of Gap Junction Structure,** *Lee Makowski, Columbia University School of Medicine.* **Processing of Endocytosed Material,** *Robert F. Murphy, Carneige Mellon University.* **The Cytoskeleton of the Blood Platelet: A Dynamic Structure,** *Vivian Nachmias and Ken-ichi Yoshida, University of Pennsylvania.* **Physiological Electric Fields Can Influence Cell Mobility, Growth, and Polarity,** *Richard Nuccitelli, University of California, Davis.* **Mapping Detailed Shape and Specific Loci of Macromolecules by Brightfield, Darkfield and Immunoelectron Microscopy,** *Henry S. Slayter, Dana-Farber Cancer Institute.* **Recent Advances in the Study of Mitochrondria in Living Cells,** *James R. Wong and Lan Bo Chen, Dana-Farber Cancer Institute and Harvard Medical School.* **DNA-Protein Interactions at Telomeres in Ciliated Protozoans,** *Daniel E. Gottschling and Virginia Zakian, Fred Hutchinson Cancer Research Center.*

Volume 3, 1990, 275 pp. $90.25
ISBN 1-55938-013-6

CONTENTS: The *Dictyostelium Discoideum* Plasma Membrane: A Model System for the Study of Actin-Membrane Interactions, *Elizabeth J. Luna, Linda J. Wuestehube, Hilary M. Ingalls and Catherine P. Chia, Worcester Foundation for Experimental Biology. Tektins and Microtubules, R.W. Linck, University of Minnesota Medical School.* **The Elusive Organization of the Spindle and the Kinetochore Fiber: A Conceptual Retrospective,** *Andrew S. Bajer, University of Oregon.* **Talin: Biochemistry and Cell Biology,** *Keith Burridge and Leslie Molony, University of North Carolina, Medical School.* **Digital Imaging Flourescence Microscopy: Statistical Analysis of Photobleaching and Passive Cellular Uptake Processing,** *Zeljko Jericevic, B. Wiese, R. Homan, J. Bryan and L.C. Smith, Baylor College of Medicine.* **Clathrin Assembly Proteins and the Organization of the Coated Membrane,** *James H. Keen, Temple University of Medicine.* **Control of the Mast Cell Secretion,** *David Lagunoff, St. Louis University Medical School.* **Pattern Formation: The Differentiation of Pigment Cells from Embryonic Neural Cells,** *Sally K. Frost, University of Kansas.* **Experimental Analysis of Centrosome Reproduction in Echinoderm Eggs,** *Greenfield Sluder, Worcester Foundation for Experimental Biology.* **Multiple Pathways of Protein Secretion in Exocrine Cells,** *J. David Castle, University of Virginia Medical School.*

Please note: Volumes 1-3 published as Advances in Cell Biology and edited by Kenneth R. Miller, Division of Biology and Medicine, Brown University

Volume 4, 1992, 280 pp. $90.25
ISBN 1-55938-209-0

CONTENTS: Preface, *E. Edward Bittar.* **The Centromere,** *Jerome B. Rattner, University of Calgary, Calgary.* **The Nuclear Matrix,** *Ronald Berezney, SUNY at Buffalo.* **The Relation between Scaffolds and Genome Function,** *Susan M. Gasser, ISREC, Epalinges, Switzerland.* **Signal Transduction in the Nucleus,** *Eric A. Nigg, ISREC, Epalinges, Switzerland.* **The Peroxisome,** *Colin Masters and Dennis Crane, Griffith University, Nathan, Australia.* **The Endoplasmic Reticulum,** *Gordon Koch, University of Cambridge, UK.* **The Golgi Complex,** *Brian Storrie, VPI and State University, Blacksburg, Virginia.* **The Lysosome,** *Glenn Mortimore, Pennsylvania State University, Hershey, Pennsylvania.* **Lysosomal Acidity,** *Donald L. Schneider and Jean Chin, National Institutes of Health, Bethesda.* **The Ribosome,** *Richard Brimacombe, Max Planck Institut, Berlin.* **Subject Index.**

Volume 5, Molecular Immunology
1992, 243 pp. $90.25
ISBN 1-55938-517-0

Edited by **Jacques F.A.P. Miller,** *The Walter and Eliza Hall Institute of Medical Research, Royal Melbourne Hospital, Victoria, Australia*

CONTENTS: Preface, *Jacques F.A.P. Miller.* **Long-Term Human Hematopoiesis *in Vitro* Using Cloned Stromal Cell Lines and Highly Purified Progenitor Cells,** *Flavia M. Cicuttiin, Michael Martin, Darryl Maher, and Andrew W. Boyd.* **Multiple Routes for Late Intrathymic Precursor to Generate $CD4^+CD8^+$ Thymocytes,** *Patrice Hugo and Howard T. Petrie.* **Immunity Versus Tolerance: The Cell Biology of Positive and Negative Signaling of B Lymphocytes,** *G.J.V. Nossal.* **Self-Tolerance in the T Cell Repertoire,** *Jacques F.A.P. Miller and Grant Morahan.* **Coordinate and Differential Regulation of GM-CSF and IL-3 Synthesis in Murine T Lymphocytes,** *Antony B. Troutt, Nikki Tsoudis, and Anne Kelso.* **Host-Parasite Interactions in Leishmaniasis,** *Emanuela Handman.* **The Nonobese Diabetic (NOD) Mouse: A Model for the Study of the Cell Biology of the Pathogenesis of an Organ-Specific Autoimmune Disease,** *T.E. Mandal.* **Epigenetic Regulation of the Early Development of the Nervous System,** *Perry F. Bartlett and Mark Murphy.* **Subject Index.**

JAI PRESS